The Polysiloxanes

The Polysiloxanes

James E. Mark, Dale W. Schaefer, and Gui Lin

University of Cincinnati

OXFORD
UNIVERSITY PRESS

OXFORD
UNIVERSITY PRESS

Oxford University Press is a department of the University of Oxford.
It furthers the University's objective of excellence in research, scholarship,
and education by publishing worldwide.

Oxford New York
Auckland Cape Town Dar es Salaam Hong Kong Karachi
Kuala Lumpur Madrid Melbourne Mexico City Nairobi
New Delhi Shanghai Taipei Toronto

With offices in
Argentina Austria Brazil Chile Czech Republic France Greece
Guatemala Hungary Italy Japan Poland Portugal Singapore
South Korea Switzerland Thailand Turkey Ukraine Vietnam

Oxford is a registered trade mark of Oxford University Press
in the UK and certain other countries.

Published in the United States of America by
Oxford University Press
198 Madison Avenue, New York, NY 10016

Library of Congress Cataloging-in-Publication Data
Mark, James E., 1934–
The polysiloxanes / James E. Mark, Dale W. Schaefer, and Gui Lin.
p. cm.
Includes bibliographical references and index.
ISBN 978-0-19-518173-9 (alk. paper)
1. Silicones. I. Schaefer, Dale W. II. Lin, Gui (Scientist) III. Title.
QD383.S54M37 2015
668.4'227—dc23
2014030484

9 8 7 6 5 4 3 2 1

Printed in the United States of America on acid-free paper

CONTENTS

PREFACE

The polysiloxanes are by far the most important polymers in the category of inorganic or semi-inorganic polymers, with a large industry devoted to them in numerous countries in the industrialized world. This book attempts to give an overview of these polymers, and to describe some unsolved problems in this area.

The material presented here is an updating and significant expansion of the material on polysiloxanes in the more general book, Mark, J. E.; Allcock, H. R.; West, R. *Inorganic Polymers*, 2nd ed. Oxford University Press: New York, 2004.

It is hoped that the present book will be useful to polymer chemists and physicists, inorganic chemists, chemical engineers, and materials scientists. The authors also hope that readers will join in studying the synthesis, characterization, and applications of these fascinating polymeric materials.

The Polysiloxanes

CHAPTER 1
Introduction

1.1 BACKGROUND

Polysiloxanes are unique among inorganic and semi-inorganic polymers; they are also the most studied and the most important with regard to commercial applications. Thus, it's not surprising that there is an extensive literature describing the synthesis, properties, and applications of these materials, including books,[1–10] proceedings books,[11–14] sections of books or encyclopedias,[15–58] review articles,[59–64] and historical articles.[65–74] The purpose of this volume is not to give a comprehensive overview of these polymers but rather to focus on some novel and interesting aspects of polysiloxane science and engineering, including properties, work in progress, and important unsolved problems.

The Si–O backbone endows polysiloxanes with a variety of intriguing properties. The strength of the Si–O bond, for example, imparts considerable thermal stability, which is important for high-temperature applications (e.g., as heat-transfer agents and high-performance elastomers). The nature of the bonding and the chemical characteristics of typical side groups impart low surface free energy and therefore desirable surface properties. Polysiloxanes, for example, are used as mold-release agents, waterproofing sprays, and biomedical materials.

Structural features of the chains give rise to physical properties that are also of considerable scientific interest. For example, the substituted Si atom and the unsubstituted O atom differ greatly in size, giving the chain a nonuniform cross section. This characteristic affects the way the chains pack in the bulk, amorphous state, which explains the unusual equation-of-state properties (such as compressibility). Also, the bond angles around the O atom are much larger than those around the Si, which makes the planar all-*trans* form of the chain approximate a series of closed polygons, as illustrated in figure 1.1.[75] As a result, siloxane chains exhibit a number of interesting configurational characteristics that impact properties and associated applications.

The major categories of homopolymers and copolymers to be discussed are[11] (i) linear siloxane polymers [–SiRR'O–] (with various alkyl and aryl R,R' side groups), (ii) sesquisiloxane polymers possibly having a ladder structure, (iii) siloxane-silarylene polymers [$-Si(CH_3)_2OSi(CH_3)_2(C_6H_4)_m-$] (where the skeletal phenylene units are either *meta* or *para*), (iv) silalkylene polymers [$-Si(CH_3)_2(CH_2)_m-$], and (v) random and block copolymers, and blends of some of the above. Table 1.1 provides some illustrative examples. Important topics include the structure, flexibility, transition temperatures, permeability, and other physical properties. Applications include high-performance fluids, elastomers, and coatings; surface modifiers; gas separation membranes; photoresists; soft contact lenses; body implants; and controlled-release systems. Also of interest are the use of sol-gel techniques to convert silicon-containing materials to novel reinforcing fillers or to polymer-modified ceramics, and the use of pyrolysis to form high-performance fibers.

1.2 HISTORY

The first reaction of relevance in silicon chemistry was the conversion of elemental silicon to silicontetrachloride, $SiCl_4$, and to trichlorosilane, $SiHCl_3$. These purely inorganic substances were then converted into organometallic species such as $RSiX_3$, by reaction with diethyl zinc and related compounds.[1, 16, 23, 29, 33]

Organosilicon chemistry blossomed with Kipping's preparation of such compounds by the more convenient Grignard process. The resulting silanes turned out to be of paramount importance since they hydrolyzed readily to form compounds containing Si–O bonds, both linear and cyclic.[4, 29, 33] These new materials were first called silicoketones or "silicones" by analogy with ketones R–(C = O)–R' in the organic area. Structural studies, however, showed that they did not contain the Si = O double bond. Thus, the silicone name is a misnomer, but it has persisted, at least in casual usage. However, those in the field prefer the terms "siloxanes" and "polysiloxanes."

Kipping [68] prepared the earliest hybrid inorganic/organic material containing Si–C bonds and reported in a series of papers over the period 1899–1944. Around 1930, the Corning Glass Company prepared an

Table 1.1. BACKBONE AND SIDE CHAINS ON SOME ILLUSTRATIVE POLYSILOXANES

Polymer	Backbone	Side chains	Characteristics
Poly(dimethylsiloxane)	Si–O–	CH_3 CH_3	Most extensively used polysiloxane
Poly(phenylmethylsiloxane)	Si–O–	C_6H_5 CH_3	Suppressed crystallinity and improved radiation resistance
Poly(vinylmethylsiloxane)	Si–O–	C_2H_3 CH_3	Unsaturation sites simplify cross linking
Siloxane-silarylene polymers	Si–O– and Si–(*m* or *p* *phenylene*)	CH_3 CH_3	Stiffer backbones useful in some applications
Siloxane-silalkylene polymers	Si–O– and Si–C–	CH_3 CH_3	Reduced thermal stability

apparently largely uncharacterized material of this type. This latter work resulted in a number of collaborations with the General Electric Company and the Mellon Institute focused on siloxane-based, thermally resistant coatings. Possible use in fiberglass insulation in submarine motors helped accelerate work in this area, including construction of a siloxane pilot plant based on Grignard reactions at the Mellon Institute in 1940.

In the case of the direct process, only one of the resulting products can be used to obtain siloxane polymers.[28] Table 1.1 describes some of these products.

1.3 NOMENCLATURE

In most of the literature, siloxane terminology consists simply of specifying the side groups and then the backbone. For example, the polymer having the repeat unit [–Si(CH_3)$_2$O–] is called poly(dimethylsiloxane), and that having the repeat unit [–Si(CH_3) (C_6H_5) O–] is called poly(phenylmethylsiloxane). Closely related polymers are the poly(silmethylenes) of repeat unit [–SiRR'CH_2–] and the poly(siloxane-silphenylenes) of repeat unit [–SiRR'OSiRR'C_6H_4–], in which the second oxygen atom in the doubled repeat unit is replaced by a phenylene group.

Because certain structures and structural segments appear over and over again in the siloxane area, several abbreviations are used in specialized areas of the literature.[28, 29] The monofunctional unit $R_3SiO_{0.5}$ is designated

Table 1.2. NOTATIONS FOR VARIOUS POLYSILOXANE STRUCTURAL UNITS

Notation	Structure	Examples	
M	$R_3SiO_{0.5}$	MM	Dimer $(CH_3)_3SiOSi(CH_3)_3$
D	R_2SiO	MD_6M	Oligomer $(CH_3)_3Si[OSi(CH_3)_2]_6OSi(CH_3)_3$
D	R_2SiO	D_3	Cyclic trimer $[Si(CH_3)_2O]_3$
T	$RSiO_{1.5}$	TT	Two trifunctional cross links $RSi(O_2)OSi(O_2)R$
Q	SiO_2	Q	Silica

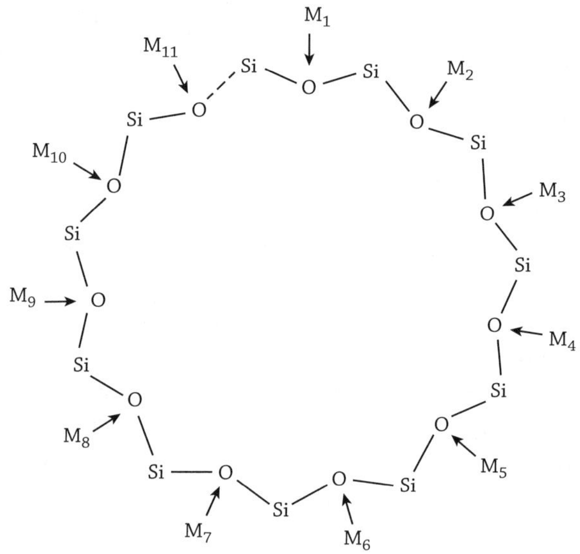

Figure 1.1:
The polysiloxane chain backbone in the low-energy, planar, all-*trans* conformation.[75] Arrows represent group dipoles *M* for each Si–O–Si pair of consecutive skeletal bonds. Because of the cancelation of dipoles, the dipole moment of oligomers depends on molecular weight.

"M," the difunctional unit R_2SiO "D," the trifunctional unit RSiO1.5 "T," and the quadrifunctional SiO_2 "Q." For example, the dimer $(CH_3)_3SiOSi(CH_3)_3$ is termed "MM," the oligomer $(CH_3)_3Si[OSi(CH_3)_2]_{10}OSi(CH_3)_3$ is termed "$MD_{10}M$," and the cyclic trimer $[Si(CH_3)_2O]_3$ is called "D_3." Table 1.2 summarizes this nomenclature. Unprimed abbreviations are taken to mean the R substituents are methyl groups, since these are the most common and the most important. Primed abbreviations are used for other substituents, the most important of which is probably the phenyl group.

REFERENCES

1. Noll, W., *Chemistry and Technology of Silicones*. Academic Press: New York, 1968.
2. Bobear, W. J., Silicone Rubber. In *Rubber Technology*, Morton, M., Ed. Van Nostrand Reinhold: New York, 1973; pp 368–406.
3. Smith, A. L., *Analysis of Silicones*. Krieger Publishing Co.: Malabar, FL, 1983.
4. Rochow, E. G., *Silicon and Silicones*. Springer-Verlag: Berlin, 1987.
5. Warrick, E. L., *Forty Years of Firsts. The Recollections of a Dow Corning Pioneer.* McGraw-Hill: New York, 1990.
6. Clarson, S. J.; Semlyen, J. A., *Siloxane Polymers*. Prentice Hall: Englewood Cliffs, N J, 1991.
7. Smith, A. L., *The Analytical Chemistry of Silicones*. John Wiley & Sons: New York, 1991.
8. Koerner, G.; Schulze, M.; Weis, J., *Silicones: Chemistry and Technology*. CRC Press: Boca Raton, FL, 1992.
9. Clarson, S. J.; Semlyen, J. A., *Siloxane Polymers*. Prentice Hall: Englewood Cliffs, NJ, 1993.
10. Cazacu, M., *Recent Developments in Silicone-Based Materials*. Nova Science: New York, 2010.
11. Zeigler, J. M.; Fearon, F. W. G., *Silicon-Based Polymer Science. A Comprehensive Resource*. American Chemical Society: Washington, DC, 1990; Vol. 224.
12. Clarson, S. J.; Fitzgerald, J. J.; Owen, M. J.; Smith, S. D., *Silicones and Silicone-Modified Materials*. American Chemical Society: Washington, DC, 2000; Vol. 729.
13. Clarson, S. J.; Fitzgerald, J. J.; Owen, M. J.; Smith, S. D.; van Dyke, M. E., *Synthesis and Properties of Silicones and Silicone-Modified Materials*. American Chemical Society: Washington, DC, 2003; Vol. 838.
14. Clarson, S. J.; Fitzgerald, J. J.; Owen, M. J.; Smith, S. S.; Dyke, M. E. V., Eds., *The Science and Technology of Silicones and Silicone-Modified Materials*, ACS Symposium Series, Vol. 964, American Chemical Society: Washington, DC, 2007.
15. Allcock, H. R., *Heteroatom Ring Systems and Polymers*. Academic Press: New York, 1967.
16. Meals, R., Silicones. In *Kirk-Othmer Encyclopedia of Chemical Technology*. Wiley: New York, 1969; Vol. 22.
17. Borisov, S. N.; Voronkov, M. G.; Lukevits, E. Y., *Organosilicon Heteropolymers and Heterocompounds*. Plenum Press: New York, 1970.
18. Allcock, H. R., Inorganic Polymers. *Sci. Am.* **1974**, *230*(3), 66–74.
19. Elias, H. G., *Macromolecules: Synthesis and Materials*. Plenum Press: New York, 1977; Vol. 2.
20. Noshay, A.; McGrath, J. E., *Block Copolymers: Overview and Critical Survey*. Academic Press: New York, 1977.
21. Roy, A. K., Inorganic High Polymers. In *Kirk Othmer Encyclopedia of Chemical Technology*, John Wiley and Sons: New York, 2009. pp 1–22.
22. Carraher, C. E., Jr.; Sheats, J. E.; Pittman, C. U., *Advances in Organometallic and Inorganic Polymers*. Marcel Dekker: New York, 1982.
23. Stark, F. O.; Falender, J. R.; Wright, A. P., Silicones. In *Comprehensive Organometallic Chemistry*, Wilkinson, G., Ed. Pergamon Press: Oxford, 1982; Vol. 2.
24. Allcock, H. R., Developments at the Interface of Inorganic, Organic, and Polymer Chemistry. *Chem. Eng. News* **1985**, *63*(11), 22–36.
25. Archer, R. D., Inorganic Polymers. In *Encyclopedia of Materials Science and Engineering*, Bever, M. B., Ed. Pergamon Press: Oxford, 1986; Vol. 3, pp 2335–2339.

26. Zeldin, M., Silicones. In *Encyclopedia of Materials Science and Engineering*, Bever, M. B., Ed. Pergamon Press: Oxford, 1986; pp 4440–4443.
27. Rheingold, A., Inorganic Polymers. In *Encyclopedia of Polymer Science and Engineering*, 2nd ed. Wiley-Interscience: New York, 1987.
28. Caprino, J. C.; Macander, R. F., Silicone Rubber. In *Rubber Technology*, 3rd ed.; Morton, M., Ed. Van Nostrand Reinhold: New York, 1987.
29. Hardman, B.; Torkelson, A., Silicones. In *Encyclopedia of Polymer Science and Engineering*, 2nd ed. Koschwitz, J. I., Ed. Wiley-Interscience: New York, 1989; pp 204–308.
30. Zeldin, M.; Wynne, K. J.; Allcock, H. R., *Inorganic and Organometallic Polymers. Macromolecules Containing Silicon, Phosphorous, and Other Inorganic Elements.* American Chemical Society: Washington, DC, 1988; Vol. 360.
31. Patai, S.; Rappoport, Z., *The Chemistry of Organic Silicon Compounds. Parts 1 and 2.* John Wiley & Sons: New York, 1989.
32. Lazar, M.; Bleha, T.; Rychly, J., *Chemical Reactions of Natural and Synthetic Polymers.* Wiley: New York, 1989.
33. Kendrick, T. C.; Parbhoo, B.; White, J. W., Siloxane Polymers and Copolymers. In *The Chemistry of Organic Silicon Compounds*, Patai, S.; Rappoport, Z., Eds. John Wiley & Sons, Ltd, Chichester, UK, 1989; pp 1289–1361.
34. Lukevics, E.; Pudova, O.; Sturkovich, R., *Molecular Structure of Organosilicon Compounds.* John Wiley & Sons: New York, 1989.
35. Barton, T. J.; Boudjouk, P., Organosilicon Chemistry: A Brief Overview. In *Silicon-Based Polymer Science. A Comprehensive Resource*, Zeigler, J. M.; Fearon, F. W. G., Eds. American Chemical Society: Washington, DC, 1990; Vol. 224.
36. Mark, J. E., Silicon-Containing Polymers. In *Silicon-Based Polymer Science. A Comprehensive Resource*, Zeigler, J. M.; Fearon, F. W. G., Eds. American Chemical Society: Washington, DC, 1990; Vol. 224, pp 47–68.
37. Weyenberg, D. R.; Lane, T. H., Future Directions for Silicon-Based Polymer Science. In *Silicon-Based Polymer Science. A Comprehensive Resource*, Zeigler, J. M.; Fearon, F. W. G., Eds. American Chemical Society: Washington, DC, 1990; Vol. 224, pp 753–764.
38. Sheats, J.; Carraher, C. E., Jr.; Zeldin, M.; Currell, B.; Pittman, C. U., Jr., *Inorganic and Metal-Containing Polymeric Materials.* Plenum Press: New York, 1991.
39. Auner, N.; Weis, J., *Organosilicon Chemistry: From Molecules to Materials.* Wiley-VCH: Weinheim, 1994.
40. Jones, R. G., *Silicon-Containing Polymers.* Royal Society of Chemistry: Cambridge, UK, 1995.
41. Manners, I., Polymers and the Periodic Table: Recent Developments in Inorganic Polymer Science. *Angew. Chem. Int. Ed. Engl.* **1996**, *35*, 1602–1621.
42. Simionescu, B. C.; Harabagiu, V.; Simionescu, C. I., Siloxane-Containing Polymers. In *Polymeric Materials Encyclopedia*, Salamone, J. C., Ed. CRC Press: Boca Raton, 1996; Vol. 10, pp 7751–7759.
43. Folland, P. D.; Matisons, J. G., Siloxane Elastomers. In *Polymeric Materials Encyclopedia*, Salamone, J. C., Ed. CRC Press: Boca Raton, 1996; Vol. 10, pp 7759–7768.
44. Rappoport, Z.; Apeloig, Y., *The Chemistry of Organic Silicon Compounds. Parts 1–3.* John Wiley & Sons: New York, 1998; Vol. 2.
45. Auner, N.; Weis, J., *Organosilicon Chemistry III. From Molecules to Materials.* Wiley-VCH: Weinheim, 1998.
46. Mark, J. E., *Polymer Data Handbook.* Oxford University Press: New York, 1999.

47. Brook, M. A., *Silicon in Organic, Organometallic, and Polymer Chemistry*. John Wiley & Sons: New York, 2000.
48. Jones, R. G.; Ando, W.; Chojnowski, J., *Silicon-Containing Polymers. The Science and Technology of Their Synthesis and Applications*. Kluwer Academic Publishers: Dordrecht, 2000.
49. Archer, R. D., *Inorganic and Organometallic Polymers*. Wiley-VCH: New York, 2001.
50. Allcock, H. R.; Lampe, F. W.; Mark, J. E., *Contemporary Polymer Chemistry*, 3rd ed. Prentice Hall: Englewood Cliffs, NJ, 2003.
51. Mark, J. E., Some Unusual Elastomers and Experiments on Rubberlike Elasticity. *Prog. Polym. Sci.* **2003**, *28*, 1205–1221.
52. Schubert, U.; Husing, N., *Synthesis of Inorganic Materials*, 2nd ed. Wiley-VCH: Weinheim, 2005.
53. Chandrasekhar, V., *Inorganic and Organometallic Polymers*. Springer Verlag: Heidelberg, 2005.
54. Mark, J. E.; Allcock, H. R.; West, R., *Inorganic Polymers*, 2nd ed. Oxford University Press: New York, 2005.
55. Cypryk, M., General Review of Polysiloxane Synthesis. In *Silicon-Based Inorganic Polymers*, De Jaeger, R.; Gleria, M., Eds. Nova Science: New York, 2008; pp 1–59.
56. Cazacu, M., Polymers Containing Si, O, and Other Elements Within Backbone. In *Advances in Functional Heterochain Polymers*, Cazacu, M., Ed. Nova Science: New York, 2008; pp 1–32.
57. Pagliaro, M., *Silica-Based Materials for Advanced Chemical Applications*. RSC Publishing: London, 2009.
58. Dvornic, P. R.; Owen, M. J., *Silicon-Containing Dendritic Polymers*. Springer: Berlin, 2009.
59. Mark, J. E., Conformations and Spatial Configurations of Inorganic Polymers. *Macromolecules* **1978**, *11*, 627–633.
60. Warrick, E. L.; Pierce, O. R.; Polmanteer, K. E.; Saam, J. C., Silicone Elastomer Developments 1967–1977. *Rubber Chem. Technol.* **1979**, *52*, 437–525.
61. Owen, M. J., Say Yes to Silicones! [Si to Si!]. In *Science and Technology of Silicones and Silicone-Modified Materials*, Clarson, S. J.; Fitzgerald, J. J.; Owen, M. J.; Smith, S. D.; Van Dyke, M. E., Eds. American Chemical Society: Washington, DC, 2007; Vol. 964, pp 3–8.
62. Wu, S.; Mark, J. E., Some Simulations and Theoretical Studies on Poly(dimethylsiloxane). *J. Macromol. Sci., Polym. Rev.* **2007**, *47*, 463–485.
63. Bokobza, L., Some Issues in Rubber Nanocomposites: New Opportunities for Silicone Materials. *Silicon* **2009**, *1*, 141–145.
64. Willoughby, J. A., Silicones Today. *Silicon* **2011**, *3*, 153–159.
65. Clarson, S. J., Some Reflections on the Element Silicon. *Silicon* **2009**, *1*, 1–2.
66. Clarson, S. J., Charles Darwin and Silicon. *Silicon* **2009**, *1*, 59–63.
67. Owen, M. J., Tom Lane: Taking Silicon to the Forefront. *Silicon* **2009**, *1*, 137–139.
68. Thomas, N. R., Frederic Stanley Kipping—Pioneer in Silicon Chemistry: His Life and Legacy. *Silicon* **2010**, *2*, 187–193.
69. Clarson, S. J., SILICON Highlights. *Silicon* **2010**, (2), 113–115.
70. Owen, M. J., Why Silicon? *Silicon* **2011**, *3*, 6–7.
71. Clarson, S. J., On Biological and Bioactive Silicon Systems. *Silicon* **2011**, *3*, 3–4.

72. Clarson, S. J., A Personal Commentary on Bioiogical and Bioactive Silicon Systems. *Silicon* **2011**, *3*, 5–6.
73. Owen, M. J., Midland Silicones Memories. *Silicon* **2011**, *3*, 53–55.
74. Mark, J. E., The Use of Polysiloxanes to Elucidate Molecular Aspects of Rubberlike Elasticity. *J. Inorg. Organomet. Polym. Mater.* **2012**, *22*, 560–563.
75. Mark, J. E., Dipole Moments of Dimethylsiloxane Chains. *J. Chem. Phys.* **1968**, *49*, 1398–1402.

CHAPTER 2

Preparation, Analysis, and Degradation

2.1 PREPARATION OF MONOMERS

Elemental silicon on which the entire technology is based is typically obtained by reduction of the mineral silica with carbon at high temperatures[1–3]:

$$SiO_2 + 2C \rightarrow Si + 2CO \tag{2.1}$$

The silicon is then converted directly to tetrachlorosilane by the reaction

$$Si + 2Cl_2 \rightarrow SiCl_4 \tag{2.2}$$

Tetrachlorosilane can be used to form an organosilane by the Grignard reaction

$$SiCl_4 + 2\,RMgX \rightarrow R_2SiCl_2 + 2\,MgClX \tag{2.3}$$

This relatively complicated reaction has been replaced by the so-called Direct Process or Rochow Process,[2, 4] which starts from elemental silicon as is illustrated by the reaction

$$Si + 2\,RCl \rightarrow R_2SiCl_2 \tag{2.4}$$

This process also yields $RSiCl_3$ and R_3SiCl, which can be removed by distillation. Compounds of formula R_2SiCl_2 are extremely important as intermediates to a variety of substances having both organic and inorganic character.[5–8] Hydrolysis gives dihydroxy structures, which can condense to give the basic [–SiR_2O–] repeat unit. The nature of the product obtained depends greatly on the reaction conditions.[8] Basic catalysts and higher temperatures favor higher molecular weight linear polymers. Acidic catalysts tend to produce cyclic small molecules or low molecular weight polymers.

The hydrolysis approach to polysiloxane synthesis has been largely replaced by ring-opening polymerization[2, 8–14] of organosilicon cyclic trimers and tetramers, with ionic initiation. These cyclic monomers are produced by the hydrolysis of dimethyldichlorosilane. Under the right conditions, at least 50 wt % of the products are cyclic oligomers. The desired cyclic species are separated from the mixture for use in ring-opening polymerizations such as those described in the following section.

In addition, "click" chemistry has been developed for new synthesis techniques in general,[15–18] and polymerizations in particular.[19] These approaches have been used to prepare polysiloxane elastomers[20] and polydimethylsiloxane (PDMS) copolymers that can function as thermoplastic elastomers.[21] New synthetic strategies for structured silicones, based on $B(C_6F_5)_3$ have also been developed.[22] Another new approach involves enzymes,[23] such as the lipase enzymatically catalyzed synthesis of silicone aromatic polyesters and silicone aromatic polyamides.[24]

2.2 RING-OPENING POLYMERIZATIONS

Cyclic siloxanes can undergo a ring-opening chain-growth polymerization. Free radicals are not useful as initiator species, because of the nature of the siloxane bond, but cationic initiators are very effective. The reaction is illustrated using the most common cyclic oligomers, the timer hexamethylcyclotrisiloxane, or the tetramer octamethylcyclotetrasiloxane[9–10]:

$$(SiR_2O)_{3,4} \rightarrow [-SiR_2O-]_x \tag{2.5}$$

where R can be alkyl or aryl and x is the degree of polymerization. In principle, the reaction is reversible, but in practice it is made essentially irreversible by the choice of monomer, initiator, and polymerization conditions. Because of this potential reversibility however it is important to remove all initiators from the finished product, typically done by neutralization of the acidic or basic terminal chain residues. Alternatively, some initiators can be removed by volatilization or thermolysis. End blocking is frequently used to place groups on chain ends that increase thermal stability and prevent re-equilibration.

Because it is frequently impossible to remove all traces of active species, some reorganization is almost inevitable. In these processes, siloxane bonds interchange to bring about variation in both molecular weight and in the relative amounts of cyclic and linear species. At equilibrium, a Gaussian distribution of molecular weight exists. The cyclic oligomers that occur most frequently are D4 through D6. The amount obtained depends greatly on the "monomer" and on the polymerization conditions. D4 through D6 are typically present to the extent of 10–15 wt %. The lower molecular weight products are generally removed from the polymer before it is used in a commercial application.

The low molecular weight products are also of interest from a fundamental scientific point of view in two respects. First, the extent to which such products are formed can be used as a measure of chain flexibility.[25–27] Second, the various cyclic species formed can be purified using standard separation techniques, and then used to test theoretical predictions regarding the differences between otherwise identical cyclic and linear molecules.[28–33]

For anionic equilibrations, typical catalysts are alkali metal oxides and hydroxides, and bases in general.[34–40] Initiation and propagation involve nucleophilic attack on the monomer, causing opening of the ring followed by chain extension. As is frequently the case in ionic polymerizations, the nature of the counter cation, particularly its size, can have a large effect on the reaction. This polymerization is very different from most others,[8] which are driven by a decrease in enthalpy. That is, the decrease in entropy that accompanies the linkage of monomers into a chain-like structure is counteracted and overcome by the decrease in energy generated by the formation of new chemical bonds. In the siloxane case however the bonds linking the monomer units into the chain are similar in energy to those found in siloxane rings, and the net energy change is very small. There is an increase in entropy, presumably from increased internal molecular freedom of the siloxane segments in going from the cyclic structures to the linear chains. It is this increase in entropy that drives the polymerization reaction. Ring opening has also been induced by hydride transfer reactions.[41]

A new, low-Tg siloxane thermoplastic elastomer with a functionalizable backbone was recently synthesized via sequential anionic ring-opening polymerization and coupling.[42] The attachment of a photoresponsive liquid crystal led to a rapid thermoplastic photoactuator. Polystyrene was used as a hard glassy end block, and poly(vinylmethylsiloxane) served as the soft middle segment in this polystyrene-b-poly(vinylmethylsiloxane)-b-polystyrene ABA triblock copolymer.

Cationically catalyzed polymerizations[12, 43–45] have not received as much attention as the anionic variety. Typical cationic (acidic) catalysts in this case are Lewis acids. Yields and proportions of the various species are generally very similar to those obtained in anionic polymerizations, although the mechanism is very different. The reaction is thought to proceed through a tertiary oxonium ion formed by addition of a proton to one of the O atoms of the cyclic siloxane. The mechanism may involve step growth, as well as the expected chain growth. Examples of other catalysts used include metal sulfonate/acid chloride combinations.[46] In the case of step-growth polymerizations, tin catalysts are very effective.[47]

Polymerization of nonsymmetrical cyclic siloxanes gives stereochemically variable polymers [–SiRR'O–] that are analogous to the totally organic vinyl and vinylidene polymers [–CRR'CH_2–]. In principle, it should be possible to prepare siloxanes in the same stereoregular forms (isotactic and syndiotactic) that have been achieved for some of their organic counterparts,[8, 48–49] as mentioned in chapter 1. This goal has been accomplished to some extent (see chapter 5). The major advantage is the crystallizability generally observed for isotactic and syndiotactic stereoregular forms of the polymer, whereas the stereoirregular (atactic) modification is inherently noncrystallizable. Strictly alternating copolymers have recently been reported.[50]

In some cases, an end blocker such as YR'SiR_2OSiR_2R'Y is used to form reactive -OSiR_2R'Y chain ends.[51–52] Interesting examples include triarylamines,[53] nitrobenzoxadiazole fluorescent groups via thiol-ene coupling.[54] Homopolymerizations of this type are discussed in detail elsewhere.[8, 10]

Some studies have focused on the preparation of porous polysiloxane materials[55–60] such as low-density aerogels.[61] Mesoporous and ultra-large pore siloxane structures can be prepared by condensation of tetraethylorthosilicate (TEOS) and other silica precursors. These materials show porosity, sometimes ordered, with pore sizes up to 30 nm.[62–66] Hollow nano/microstructures have also been prepared, by ionic polymerization.[67] In a reversal of roles, siloxane chains have been substituted into poly(*p*-xylylene).[68]

It has become important to develop environmentally-friendly methods for preparing any of these materials.[69]

2.3 OTHER APPROACHES AND COPOLYMERIZATIONS

Atom transfer radical polymerization (ATRP) has been used in the area of polysiloxane[70] and there have been a number of studies on controlling the stereochemical structures of polysiloxanes.[71–74] Templated syntheses of ladder-like siloxane structures[75] and the placement of silsesquioxane units in siloxane backbones[76] have also been explored. Quintana et al.

demonstrated robust, antifouling, hydrophilic polysulfobetaine-based brushes with diblock architecture fabricated by ATRP using initiator-modified surfaces.[77]

Emulsions play a role in some aspects of polysiloxane technology. Polymerization has been carried out in emulsions,[78–85] and room temperature–cured PDMS elastomers have been obtained from aqueous dispersions.[86] Poly(vinyl pyrrolidone) dispersions have been prepared in reactive PDMS media.[87] Microemulsions containing PDMS also find use in applications such as cosmetics. One novel study involved a surfactant solution in a polysiloxane, droplets of which bounced on a vertically vibrated liquid surface.[88] When the amplitude of the vibration exceeded a threshold value, a polysiloxane-water-polysiloxane *double* emulsion was formed!

Polymerizations have also been carried out in the vapor phase,[89] in plasmas,[90] enzymatically,[24, 91–94] and in compressed carbon dioxide.[95] Polysiloxanes have also been obtained by acid leachate from chrysotile asbestos.[96]

Polymerization of mixtures of monomers, such as $(SiR_2O)_m$ with $(SiR'_2O)_m$, can be used to generate random copolymers. These copolymers are generally highly irregular in structure in a chemical rather than stereochemical sense. Correspondingly, they also show little if any crystallizability. Copolymerizations of this type are discussed in detail elsewhere.[8, 10]

A higher level of structure is illustrated by the preparation of microporous hybrid polymers from functionalized cubic siloxane cages.[97]

2.4 STRUCTURAL FEATURES

Several structural features make the siloxane backbone one of the most flexible in all of polymer science.[98–99] Figure 2.1 illustrates the reasons for this extraordinary flexibility. First, because of the nature of the bonding[25–27, 100–101] the Si–O skeletal bond has a length (1.64 Å) that is significantly larger than that of the C–C bond (1.53 Å) of paramount importance in most organic polymers. As a result, steric interference and intramolecular congestion are diminished.[25–27] This circumstance is true for inorganic and semi-inorganic polymers in general. Almost any single bond between a pair of inorganic atoms (Si–Si, Si–C, Si–N, P–N, etc.) is longer than the C–C bond. Also, the oxygen skeletal atoms are unencumbered by side groups and they are as small as an atom can be and still have the divalency needed to continue a chain structure. In addition, the

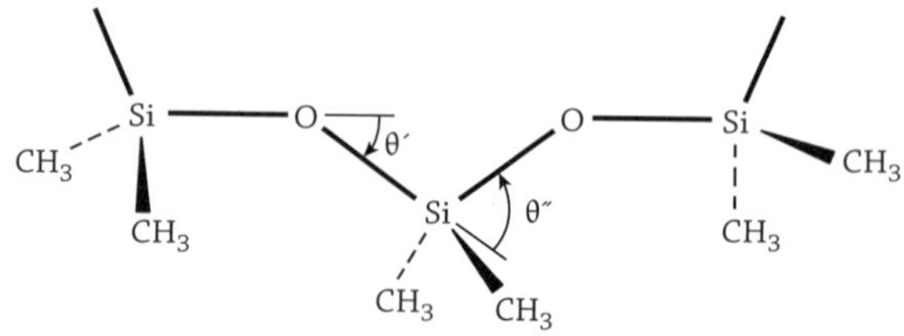

Figure 2.1:
Sketch of the PDMS chain, showing some structural information relevant to its high flexibility.[25] Because of the difference in bond angles ($\theta' = 37°$, $\theta'' = 70°$). The all-trans configuration closes to Figure 1.1 after $360°/(70°–37°) = 11$ repeat units. The torsional barrier for rotations about the skeletal bonds is very low, which accounts for the high dynamic flexibility and low glass transition temperature. Reproduced by permission of John Wiley and Sons.

Si–O–Si bond angle of ~143° is much more open than the usual tetrahedral bonding (occurring at ~110°), and can "invert"[102] through the linear (180° form) with little cost in energy. Similarly, torsional rotations can occur without serious increases in energy. These structural features have the effect of increasing the *dynamic* flexibility of the chain[25–27]; they also increase its *equilibrium* flexibility, which is the ability of a chain to take on a compact shape when in the form of a random coil. Flexibility is generally measured, inversely, by the mean-square end-to-end distance or radius of gyration of the chain in the absence of excluded-volume effects.

2.5 ELASTOMER TECHNOLOGY

Pure siloxane polymers are only rarely appropriate for use in technology. Numerous additives are incorporated in order to improve their properties. A typical formulation contains the siloxane polymer, plus some or all of the following ingredients: reinforcing fillers, extending (nonreinforcing) fillers, processing aids, heat-aging additives, pigments, and curing agents (e.g., end-linking agents with associated catalysts, or organic peroxides).[103–107]

The siloxane polymer usually has a rather high molecular weight, and may have reactive ends for end-linking or vinyl side chains for peroxide curing, as described in chapter 7.

The preferred reinforcing filler is high surface area silica, particularly that made by the fume process, which gives the greatest reinforcement, and, because of its high purity, yields excellent electrical insulation properties.[108–109] Silicas obtained from aqueous solutions[110] impart moderately good reinforcement but, because of the presence of water on the filler particles, can adversely affect the electrical properties of the elastomer. Ex situ generated titania particles have also been used, particularly particles derivitized with trifunctional siloxanes to make them highly hydrophobic.[111] Carbon black provides some reinforcement, but can interfere with some types of peroxide cures. Also, its electrical conductivity can severely compromise the electrical properties of the material.[104, 107] In some cases, silane coupling agents are used to improve the bonding between the reinforcing phase and the polymer.[112–114] These molecules typically have the structure X_3SiY, with X chosen to interact strongly with one phase, and Y with the other. For example, if X is an alkoxy group it can hydrolyze and react with OH groups on the surface of a filler particle.[2] Similarly, if Y is a vinyl group it can be polymerized into the organic matrix being reinforced, providing enhanced filler-matrix bonding.[2] It is also possible to form reinforcing filler particles in situ, for example by the sol-gel hydrolysis and condensation of precursors such as organosilicates.[115–116]

Not all fillers are designed to improve mechanical properties. Extending fillers, for example, reduce the cost of the compounded elastomer. Nonreinforcing fillers are exemplified by kaolin, diatomaceous earth, and minerals such as calcium carbonate. Coloring agents can be either organic or inorganic, but the former can adversely affect heat stability. Examples of suitable inorganic colorants are oxides and salts of iron, chromium, cobalt, titanium, and cadmium. Some not only provide color but can also have some beneficial heat-aging effects.

Processing aids are particularly important in the case of elastomers that contain highly reinforcing silicas, since these fillers adsorb polymer chains so strongly to their surfaces that premature gelation can occur. These additives have a softening or plasticizing effect, thus ameliorating the occurrence of this complication.[104, 107]

The nature of the curing (cross-linking) agents introduced depends on the particular chemical reaction chosen for generating the cross links.[117–121] In the case of end-linking reactions, the end groups are generally either hydroxyl or vinyl units. In the former case, the end-linking agent may be TEOS [$Si(OC_2H_5)_4$], which reacts by a condensation reaction, with a stannous salt used as catalyst. In the latter case, the end-linking agent can be an oligomeric siloxane that contains reactive Si–H groups, with the H atoms adding to the double bonds in the polymeric siloxane. Platinum salts are catalysts for this type of addition curing reaction.[117–120, 122] Other functional groups at the ends or along the chains can serve the same purpose.[17, 123–134] In an interesting reversal of roles, a polysiloxane was used to cross link cellulose acetate.[135]

Aliphatic or aromatic peroxide curing agents can also be used, by reactions with vinyl side chains or even saturated alkyl groups.[136–137] Specific peroxides are chosen on the basis of their decomposition temperatures, and the reaction products they leave behind after the curing process is complete. Some peroxides used are bis(2,4-dichlorobenzoyl)peroxide, benzoyl peroxide, dicumyl peroxide, and di-*t*-butyl peroxide.[117–120, 138]

PDMS has also been cured using UV (ultraviolet) radiation,[139–142] gamma or electron beams,[143–150] and laser irradiation.[151] Thermal cures are also available.[20, 152–155] Also relevant here are physically cross-linked fluorosilicone elastomers obtained by self-assembly and template polycondensation of tailored building blocks.[156]

Evaluation of the results of any of these curing processes is important, including rheological characterization.[157] Control of the material during cross linking is quite important since movement of the polymer during the curing process can affect the mechanical properties of the resulting elastomer.[158] Some modeling of the vulcanization process has been carried out. Additional work could be very useful for optimizing the properties of the resulting elastomers.[159]

Using cyclic polystyrene and cyclic PDMS can give PDMS networks with movable cross-link sites[160] and some chemistries can yield elastomers in which the cross linking is reversible.[161] Also, including the proper functional groups on the chains can make PDMS elastomers self-healing after rupture.[162–163] In an extension of these ideas, the cross linking of encapsulated PDMS resins has been used for the self healing of high-temperature cured epoxies.[164]

There has been some interest in cross-linking in solution, since the network chains in the subsequently dried elastomer are "super compressed." These materials can have some unusual properties, including unusually high extensibility.[165–167]

Mechanical property measurements are the most common way to characterize cross linking.[115, 168] Different curing methods can give different mechanical properties.[169] New characterization methods are being developed, including fluorescence,[170] small-angle neutron scattering,[171] and multiple quantum ^{1}H NMR.

The compounding process can be very complicated.[104, 107] The amount of some ingredients used is fixed by the stoichiometry of the reaction in which they participate. The end-functionalized polymer and the end-linking agent are in this category. The relative amounts of other ingredients are often chosen by experience or by trial and error. After all the amounts have been selected, they are mixed ("compounded") in conventional equipment (e.g., a Banbury mixer). Although the resulting mixture can be fabricated immediately, it is common practice to permit the material to "recover" or "age" for a few days. If hardening occurs during this period, some remilling ("refreshening") will be required.

A variety of processing steps may then be carried out. Examples are compression molding, injection molding, transfer molding, extrusion, calendering, dispersion coating, and blowing into foams.[104, 107]

Although some cures can take place at room temperature, most are carried out at elevated temperatures.[104, 117–120] Conventional electric, gas or forced-air ovens are used, but presumably microwave heating could be used as well. There has been some interest in developing polysiloxane thermoplastic elastomers.[172] Attempts have also been made to model the processing various types of polysiloxane materials.[173]

2.6 ANALYSIS AND TESTING

Infrared and UV spectroscopy are often used to determine the composition of siloxane copolymers and of mixtures of siloxanes and silicates with other species.[1, 174] These spectroscopies can also be used to monitor (i) vinyl groups introduced to facilitate cross-linking, (ii) phenyl groups to suppress crystallization or to improve radiation resistance, or (iii) silanol end groups introduced during polymerization and used to determine number-average molecular weights, or for chemical reactions such as end linking. Some important absorbances are those for Si–O–Si groups at 1010 cm^{-1}, $Si(CH_3)_2$ groups at 800 cm^{-1}, $SiCH_3$ groups at 1260 cm^{-1}, and SiH groups at 2200 cm^{-1}. Not surprisingly, these methods are used for quality control in the commercial production of siloxane products.[1, 174]

NMR (nuclear magnetic resonance) is used for a variety of purposes, most of which parallel those used to characterize small-molecule systems.[1, 174] In addition to ^{1}H and ^{13}C NMR, ^{29}Si NMR is frequently employed. These methods are used to characterize chemical composition, structural features, and conformational preferences. NMR is also used to characterize hybrid inorganic composites, silica-type ceramics, and siloxane films.[175–176]

Specific functional groups are also analyzed by chemical methods. For example, the various chlorosilanes can be hydrolyzed and the resulting chloride ions determined by titration with silver salts. Si–H groups can be determined by measurement of the amount of hydrogen gas evolved during hydrolyses. As a final example, silanol groups can be monitored through measurement of the amount of methane gas evolved when they react with methyl Grignard reagent.[1, 174]

It is possible to determine the amount of silicon present by pyrolysis to silica, followed by atomic absorption methods.

Both mass spectrometry and gas chromatography are used to identify and determine the amounts of volatile siloxane-type materials.[1, 174] The more important nonvolatile materials, including polymers, can be characterized by liquid chromatography and by gel permeation chromatography (GPC). When used analytically, these techniques give molecular weight distributions; used preoperatively, they yield narrow molecular weight distribution fractions that are suitable for determining structure-property relationships. Average molecular weights can be determined by a variety of techniques, including dilute solution viscometry, osmometry, ultracentrifugation, and light scattering intensity measurements. Other techniques for determining molecular weight distributions include fractional precipitation and gradient elution. Extractions with supercritical fluids seem particularly promising in this regard.[177]

Thermal properties are measured and evaluated by common thermal analysis instruments. For identification of transition temperatures, heat of fusion, differential thermal analysis (DTA), and differential scanning calorimetry (DSC) are available. Thermal stability is measured by thermogravimetric analysis (TGA), although this technique can give overly optimistic results unless used with great care.

Rheological measurements are of central importance in the processing of siloxane polymers. Typical studies include determination of the dependence of the bulk viscosity on the average molecular weight, molecular weight distribution, and rate of shear. Characterization of the effects of branched chains or reinforcing fillers present is also of great importance.[2]

Most siloxane polymers are excellent insulators so electrical properties are important for many applications. Such properties include resistivity, dielectric constant, dielectric losses, dielectric strength (resistance to electrical breakdown), and power factors.[2]

The use of siloxane polymers in applications such as separation membranes, drug release systems[178–179] and blood oxygenators requires extensive permeability studies. These applications also involve measurements of diffusivity and solubility.[2]

For some specialized applications, optical properties can be of crucial importance. Two examples are contact lenses and interlayers for glass windshields. Here, transparency is of primary importance, but index of refraction (n) is important for matching values of n for polymers and fillers.[2, 180–181]

Applications in the biomedical area require extensive testing of biocompatibility,[182] including acute, dietary and implant testing, and monitoring toxicological effects such as carcinogenicity, mutagenicity, teratogenicity, and bacterial or fungal colonization.[2]

In the case of siloxane elastomers, the testing of mechanical properties is of particular importance. Elongation or tensile measurements are used almost to the exclusion of other types of mechanical tests, probably because of their simplicity. In this way, structural information is obtained about the networks, such as their degrees of cross linking.[168] Measurements of the ultimate strength (modulus at rupture), and the maximum extensibility (elongation at rupture) are also important, as is recovery after compression.[183] Relating such properties to the chemical nature of the siloxane polymer, to the curing conditions, and to the nature and amount of any reinforcing filler is a task of paramount importance in the area of elastomeric applications.[168]

Many applications of siloxane materials involve such a complicated array of properties that the ultimate evaluation has to involve a "use" test. Many surface applications, such as release coatings, are in this category. In this approach, the prospective material is tested directly under standard conditions chosen to mimic those under which the material would actually be employed.[1–2, 174] The advantage of such a test is its direct connection with the desired application. The major disadvantage results from the fact that the underlying reasons why a material fails are not uncovered in a global test. One purpose of such analysis is determination of the extent to which a polymer has degraded under a specific set of conditions.

2.7 DEGRADATION

Polysiloxanes are known to degrade in acids and in bases,[33] in the presence of clay-like materials,[184] and even in the presence of minerals such as aluminum oxide.[185]

Degradation of PDMS at elevated temperatures,[186–191] and at various levels of humidity[192] is relevant to many applications. Decomposition products have also been reported after siloxane pyrolysis.[193–194] Simulations have been carried out on the thermal decomposition of PDMS using a reactive force field,[195] or density functional theory.[196] Other studies have focused on the effects of photons,[197] high-energy protons[198] or ^{4}He ion beams and flammability.[199–202] Ablation and deposition of PDMS with x-rays have also been reported.[203] Degradation of biodegradable polysiloxanes is of particular importance,[204] as is their behavior in biomedical applications in general.[205]

REFERENCES

1. Hardman, B.; Torkelson, A., Silicones. In *Encyclopedia of Polymer Science and Engineering*, 2nd ed. Koschwitz, J. I., Ed. Wiley-Interscience: New York, 1989; pp 204–308.
2. Stark, F. O.; Falender, J. R.; Wright, A. P., Silicones. In *Comprehensive Organometallic Chemistry*, Wilkinson, G., Ed. Pergamon Press: Oxford, 1982; Vol. 2, pp 305–363.
3. Lewis, L. N., From Sand to Silicones: An Overview of the Chemistry of Silicones. In *Silicones and Silicone-Modified Materials*, Clarson, S. J.; Fitzgerald, J. J.; Owen, M. J.; Smith, S. D., Eds. American Chemical Society: Washington, DC, 2000; Vol. 729, pp 11–19.
4. Rochow, E. G., *Silicon and Silicones*. Springer-Verlag: Berlin, 1987.
5. West, R.; Barton, T. J., Organosilicon Chemistry. Part I. *J. Chem. Educ.* **1980**, *57*, 165–169.
6. Bock, H., Fundamentals of Silicon Chemistry: Molecular States of Silicon-Containing Compounds. *Angew. Chem. Int. Ed. Engl.* **1989**, *28*(12), 1627–1650.
7. Goodwin, G. B.; Kenney, M. E., A Silicate Substitution Route to Organosilicon Compounds. In *Silicon-Based Polymer Science. A Comprehensive Resource*, Zeigler, J. M.; Fearon, F. W. G., Eds. American Chemical Society: Washington, DC, 1990; Vol. 224, pp 251–263.
8. Odian, G., *Principles of Polymerization*, 4th ed. Wiley-Interscience: New York, 2004.
9. Frisch, K. C.; Reegen, S. L., *Ring-Opening Polymerization*. Marcel Dekker: New York, 1969.
10. McGrath, J. E.; Riffle, J. S.; Banthia, A. K.; Yilgor, I.; Wilkes, G. L., An Overview of the Polymerization of Cyclosiloxanes. In *Initiation of Polymerization*, Bailey, F. E., Jr., Ed. American Chemical Society: Washington, 1983; pp 145–172.
11. McGrath, J. E., Ring-Opening Polymerization: Introduction. In *Ring-Opening Polymerization*, McGrath, J. E., Ed. American Chemical Society: Washington, DC, 1985; Vol. 286, pp 1–22.
12. Kendrick, T. C.; Parbhoo, B.; White, J. W., Siloxane Polymers and Copolymers. In *The Chemistry of Organic Silicon Compounds*, Patai, S.; Rappoport, Z., Eds. Wiley: New York, 1989; pp 1289–1361.
13. Saam, J. C., Formation of Linear Siloxane Polymers. In *Silicon-Based Polymer Science. A Comprehensive Resource*, Zeigler, J. M.; Fearon, F. W. G., Eds. American. Chemical Society: Washington, DC, 1990; Vol. 224, pp 71–90.
14. Chojnowski, J., Polymerization. In *Siloxane Polymers*, Clarson, S. J.; Semlyen, J. A., Eds. Prentice Hall: Englewood Cliffs, NJ, 1993; pp 1–71.
15. Lahann, J., *Click Chemistry for Biotechnology and Materials Science*. John Wiley & Sons: Hoboken, NJ, 2009.
16. Campos, L. M.; Meinel, I.; Guino, R. G.; Schierhorn, M.; Gupta, N.; Stucky, G. D.; Hawker, C. J., Highly Versatile and Robust Materials for Soft Imprint Lithography Based on Thiolene Click Chemistry. *Adv. Mater.* **2008**, *20*, 3728–3733.
17. Hoyle, C. E.; Bowman, C. N., Thiol-Ene Click Chemistry. *Angew. Chem. Int. Ed. Engl.* **2010**, *49*, 1540–1573.
18. Lowe, A. B.; Hoyle, C. E.; Bowman, C. N., Thiol-yne Click Chemistry: A Powerful and Versatile Methodology for Materials Synthesis. *J. Mater. Chem.* **2010**, *20*, 4745–4750.

19. Qin, A.; Lam, J. W. Y.; Tang, B. Z., Click Polymerization: Progress, Challenges, and Opportunities. *Macromolecules* **2010**, *43*, 8693–8702.
20. Gonzaga, F.; Yu, G.; Brook, M. A., Polysiloxane Elastomers via Room Temperature, Metal-Free Click Chemistry. *Macromolecules* **2009**, *42*, 9220–9224.
21. Schmidt, U.; Zehetmaier, P. C.; Rieger, B., Direct Synthesis of Poly(dimethylsiloxane) Copolymers with TPE-Properties via CuAAC (Click Chemistry). *Macromol. Rapid Commun.* **2010**, *31*, 545–548.
22. Brook, M. A.; Grande, J. B.; Ganachaud, F., New Synthetic Strategies for Structured Silicones Using $B(C_6F_5)_3$. *Adv. Polym. Sci.* **2011**, *235*, 161–184.
23. Zhou, Y.; Shimizu, K.; Cha, J. N.; Stucky, G. D.; Morse, D. E., Efficient Catalysis of Polysiloxane Synthesis by Silicatein *a* Requires Specific Hydroxy and Imidazole Functionalities. *Angew. Chem. Int. Ed. Engl.* **1999**, *38*, 780–782.
24. Poojari, Y.; Clarson, S. J., Lipase-Catalyzed Synthesis and Properties of Silicone Aromatic Polyesters and Silicone Aromatic Polyamides. *Macromolecules* **2010**, *43*, 4616–4622.
25. Flory, P. J., *Statistical Mechanics of Chain Molecules*. Interscience: New York, 1969.
26. Mattice, W. L.; Suter, U. W., *Conformational Theory of Large Molecules. The Rotational Isomeric State Model in Macromolecular Systems*. Wiley: New York, 1994.
27. Rehahn, M.; Mattice, W. L.; Suter, U. W., Rotational Isomeric State Models in Macromolecular Systems. *Adv. Polym. Sci.* **1997**, *131*, 1–18.
28. Semlyen, J. A., Cyclic Polymers. *Pure Appl. Chem.* **1981**, *53*(9), 1797–1804.
29. Semlyen, J. A., *Cyclic Polymers*. Elsevier: London, 1986.
30. Clarson, S. J.; Semlyen, J. A., *Siloxane Polymers*. Prentice Hall: Englewood Cliffs, NJ, 1993.
31. Semlyen, J. A.; Wood, B. R.; Hodge, P., Cyclic Polymers: Past, Present and Future. *Polym. Adv. Technol.* **1994**, *5*, 473–478.
32. Dagger, A. C.; Semlyen, J. A., Studies of Cyclic and Linear Poly(dimethylsiloxanes): 34. Preparation, Fractionation and Characterization of the First Per-Deuterated Macrocyclic Poly(dimethylsiloxane). *Polym. Commun.* **1999**, *40*, 3243–3245.
33. Clarson, S. J.; Fitzgerald, J. J.; Owen, M. J.; Smith, S. D.; van Dyke, M. E., *Synthesis and Properties of Silicones and Silicone-Modified Materials*. American Chemical Society: Washington, DC, 2003; Vol. 838.
34. Andrianov, K. A.; Vardosanidze, T. N.; Nogaideli, A. I.; Yakushkina, S. Y., Anionic Polymerization of Methylphenylcyclosiloxanes. *Polym. Sci. U.S.S.R.* **1968**, *10*, 301–307.
35. Molenberg, A.; Moller, M., A Fast Catalyst System for the Ring-Opening Polymerization of Cyclosiloxanes. *Macromol. Rapid Commun.* **1995**, *16*, 449–453.
36. Kress, J. D.; Leung, P. C.; Tawa, G. J.; Hay, P. J., Calculation of a Reaction Path for KOH Catalyzed Ring-Opening Polymerization of Hexamethylcyclotrisiloxane. *J. Am. Chem. Soc.* **1997**, *119*, 1954–1960.
37. Brook, M. A., *Silicon in Organic, Organometallic, and Polymer Chemistry*. John Wiley & Sons: New York, 2000.
38. Kress, J. D.; Leung, P. C.; Tawa, G. J.; Hay, P. J., Calculation of a Reaction Path for KOH-Catalyzed Ring-Opening Polymerization of Cyclic Siloxanes. In *Silicones and Silicone-Modified Materials*, Clarson, S. J.; Fitzgerald, J. J.; Owen, M. J.; Smith, S. D., Eds. American Chemical Society: Washington, DC, 2000; Vol. 729, pp 81–97.

39. Dubois, P.; Coulembier, O.; Raquez, J.-M., *Handbook of Ring-Opening Polymerizations*. Wiley-VCH: Weinheim, 2009.
40. Ninago, M. D.; Satti, A. J.; Ressia, J. A.; Ciolino, A. E.; Villar, M. A.; Valles, E. M., Controlled Synthesis of Poly(dimethylsiloxane) Homopolymers Using High-Vacuum Anionic Polymerization Techniques. *J. Polym. Sci., Part A: Polym. Chem.* **2009**, *47*, 4774–4783.
41. Chojnowski, J.; Kurjata, J.; Fortuniak, W.; Rubinsztasjn, S.; Trzebicka, B., Hydride Transfer Ring-Opening Polymerization of a Cyclic Oligomethylhydrosiloxane. Route to a Polymer of Closed Multicyclic Structure. *Macromolecules* **2012**, *45*, 2654–2661.
42. Petr, M.; Katzman, B. A.; DiNatale, W.; Hammond, P. T., Synthesis of a New, Low-T_g Siloxane Thermoplastic Elastomer with a Functionalizable Backbone and Its Use as a Rapid, Room Temperature Photoactuator. *Macromolecules* **2013**, *46*(7), 2823–2832.
43. Wilczek, L.; Rubinsztajn, S.; Chojnowski, J., Comparison of the Cationic Polymerization of Octamethylcyclotetrasiloxane and Hexamethylcyclotrisiloxane. *Makromol. Chem.* **1986**, *187*(1), 39–51.
44. Crivello, J. V.; Malik, R., Synthesis and Photoinitiated Cationic Polymerization of Monomers Containing the Silsesquioxane Core. In *Silicones and Silicone-Modified Materials*, Clarson, S. J.; Fitzgerald, J. J.; Owen, M. J.; Smith, S. D., Eds. American Chemical Society: Washington, DC, 2000; Vol. 729, pp 284–295.
45. Apedaile, A.; Liggat, J.; Parkinson, J.; Nikiforidis, G.; Berlouis, L.; Patel, M., Lewis-Acid Mediated Polymerization of Poly(dimethylsiloxane) Polymers: Investigating Reaction Kinetics Using Both NMR Spectroscopy and Cyclic Voltammetry. *J. Appl. Polym. Sci.* **2012**, *123*, 2601–2608.
46. Qayouh, H.; Lahcini, M.; Six, J.-L.; Kricheldorf, H. R., Polymerizations of Hexamethylcyclosiloxane Catalyzed by Metal Sulfoniate/Acid Chloride Combinations. *J. Appl. Polym. Sci.* **2012**, *124*, 4114–4120.
47. Kuo, C. M.; Clarson, S. J., On the Role of Tin Catalysts in Step-Growth Polymerization Reactions. *J. Inorg. Organomet. Polym. Mater.* **2012**, *22*, 577–587.
48. Elias, H. G., *Macromolecules: Synthesis and Materials*. Plenum Press: New York, 1977; Vol. 2.
49. Hreczycho, G.; Pawluc, P.; Marciniec, B., A New Selective Approach to Unsymmetrical Siloxanes and Germasiloxanes *via* O-Metalation of Silanols with 2-Methylallylsilanes and 2-Methylallylgermanes. *New J. Chem.* **2011**, *35*, 2743–2746.
50. Nguyen, K. A. T.; Clarke, S. R.; Matisons, J.; Skelton, B. W.; White, A. H.; Markovic, E., Synthesis of Strictly Alternating Siloxane Copolymers and Cyclics. *Silicon* **2014**, *6*(1), 21–26.
51. Yilgor, I.; Riffle, J. S.; McGrath, J. E., Reactive Siloxane Oligomers. In *Reactive Oligomers*, Harris, F. W.; Spinelli, H. J., Eds. American Chemical Society: Washington, DC, 1985; Vol. 282, pp 161–174.
52. McGrath, J. E.; Sormani, P. M.; Elsbernd, C. S.; Kilic, S., Kinetics, Mechanisms, and Synthesis Studies of Difunctional Aminopropyl Terminated Polydimethylsiloxane Oligomers. *Macromol. Chem., Macromol. Symp.* **1986**, *6*, 67–80.
53. Gretton, M. J.; Kamino, B. A.; Brook, M. A.; Bender, T. P., The Use of Piers-Rubinsztajn Conditions for the Placement of Triarylamines Pendant to Silicone Polymers. *Macromolecules* **2012**, *45*, 723–728.

54. Cohen, C.; Damiron, D.; Dkhil, S. B.; Drockenmuller, E., Synthesis of Well-Defined Poly(dimethylsiloxane) Telechelics Having Nitrobenzoxadiazole Fluorescent Chain-Ends via Thiol-Ene Coupling. *J. Polym. Sci., Part A: Polym. Chem.* **2012**, *50*, 1827–1833.
55. Stevens, N. S. M.; Rezac, M. E., Nanoporous Organic/Inorganic Hybrid Materials Produced from Poly(dimethylsiloxane). *Polymer* **1999**, *40*, 4289–4298.
56. Ndoni, S.; Vigild, M. E.; Berg, R. H., Nanoporous Materials with Spherical and Gyroid Cavities Created by Quantitative Etching of Polydimethylsiloxane in Polystyrene-Polydimethylsiloxane Block Copolymers. *J. Am. Chem. Soc.* **2003**, *125*, 13366–13367.
57. Gunji, T.; Shimano, R.; Arimitsu, K.; Abe, Y., Preparation and Characterization of Mesoporous Silica Thin Films from Polyethoxysiloxanes. *J. Polym. Sci., Part B: Polym. Phys.* **2006**, *44*, 2542–2550.
58. Shibuya, M.; Takahashi, T.; Koyama, K., Microcellular Ceramics by Using Silicone Preceramic Polymer and PMMA Polymer Sacrificial Microbeads. *Compos. Sci. Tech.* 2007, *67*, 119–124.
59. Grosse, M.-T.; Lamotte, M.; Birot, M.; Deleuze, H., Preparation of Microcellular Polysiloxane Monoliths. *J. Polym. Sci., Part A: Polym. Chem.* **2007**, *46*, 21–32.
60. Wong, K. H.; Stenzel, M. H.; Duvall, S.; Ladouceur, F., Exploitable Flexible Honeycomb Structured Porous Films from Sol-Gel Cross-Linkable Silicone Based Random Branched Copolymers. *Chem. Mater.* **2010**, *22*, 1878–1891.
61. Soraru, G. D.; Dalcanale, F.; Campostrini, R.; Gaston, A.; Blum, Y.; Carturan, S.; Aravindi, P. R., Novel Polysiloxane and Polycarbosilane Aerogels *via* Hydrosilylation of Preceramic Polymers. *J. Mater. Chem.* **2012**, *22*, 7676–7680.
62. Huerta, L.; Guillem, C.; Latorre, J.; Beltrán, A.; Martínez-Máñez, R.; Marcos, M. D.; Beltrán, D.; Amorós, P., Bases for the Synthesis of Nanoparticulated Silicas with Bimodal Hierarchical Porosity. *Solid State Sciences* **2006**, *8*(8), 940–951.
63. Stebe, M. J.; Emo, M.; Forny-Le Follotec, A.; Medas-Komunjer, L.; Pezron, I.; Blin, J. L., Triblock Siloxane Copolymer Surfactant: Template for Spherical Mesoporous Silica with a Hexagonal Pore Ordering. *Langmuir* **2013**, *29*(5), 1618–1626.
64. Lettow, J. S.; Han, Y. J.; Schmidt-Winkel, P.; Yang, P. D.; Zhao, D. Y.; Stucky, G. D.; Ying, J. Y., Hexagonal to Mesocellular Foam Phase Transition in Polymer-Templated Mesoporous Silicas. *Langmuir* **2000**, *16*(22), 8291–8295.
65. Schmidt-Winkel, P.; Lukens Wayne, W.; Yang, P.; Margolese, D. I.; Lettow, J. S.; Ying, J. Y.; Stucky, G. D., Microemulsion Templating of Siliceous Mesostructured Cellular Foams with Well-Defined Ultralarge Mesopores. *Chem. Mater.* **2000**, *12*, 686–696.
66. Goltner, C. G.; Smarsly, B.; Berton, B.; Antonietti, M., On the Microporous Nature of Mesoporous Molecular Sieves. *Chem. Mater.* **2001**, *13* (5), 1617–1624.
67. Kuzmicz, D.; Kepczynski, M.; Bednar, J.; Mistrikova, V.; Jachimsko, B.; Nowakowska, M., Silicone Nano/Microstructures Obtained in Ionic Polymerization. *Macromol. Symp.* **2011**, *308*, 43–48.
68. Bier, A. K.; Bognitzki, M.; Schmidt, A.; Greiner, A.; Gallo, E.; Klack, P.; Schartel, B., Synthesis, Properties and Processing of New Siloxane-Substituted Poly(*p*-xylene) via CVD. *Macromolecules* **2012**, *45*, 633–639.

69. Cheng, H. N.; Gross, R. A., *Green Polymer Chemistry: Biocatalysis and Biomaterials*. American Chemical Society: Washington, D C, 2010; Vol. 1043.
70. Matyjaszewski, K.; Miller, P. J.; Kickelbick, G.; Nakagawa, Y.; Diamanti, S.; Pacis, C., Organic-Inorganic Hybrid Polymers from Atom Transfer Radical Polymerization and Poly(dimethylsiloxane). In *Silicones and Silicone-Modified Materials*, Clarson, S. J.; Fitzgerald, J. J.; Owen, M. J.; Smith, S. D., Eds. American Chemical Society: Washington, DC, 2000; Vol. 729, pp 270–283.
71. Kuo, C.-M.; Battjes, K. J.; Miller, R. L.; Saam, J. C., Strain-Induced Crystallization in Stereoregular Poly[(methyl(3,3,3-trifluoropropyl)siloxane] Networks. *Rubber Chem. Technol.* **1997**, *70*, 769–780.
72. Saam, J. C., Stereoregular Polysiloxanes via Ring-Opening Polymerization, a Review. *J. Inorg. Organomet. Polym.* **1999**, *9*, 1–16.
73. Kawakami, Y.; Imae, I., Functionality and Stereochemical Design of Silicon-Containing Polymers. *Macromol. Chem. Phys.* **1999**, *200*, 1245–1256.
74. Kawakami, Y., Control of the Stereochemical structure of Silicon-Containing Polymers. *Macromol. Symp.* **2001**, *174*, 145–153.
75. Tang, H.; Sun, J.; Zhou, X.; Fu, P.; Xie, P.; Zhang, R., Hydrogen-Bonding-Directed Template Synthesis of Novel Stereo-Regular Organo-Bridged Ladder-Like Polymethylsiloxane. *Macromol. Chem. Phys.* **2003**, *204*, 155–163.
76. Tebeneva, N. A.; Rebrov, E. A.; Muzafarov, A. M., Synthesis of Organosilicon Linear Polymer with Organooctasilsesquioxanes Structural Units in Main Chain. In *Silicones and Silicone-Modified Materials*, Clarson, S. J.; Fitzgerald, J. J.; Owen, M. J.; Smith, S. D., Eds. American Chemical Society: Washington, DC, 2000; Vol. 729, pp 214–225.
77. Quintana, R.; Gosa, M.; Janczewski, D.; Kutnyanszky, E.; Vancso, G. J., Enhanced Stability of Low Fouling Zwitterionic Polymer Brushes in Seawater with Diblock Architecture. *Langmuir* **2013**, *29*(34), 10859–10867.
78. Gravier, D.; Huebner, D. J.; Saam, J. C., Emulsion Polymerized Polydimethylsiloxane. *Rubber Chem. Technol.* **1983**, *56*, 918–926.
79. Barrere, M.; De Silva, S. C.; Balic, R.; Ganachaud, F., Synthesis of Monodisperse Poly(dimethylsiloxane) Micro- and Macroemulsions. *Langmuir* **2002**, *18*, 941–944.
80. Landfester, K.; Pawelzik, U.; Antonietti, M., Polydimethylsiloxane Latexes and Copolymers by Polymerization and Polyaddition in Miniemulsion. *Polymer* **2005**, *46*, 9892–9898.
81. Zhang, D.; Jiang, X.; Yang, C., Preparation of Polydimethylsiloxane Nanolatices by Emulsion Polymerization in a Water-Aminoethanol System. *J. Appl. Polym. Sci.* **2005**, *98*, 347–352.
82. Sharma, S. C.; Tsuchiya, K.; Sakai, K.; Sakai, H.; Abe, M.; Komura, S.; Sakamoto, K.; Miyahara, R., Formation and Characterization of Microemulsions Containing Silicone. *Langmuir* **2008**, *24*, 7658–7662.
83. Jiang, S.; Qiu, T.; Li, X., Kinetic Study on the Ring-Opening Polymerizatioin of Octamethylcyclotetrasiloxane (D4) in Miniemulsion. *Polymer* **2010**, *51*, 4087–4094.
84. Mohoric, I.; Sebenik, L., Anionic Ring-Opening Polymerization of Octamethylcyclotetrasiloxane in Emulsion Above Critical Micelle Concentration. *Polymer* **2011**, *52*, 1234–1240.
85. Mohoric, I.; Sebenik, U., Semibatch Anionic Ring-Opening Polymerization of Octamethylcyclotetrasiloxane in Emulsion. *Polymer* **2011**, *52*, 4421–4428.

86. Saam, J. C.; Gravier, D.; Baile, M., Room-Temperature-Cured Polydimethylsiloxane Elastomers from Aqueous Dispersion. *Rubber Chem. Technol.* **1981**, *54*, 976–987.
87. Hariri, K.; Akhrass, S. A.; Delaite, C.; Moireau, P.; Riess, G., Polymerizable Oil-in-Oil Emulsions: Poly(vinyl pyrolidone) Dispersions in Reactive PDMS Medium. *Polym. Int.* **2007**, *56*, 1200–1205.
88. Terwagne, D.; Gilet, T.; Vandewalle, N.; Dorbolo, S., From a Bouncing Compound Drop to a Double Emulsion. *Langmuir* **2010**, *26*, 11680–11685.
89. Rollings, D.-A.; Veinot, J. G. C., Polysiloxane Nanofibers via Surface Initiated Polymerization of Vapor Phase Reagents: A Mechanism of Formation and Variable Wettability of Fiber-Bearing Substrates. *Langmuir* **2008**, *24*, 13653–13662.
90. Oka, T.; Ito, K.; Muramatsu, M.; Ohdaira, T.; Suzuki, R.; Kobayashi, Y., Porogen Approach for the Fabrication of Plasma-Polymerized Nanoporous Polysiloxane Films. *J. Phys. Chem. B* **2006**, *110*, 20172–20176.
91. Poojari, Y.; Palsule, A. S.; Cai, M.; Clarson, S. J.; Gross, R. A., Synthesis of Organosiloxane Copolymers Using Enzymatic Polyesterification. *Eur. Polym. J.* **2008**, *44*, 4139–4145.
92. Kobayashi, S.; Makino, A., Enzymatic Polymer Synthesis: An Opportunity for Green Polymer Chemistry. *Chem. Rev.* **2009**, *109*, 5288–5353.
93. Poojari, Y.; Clarson, S. J., Lipase Catalyzed Synthesis of Silicone Polyesters. *Chem. Commun.* **2009**, *44*, 6834–6835.
94. Poojari, Y.; Clarson, S. J., Lipase Catalyzed Synthesis and Thermal Properties of Poly(dimethylsiloxane)-Poly(ethylene glycol) Amphiphilic Block Copolymers. *J. Inorg. Organomet. Polym.* **2010**, *20*, 46–52.
95. Folk, S. L.; Desimone, J. M., Polysiloxanes in Compressed Carbon Dioxide. In *Synthesis and Properties of Silicones and Silicone-Modified Materials*, Clarson, S. J.; Fitzgerald, J. J.; Owen, M. J.; Smith, S. D.; Van Dyke, M. E., Eds. American Chemical Society: Washington, DC, 2003; Vol. 838, pp 79–93.
96. Habaue, S.; Sato, K.; Yamashita, K.; Shimamura, T.; Kaito, M.; Masuda, T.; Kajiwara, M., Polysiloxanes Derived from Chrysotile Asbestos via Acid-Leaching and Silylation Processes. *J. Appl. Polym. Sci.* **2008**, *110*, 2891–2897.
97. Chaikittisilp, W.; Sugawara, A.; Shimojima, A.; Okubo, T., Microporous Hybrid Polymer with a Certain Crystallinity Built from Functionalized Cubic Siloxane Cages as a Singular Building Unit. *Chem. Mater.* **2010**, *22*, 4841–4843.
98. Mark, J. E., Silicon-Containing Polymers. In *Silicon-Based Polymer Science. A Comprehensive Resource*, Zeigler, J. M.; Fearon, F. W. G., Eds. American Chemical Society: Washington, DC, 1990; Vol. 224, pp 47–68.
99. Dvornic, P. R.; Jovanovic, J. D.; Govedarica, M. N., On the Critical Molecular Chain Length of Polydimethylsiloxane. *J. Appl. Polym. Sci.* **1993**, *49*, 1497–1507.
100. Oberhammer, H.; Boggs, J. E., Importance of (p-d)π Bonding in the Siloxane Bond. *J. Am. Chem. Soc.* **1980**, *102*, 7241–7244.
101. Lukevics, E.; Pudova, O.; Sturkovich, R., *Molecular Structure of Organosilicon Compounds.* John Wiley & Sons: New York, 1989.
102. DeBolt, L. C.; Mark, J. E., Effects of Bond-Angle Inversion on the Statistical Properties of Poly(dimethylsiloxane). *J. Polym. Sci., Polym. Phys. Ed.* **1988**, *26*, 989–995.
103. Hartung, H. A.; Berger, S. E., Crosslinkinig of Vinylsilicone Resins by Dicumyl Peroxide. *J. Appl. Polym. Sci.* **1962**, *6*, 474–479.

104. Bobear, W. J., Silicone Rubber. In *Rubber Technology*, Morton, M., Ed. Van Nostrand Reinhold: New York, 1973; pp 368–406.
105. James, P. M.; Barrall, E. M.; Dawson, B.; Logan, J. A., Crosslinking of Methyl Silicone Rubbers. 2. Analysis of Extractables from Samples Crosslinked under Various Conditions. *J. Macromol. Sci., Part A: Chem.* **1974,** *A 8* (1), 135–155.
106. Barrall II, E. M.; Flandera, M. A.; Logan, J. A., A Thermodynamic Study of the Crosslinking of Methyl Silicone Rubber. *Thermochim. Acta* **1973,** *5,* 415–432.
107. Studebaker, M. L.; Beatty, J. R., The Rubber Compound and Its Composition. In *Science and Technology of Rubber,* Eirich, F. R., Ed. Academic Press: New York, 1978, pp 366–418.
108. Martin, J. E.; Schaefer, D. W.; Hurd, A. J., Fractal Geometry of Vapor-Phase Aggregates. *Physical Rev. A* **1986,** *33*(5), 3540–3543.
109. Schaefer, D. W.; Hurd, A. J., Growth and Structure of Combustion Aerosols: Fumed Silica. *Aerosol Sci. Tech.* **1990,** *12*(4), 876–890.
110. Schaefer, D. W.; Chen, C., Structure Optimization in Colloidal Reinforcing Fillers: Precipitated Silica. *Rubber Chem. Technol.* **2002**, *75*(5), 773–793.
111. Milanesi, F.; Cappelletti, G.; Annunziata, R.; Bianchi, C. L.; Meroni, D.; Ardizzone, S., Siloxane-TiO_2 Hybrid Nanocomposites. The Structure of the Hydrophobic Layer. *J. Phys. Chem. C* **2010,** *114,* 8287–8293.
112. Plueddemann, E. D., *Silane Coupling Agents.* Plenum Press: New York, 1982.
113. Vu, B. T. N.; Mark, J. E.; Schaefer, D. W., Interfacial Modification for Controlling Silica-Polysiloxane Interactions and Bonding in Some Elastomeric Composites. *Compos. Interfaces* **2003**, *10*, 451–473.
114. Schaefer, D. W.; Vu, B. T. N.; Mark, J. E., The Effect of Interphase Coupling on the Structure and Mechanical Properties of Silica-Siloxane Composites. *Rubber Chem. Technol.* **2002**, *75*, 795–810.
115. Erman, B.; Mark, J. E., *Structures and Properties of Rubberlike Networks.* Oxford University Press: New York, 1997.
116. Schaefer, D. W.; Jian, L.; Sun, C.-C.; McCarthy, D. W.; Jiang, C.-Y.; Ning, Y.-P.; Mark, J. E.; Spooner, S., Structure-Property Relationships in Silica-Siloxane Molecular Composites. In *Ultrastructure Processing of Advanced Materials*, Uhlmann, D. R.; Ulrich, D. R., Eds. Wiley: New York, 1992; pp 361–375.
117. Coran, A. Y., Vulcanization. In *Science and Technology of Rubber, 4th Edition,* Mark, J. E.; Erman, B.; Roland, C. M. Eds. Academic: New York, 1978, pp 337–381.
118. Brydson, J. A., *Rubber Chemistry.* Applied Science Publishers: London, 1978.
119. Thomas, D. R., Cross-Linking Reactions. In *Siloxane Polymers,* Clarson, S. J.; Semlyen, J. A., Eds. Prentice Hall: Englewood Cliffs, 1993; pp 567–615.
120. Coran, A. Y., Chemistry of the vulcanization and protection of elastomers: A review of the achievements. *J. Appl. Polym. Sci.* **2003,** *87,* 24–30.
121. Damiron, D.; Okhay, N.; Akhrass, S. A.; Cassagnau, P.; Drockenmuller, E., Crosslinked PDMS Elastomers and Coatings from the Thermal Curing of Vinyl-Functionalized PDMS and a Diazide Aliphatic Crosslinker. *J. Polym. Sci., Part A: Polym. Chem.* **2012,** *50,* 98–107.
122. Steffanut, P.; Osborn, J. A.; DeCian, A.; Fisher, J., Efficient Homogeneous Hydrosilylation of Olefins by Use of Complexes of Pt^0 with Selected Electron-Deficient Olefins as Ligands. *Chem. Eur. J.* **1998**, *4*, 2008–2017.
123. De Buyl, F.; Comyn, J.; Shephard, N. E.; Subramaniam, N. P., Kinetics of Cure, Coss Link Density and Adhesion of Water-Reactive Alkoxysilicone Sealants. *Adhesion Sci. Technol.* **2002,** *16,* 1055–1071.

124. Boileau, S.; Bouteiller, L.; Kowalewska, A., Polysiloxanes Containing Cross-linkable Diacetylene Units in the Main Chain. *Polym. Int.* **2004**, *53*, 191–197.
125. Matloka, P. P.; Sworen, J. C.; Zuluaga, F.; Wagener, K. B., Chain-End and Chain-Internal Crosslinking in "Latent Reactive" Silicon Elastomers. *Macromol. Chem. Phys.* **2005**, *206*, 218–228.
126. Yin, Y.; Zhang, Y.; Chi, Z.; Xu, J., Self-Catalytic Cross-Linking Reaction and Reactive Mechanism Studies of α-Aminomethyl Triethoxysilanes for α,ω-Dihydroxy Poly(dimethylsiloxane) *Eur. Polym. J.* **2008**, *44*, 2284–2294.
127. Han, Y.; Zhang, J.; Yang, Q.; Shi, L.; Qi, S.; Jin, R., Novel Polymethoxysiloxane-Based Crosslinking Reagent and Its In-Situ Improvement for Thermal anad Mechanical Properties of Siloxane Elastomer. *J. Appl. Polym. Sci.* **2008**, *107*, 3788–3795.
128. Dai, L.; Zhang, Z.; Zhao, Y.; Xie, Z., Polymeric Curing Agent Reinforced Silicone Rubber Composites with Low Viscosity and Low Volume Shrinkage. *J. Appl. Polym. Sci.* **2008**, *110*, 1624–1631.
129. Esteves, A. C. C.; Brokken-Zijp, J.; Laven, J.; Huinink, H. P.; Reuvers, N. J. W.; Van, M. P.; With, G. D., Influence of Cross-Linker Structure Concentration on the Cross-Linking of PDMS and the Network Structures Formed. *Polymer* **2009**, *50*, 3955–3966.
130. Dai, L.; Zhang, Z.; Zhao, Y.; Liu, H.; Xie, Z., Effects of Polymeric Curing Agent Modified with Silazanes on the Mechanical Properties of Silicone Rubber. *J. Appl. Polym. Sci.* **2009**, *111*, 1057–1062.
131. Racles, C.; Cazacu, M.; Hitruc, G.; Hamaide, T., On the Feasibility of Chemical Reactions in the Presence of Siloxane-Based Surfactants. *Coll. Polym. Sci.* **2009**, *287*, 461–470.
132. Kade, M. J.; Burke, D. J.; Hawker, C. J., The Power of Thiol-ene Chemistry. *J. Polym. Sci., Part A: Polym. Chem.* **2010**, *48*, 743–750.
133. Matloka, P. R.; Kean, Z.; Wagener, K. B., Chain Internal/Chain End Latent Crosslinking in Thermoset Polymer Systems. *J. Polym. Sci., Part A: Polym. Chem.* **2010**, *48*, 1866–1877.
134. Chen, D.; Yi, S.; Wu, W.; Zhong, Y.; Liao, J.; Huang, C., Synthesis and Characterization of Novel Room Temperature Vulcanized (RTV) Silicone Rubbers Using Vinyl-POSS Derivatives as Cross Linking Agents. *Polymer* **2010**, *51*, 3867–3878.
135. Stiubianu, G.; Racles, C.; Cazacu, M.; Simionescu, B. C., Silicone-Modified Cellulose. Crosslinking of Cellulose Acetate with Poly[dimethyl(methyl-H)siloxane] by Pt-Catalyzed Dehydrogenative Coupling. *J. Mater. Sci.* **2010**, *45*, 4141–4150.
136. Baquey, G.; Moine, L.; Degueil-Castaing, M.; Lartigue, J.-C.; Maillard, B., Decomposition of Di-tert-butyl Peroxide in Siloxane: An Approach of the Free Radical Cross-Linking of Silicones. *Macromolecules* **2005**, *38*, 9571–9583.
137. Baquey, G.; Moine, L.; Babot, O.; Degueil, M.; Maillard, B., Model Study of the Crosslinking of Polydimethylsiloxanes by Peroxides. *Polymer* **2005**, *38*, 6283–6292.
138. Mani, S.; Cassagnau, P.; Bousmina, M.; Chaumont, P., Rheological Modeling of the Free-Radical Crosslinking of PDMS Rubber in the Presence of TEMPO Nitroxide. *Polymer* **2010**, *51*, 3918–3925.
139. Chu, H. K., UV Curable Silicones from Acryloxymethyldimethylacryloxysilane. In *Silicones and Silicone-Modified Materials*, Clarson, S. J.; Fitzgerald, J. J.;

Owen, M. J.; Smith, S. D., Eds. American Chemical Society: Washington, DC, 2000; Vol. 729, pp 170–179.
140. Dworak, D. P.; Soucek, M. D., Synthesis of Cycloaliphatic Substituted Silane Monomers and Polysiloxanes for Photocuring. *Macromolecules* **2004**, *37*, 9402–9417.
141. Chen, J.; Vaino, A. R.; Smith, R. L.; Collins, S. C., Photomediated Crosslinking of Cinnamated PDMS for In-Situ Direct Photopatterning. *J. Polym. Sci., Part A: Polym. Chem.* **2008**, *46*, 3482–3487.
142. Wang, S.-J.; Fan, X.-D.; Kong, J.; Lu, J.-R., Synthesis, Characterization and UV Curing Kinetics of Hyperbranched Polysiloxanes from A2 and CB2 Type Monomers. *Polymer* **2009**, *50*, 3587–3594.
143. Miller, A. A., Radiation Chemistry of Polydimethylsiloxane. I. Crosslinking and Gas Yields. *J. Am. Chem. Soc.* **1960**, *82*, 3519–3523.
144. Miller, A. A., Radiation Chemistry of Polydimethylsiloxane. II. Effects of Additives. *J. Am. Chem. Soc.* **1960**, *83*, 31–36.
145. Hill, D. T.; Preston, C. M. L.; Whittaker, A. K.; Hunt, S. M., The Radiation Chemistry of Poly(dimethylsiloxane). *Macromol. Symp.* **2000**, *156*, 95–102.
146. Ormerod, M. G.; Charlesby, A., The Radiation Chemistry of Some Polysiloxanes: An Electron Spin Resonance Study. *Polymer* **1963**, *4*, 459–470.
147. Jiang, Z.; Zhang, J.; Feng, S., Effects of Polymethylvinylsilicone Oil with Side Tetraphenylphenyl Groups on the Radiation Resistance of Addition-Type Silicone Rubber. *J. Appl. Polym. Sci.* **2007**, *104*, 4144–4148.
148. Satti, A. J.; Andreucetti, N. A.; Ressia, J. A.; Vallat, M. F.; Sarmoria, C.; Valles, E. M., Modelling Molecular Weight Changes Induced in Polydimethylsiloxane by Gamma and Electron Beam Irradation. *Eur. Polym. J.* **2008**, *44*, 1548–1555.
149. Wu, C.; Lei, W.; Gao, D.; Jia, M., Study of Resistance of Silicone Resin to Heat and Irradiation. *Polym. Plast. Technol. Eng.* **2009**, *48* (1094–1100), 158–163.
150. Palsule, A. S.; Clarson, S. J.; Widenhouse, C. W., Gamma Irradiation of Silicones. *J. Inorg. Organomet. Polym.* **2008**, *18*, 207–221.
151. Kowalewska, A.; Kupcik, J.; Pola, J.; Stanczyk, W. A., Laser Irradiation of Oligosiloxane Copolymer Thin Films Functionalized with Side Chain Bulky Carbosilane Moieties. *Polymer* **2008**, *49*, 857–866.
152. Mani, S.; Cassagnau, P.; Bousmina, M.; Chaumont, P., Cross-Linking Control of PDMS Rubber at High Temperatures Using TEMPO Nitroxide. *Macromolecules* **2009**, *42*, 8460–8467.
153. Chakraborty, R.; Soucek, M. D., Mechanical and Film Properties of Thermally Curable Polysiloxane. *J. Appl. Polym. Sci.* **2010**, *115*, 358–369.
154. Crowe-Willoughby, J. A.; Weiger, K. L.; Ozcam, A. E.; Genzer, J., Formation of Silicone Elastomer Network Films with Gradients in Modulus. *Polymer* **2010**, *51*, 763–773.
155. Corcione, C. E.; Frigione, M., Factors Influencing Photo Curing Kinetics of Novel Uv-Cured Siloxane-Modified Acrylic Coatings: Oxygen Inhibition and Composition. *Thermochim. Acta* **2012**, *534*, 21–27.
156. Longuet, C.; Ratsimihety, A.; Andre, S.; Boutevin, G.; Guida-Pietrasanta, F.; Deschamps, B.; Ramonda, M.; Joly-Duhamel, C.; Boutevin, B.; Ganachaud, F., Physically Crosslinked Fluorosilicone Elastomers Obtained by Self-Assembly and Template Polycondensation of Tailored Building blocks. *J. Mater. Chem.* **2010**, *20*, 10269–10276.
157. McConville, C.; Andrews, G. P.; Laverty, T. P.; Woolfson, A. D.; Malcolm, R. K., Rheological Evaluation of the Isothermal Cure Characteristics of Medical Grade Silicone Elastomers. *J. Appl. Polym. Sci.* **2010**, *116*, 2320–2327.

158. White, C. C.; Hunston, D. L.; Williams, R. S., Studies on the Effect of Movement During the Cure on the Mechanical Properties of a Silicone Joint Sealant. *Polym. Eng. Sci.* **2010**, *50*, 113–119.
159. Lopez, L. M.; Cosgrove, A. B.; Hernandez-Ortiz, J. P.; Osswald, T. A., Modeling the Vulcanization Reaction of Silicone Rubber. *Polym. Eng. Sci.* **2007**, *47*, 675–683.
160. Miki, K.; Inamoto, Y.; Inoue, S.; Uno, T.; Itoh, T.; Kubo, M., Preparation of Three-Dimensional Poly(dimethylsiloxane) (PDMS) with Movable Cross-Linking. *J. Polym. Sci., Part A: Polym. Chem.* **2009**, *47*, 5882–5890.
161. Yu, T.; Wakuda, K.; Blair, D. L.; Weiss, R. G., Reversible Cross-Linking Amino-Polysiloxanes by Simple Triatomic Molecules. Facile Methods for Tuning Thermal, Rheological, and Adhesive Properties. *J. Phys. Chem. C* **2009**, *113*, 11546–11553.
162. Cho, S. H.; Andersson, H. M.; White, S. R.; Sottos, N. R.; Braun, P. V., Polydimethylsiloxane-Based Self-Healing Materials. *Adv. Mater.* **2006**, *18*, 997–1000.
163. Keller, M. W.; White, S. R.; Sottos, N. R., Torsion Fatigue Response of Self-Healing Poly(dimethylsiloxane) Elastomers. *Polymer* 2008, *49*, 3136–3145.
164. Mangun, C. L.; Mader, A. C.; Sottos, N. R.; White, S. R., Self-Healing of a High Temperature Cured Epoxy Using Poly(dimethylsiloxane) Chemistry. *Polymer* **2010**, *51*, 4063–4068.
165. Kohjiya, S.; Urayama, K.; Ikeda, Y., Poly(siloxane) Network of Ultra-High Elongation. *Kautschuk Gummi Kunststoffe* **1997**, *50*, 868–872.
166. Urayama, K.; Kohjiya, S., Extensive Stretch of Polysiloxane Network Chains with Random and Super-Coiled Conformations. *Eur. Phys. J. B* **1998**, *2*, 75–78.
167. Premachandra, J.; Mark, J. E., Effects of Dilution During Cross Linking on Strain-Induced Crystallization in *cis*-1,4-Polyisoprene Networks. 1. Experimental Results. *J. Macromol. Sci., Pure Appl. Chem.* **2002**, *39*, 287–300.
168. Mark, J. E.; Erman, B., *Rubberlike Elasticity: A Molecular Primer*, 2nd ed. Cambridge University Press: Cambridge, 2007.
169. Wu, Y.; McGarry, F. J.; Zhu, B.; Keryk, J. R.; Katsoulis, D. E., Temperature Effects on Mechanical Properties of Toughened Silicone Resins. *Polym. Eng. Sci.* **2005**, *45*, 1522–1531.
170. Hotta, Y.; Komatsu, K.; Wang, F. W., Novel Fluorescence Method for Cure Monitoring of Hydrosilation-Curable Silicones. *J. Appl. Polym. Sci.* **2009**, *112*, 2441–2444.
171. Maxwell, R. S.; Chinn, S. C.; Solyom, D.; Cohenour, R., Radiation-Induced Cross-Linking in a Silica-Filled Silicone Elastomer as Investigated by Multiple Quantum ^{1}H NMR. *Macromolecules* **2005**, *38*, 7026–7032.
172. Mark, J. E., Some Interesting Things About Polysiloxanes. *Acc. Chem. Res.* **2004**, *37*, 946–953.
173. Hernandez-Ortiz, J. P.; Osswald, T. A., Modeling Processing of Silicone Rubber: Liquid Versus Hard Silicone Rubbers. *J. Appl. Polym. Sci.* **2011**, *119*, 1864–1871.
174. Smith, A. L., *Analysis of Silicones*. Krieger Publishing Co.: Malabar, FL, 1983.
175. Metroke, T. L.; Wang, Y.; Ooij, W. J.; Schaefer, D. W., Chemistry of Mixtures of Bis-[trimethoxysilylpropyl]amine and Vinyltriacetoxysilane: An NMR analysis. *J. Sol-Gel Sci. Technol.* **2009**, *51* (1), 23–31.
176. Schaefer, D. W.; Wang, Y.; Metroke, T. L., Metal-Protective Films Based on Mixed Silane Coupling Agents. In *Silanes and Other Coupling Agents*, Mittal, K. L., Ed. VSP Publishing: 2009; Vol. 5, pp 243–257.

177. Elsbernd, C. S.; Spinu, M.; Krukonis, V. J.; Gallagher, P. M.; Mohanty, D. K.; McGrath, J. E., Synthesis and Fractionation Studies of Functionalized Organosiloxanes. In *Silicon-Based Polymer Science. A Comprehensive Resource*, Zeigler, J. M.; Fearon, F. W. G., Eds. American Chemical Society: Washington, DC, 1990; Vol. 224, pp 145–164.
178. Tian, X.-H.; Wei, F.; Wang, T.-X.; Wang, P.; Lin, X.-N.; Wang, J.; Wang, D.; Ren, L., In Vitro and In Vivo Studies on Gelatin-Siloxane Nanoparticles Conjugated with Synb Peptide to Increase Drug Delivery to the Brain. *Int. J. Nanomedicine* **2012**, *7*, 1031–1041.
179. Hu, Y.; Zhuo, R., The Synthesis and Drug Release Property of Polysiloxane Bearing Pendant Long Alkyl Ether Group. *Acta Polym. Sinica* **1997**, 1, 62–67.
180. Gu, Q. G.; Zhou, Q. L., Preparation of High Strength and Optically Transparent Silicone Rubber. *Eur. Polym. J.* **1998**, *34*, 1727–1733.
181. Rajan, G. S.; Sur, G. S.; Mark, J. E.; Schaefer, D. W.; Beaucage, G., Preparation and Characterization of Some Unusually Transparent Poly(dimethylsiloxane) Nanocomposites. *J. Polym. Sci., Part B: Polym. Phys.* **2003**, *41*, 1897–1901.
182. Lin, G.; Zhang, X.; Kumar, S. R.; Mark, J. E., Modification of Polysiloxane Networks for Biocompatibility. *Mol. Cryst. Liq. Cryst.* **2010**, *521*, 56–71.
183. Katzman, H. A.; Ross, F. D.; Valenzuela, P. R., Resiliency of Silicone O-Rings. *J. Appl. Polym. Sci.* **2009**, *114*, 843–846.
184. Xu, S.; Lehmann, R. G.; Miller, J. R.; Chandra, G., Degradation of Polydimethylsiloxanes (Silicones) as Influenced by Clay Minerals. *Environ. Sci. Tech.* **1998**, *32*, 1199–1206.
185. Wen, J.; Mark, J. E., Synthesis, Structure, and Properties of Poly (dimethylsiloxane) Networks Reinforced by In Situ-Precipitated Silica-Titania, Silica-Zirconia, and Silica-Alumina Mixed Oxides. *J. Appl. Polym. Sci.* **1995**, *58*, 1135–1145.
186. Patel, M.; Skinner, A. R., The Effect of Thermal Aging on the Non-Network Species in Room Temperature Vulcanized Polysiloxane Rubbers. In *Synthesis and Properties of Silicones and Silicone-Modified Materials*, Clarson, S. J.; Fitzgerald, J. J.; Owen, M. J.; Smith, S. D.; Van Dyke, M. E., Eds. American Chemical Society: Washington, DC, 2003; Vol. 838, pp 138–150.
187. Jose, N. M.; Prado, L. A. S. D. A.; Schiavon, M. A.; Redondo, S. U. A.; Yoshida, I. V. P., Partially Pyrolyzed Poly(dimethylsiloxane)-Based Networks: Thermal Characterization and Evaluation of the Gas Permeability. *J. Polym. Sci., Polym. Phys. Ed.* **2007**, *45*, 299–309.
188. Giuliani, J. R.; Gjersing, E. L.; Chinn, S. C.; Jones, T. V.; Wilson, T. S.; Alviso, C. T.; Herberg, J. L.; Pearson, M. A.; Maxwell, R. S., Thermal Degradation in a Trimodal Poly(dimethylsiloxane) Network Studied by ^{1}H Multiple Quantum NMR. *J. Phys. Chem. B* **2007**, *111*, 12977–12984.
189. Kim, E. S.; Kim, E. J.; Shim, J. H.; Yoon, J.-S., Thermal Stability and Ablation Properties of Silicone Rubber Composites. *J. Appl. Polym. Sci.* **2008**, *110*, 1263–1270.
190. Chen, X.; Cui, Y.; Yin, G.; Lian, L., Thermooxidative Degradation Behavior of Poly(silphenylene-siloxane)s. *J. Appl. Polym. Sci.* **2010**, *117*, 926–933.
191. Chrusciel, J. J.; Lesniak, E., Preparation of Flexible, Self-Extinguishing Silicone Foams. *J. Appl. Polym. Sci.* **2011**, *119*, 1696–1703.
192. Lai, S. K.; Batra, A.; Cohen, C., Characterization of Polydimethylsiloxane Elastomer Degradation via Cross-Linker Hydrolysis. *Polymer* **2005**, *46*, 4204–4211.

193. Wild, M. J.; Buhler, P., On the Phase Composition of Polydimethylsiloxane Derived Ceramics. *J. Mater. Sci.* **1998**, *33*, 5441–5444.
194. Almond, M. A.; Becerra, R.; Bowes, S. J.; Cannady, J. P.; Ogden, J. S.; Young, N. A.; Walsh, R., A Mechanistic Study of the Low Pressure Pyrolysis of Linear Siloxanes. *Phys. Chem. Chem. Phys.* **2009**, *11*, 9259–9267.
195. Chenoweth, K.; Cheung, S.; Duin, A. C. T. V.; Goddard III, W. A.; Kober, E. M., Simulations on the Thermal Decomposition of Poly(dimethylsiloxane) Polymer Using the ReaxFF Reactive Force Field. *J. Am. Chem. Soc.* **2005**, *127*, 7192–7202.
196. Assary, R. S.; Curtiss, L. A.; Redfern, P. C.; Zhang, Z.; Amine, K., Computational Studies of Polysiloxanes: Oxidation Potentials and Decomposition Reactions. *J. Phys. Chem. C* **2011**, *115*, 12216–12223.
197. Virlogeux, F.; Bianchini, D.; Delor-Jestin, F.; Baba, M.; Lacoste, J., Evaluation of Cross-Linking After Accelerated Photo-Aging of Silicone Rubber. *Polym. Int.* **2004**, *53*, 163–168.
198. Xiao, H.; Li, C.; Yang, D.; LI, X.; He, S., Optical Degradation of Polydimethylsiloxane Under 150 keV Proton Exposure. *J. Appl. Polym. Sci.* **2008**, *109*, 4060–4064.
199. Nodera, A.; Kanai, T., Flame Retardancy of Polycarbonate-Polydimethylsiloxane Block Copolymer/Silica Nanocomposites. *J. Appl. Polym. Sci.* **2006**, *101*, 3862–3868.
200. Yang, L.; Hu, Y.; Lu, H.; Song, L., Morphology, Thermal, and Mechancal Properties of Flame-Retardant/Silicone Rubber/Montmorillonite Nanocomposites. *J. Appl. Polym. Sci.* **2006**, *99*, 3275–3280.
201. Verdejo, R.; Baroso-Bujans, F.; Rodriguez-Perez, M. A.; De Saja, J. A.; Arroyo, M.; Lopez-Manchado, M. A., Carbon Nanotubes Provide Self-Extinguishing Grade to Silicone-Based Foams. *J. Mater. Chem.* **2008**, *18*, 3933–3939.
202. Kashiwagi, T.; Mu, M.; Winey, K.; Cipriano, B.; Raghavan, S. R.; Pack, S.; Rafailovich, M.; Yang, Y.; Grulke, E.; Shields, J.; Harris, R.; Douglas, J., Relation Between the Viscoelastic and Flammability Properties of Polymer Nanocomposites. *Polymer* **2008**, *49*, 4358–4368.
203. Weon, B. M.; Kwon, Y. B.; Won, K. H.; Yoo, J.; Je, J. H.; Li, M.; Hahn, J. H., Ablation and Deposition of Poly(dimethylsiloxane) with X-Rays. *ChemPhysChem* **2010**, *11*, 115–118.
204. Granqvist, H.; Helminen, A.; Vehvilainen, M.; Aaritalo, V.; Seppala, J.; Linden, M., Biodegradable and Bioactive Hybrid Organic-Inorganic PEG-Siloxane Fibers. Preparation and Characterization. *Coll. Polym. Sci.* **2004**, *282*, 495–501.
205. Kaali, P.; Momcilovic, D.; Markstrom, A.; Aune, R.; Czel, G.; Karlsson, S., Degradation of Biomedical Polydimethylsiloxanea During Exposure to In Vivo Biofilm Environments Monitored by FE-SEM, ATR-FTIR, and MALDI-TOF MS. *J. Appl. Polym. Sci.* **2010**, *115*, 802–810.

CHAPTER 3

Types of Polysiloxanes

3.1 HOMOPOLYMERS

The polysiloxane of greatest commercial importance and scientific interest is poly(dimethylsiloxane) (PDMS), $[Si(CH_3)_2–O–]_x$, a member of the symmetrical dialkyl polysiloxanes, with repeat unit $[SiR_2–O–]_x$. This polymer is discussed extensively in the following chapters, particularly in chapter 5. Other members of this series are poly(diethylsiloxane) $[Si(C_2H_5)_2–O–]_x$, and poly(di-*n*-propylsiloxane) $[Si(C_3H_7)_2–O–]_x$. An example of an aryl member of the symmetrically substituted series is poly(diphenylsiloxane), with repeat unit $[Si(C_6H_5)_2–O–]_x$.[1–17] This polymer is unusual because of its very high melting point and the mesophase it exhibits. The closely related polymer, poly(phenyl/tolylsiloxane), has also been prepared and studied.[18]

The unsymmetrically substituted polysiloxanes have the repeat unit $[SiRR'O–]_x$, and are exemplified by poly(methylphenylsiloxane) $[Si(CH_3)(C_6H_5)–O–]_x$[19–24] and poly(methylhydrosiloxane) $[Si(CH_3)(H)–O–]_x$.[25] In some cases, one of the side chains has been unusually long, for example C_6H_{13}, $C_{16}H_{33}$, and $C_{18}H_{37}$,[26–27] including a branched side chain—$CH(CH_3)–(CH_2)_m–CH_3$.[27] Another example has methoxy-substituted aromatic fragments as one of the two side chains in the repeat unit.[28] Such chains have stereochemical variability in analogy with the vinyl polymers such as polypropylene $[CH(CH_3)–CH_2–]_x$ and vinylidene polymers such as poly(methyl methacrylate) $[C(CH_3)(C=OOCH_3)–CH_2–]_x$.[29] One can also introduce optically active groups as side chains, the simplest example being the *secondary* butyl group—$CH(CH_3)(C_2H_5)$. Another example involves redox-active dendritic wedges containing ferrocenyl and carbonylchromium moieties.[30]

Other substituents have included phenylethenyl groups,[31] cyclic siloxane groups,[32] and Cr-bound carbazole chromophores.[33] In a reversal of roles, some polymers were prepared to have PDMS side chains on a poly(phenylacetylene) main chain.[34] Siloxane-terminated solubilizing side chains are used to improve the properties of thin-film transistors.[35]

Silalkylene polymers have methylene groups replacing the oxygen atoms in the backbone. Poly(dimethylsilmethylene) is an example, $[Si(CH_3)_2-CH_2]_x$.[36–46] A variation on this theme is to include aryl groups, for example, in poly(dimethyldiphenylsilylenemethylene) $[Si(CH_3)_2-CH_2-Si(C_6H_5)_2]_x$.[16, 28, 45, 47–55] Other aryl substituents, specifically tolyl groups, have also been included as side chains.[56–57]

It is also possible to insert a silphenylene group $[Si(CH_3)_2-C_6H_4-]$ into the backbone of the polysiloxane repeat unit to give $[Si(CH_3)_2-C_6H_4-Si(CH_3)_2O-]$, in which the phenylene can be *para* or *ortho* or *meta*.[58–61] A specific example is poly(tetramethyl-*p*-silphenylene-siloxane).[13, 45, 48–49, 62–71] Similarly, it is possible to use anthrylene groups[51] or fluorene groups[72] instead of phenylene groups.

Most fluorosiloxane polymers[73] have the fluorine atoms in alkyl side chains[74] but preceded by methylene spacers in the side chain to prevent the F atoms from destabilizing the chain backbone. The most important example of this type is poly((3,3,3-trifluoropropyl)methylsiloxane),[75–79] but other examples include low molecular weight model organosiloxanes containing perfluoroether side chains.[80] Some fluorinated silsesquioxane polymers have also been prepared and characterized with regard to their surface tension.[81]

Fluorine atoms have also been placed on polysiloxanes having aromatic side chains (e.g., in poly(phenylmethylsiloxanes)). Examples are poly(4-fluorophenylmethylsiloxane),[82–83] poly(3,5-difluorophenylmethylsiloxane),[82] poly(3,5-bis(trifluoromethyl)phenylmethylsiloxane),[82] poly(pentafluorophenylmethylsiloxane),[84] and perfluorocyclobutane aromatic polyethers.[85] Another interesting example involved hexafluoroisopropanol-functionalized polysiloxanes that are being offered as new coating materials for sensors.[86] In some cases, the fluorosiloxane groups were simply used to cap chains of polybutadiene.[87]

More complicated structures have been formed by bridging single polysiloxane chains with oxygen atoms to give ladder-like polymers, the most common example being poly(phenylsilsesquioxane).[16, 52, 88–94]

3.2 REACTIVE CHAINS

In the typical ring-opening polymerization, reactive hydroxyl groups are automatically formed at the ends of the chains.[95–96] Substitution reactions carried out on these chain ends can convert them into other reactive functional groups. These functionalized polymers can undergo a variety of subsequent reactions, some of which are listed in table 3.1.[97] For example, hydroxyl-terminated chains, can undergo condensation reactions with alkoxysilanes (orthosilicates).[95, 98] A difunctional alkoxysilane leads to chain extension, and a tri- or tetrafunctional one to network formation. Corresponding addition reactions with di- or triisocyanates provide other possibilities. Hydrogen atoms on silicon atoms are sufficiently reactive to react with unsaturated groups at elevated temperatures in the presence of platinum catalysts Thus, hydrogen-terminated chains can react with molecules having unsaturated groups, or vinyl chain ends or side groups can react with the active hydrogen atoms on silanes.[95, 98–99] There are reactive hydrogen atoms in the repeat units of poly(hydrogenmethylsiloxane) [–SiH(CH_3)–O–].[25, 100] Other possibilities can be found elsewhere.[101]

A pair of vinyl or other unsaturated groups can also be linked by their direct reactions with free radicals. Similar end groups can be placed on siloxane chains by the use of an end blocker during polymerization.[102–103] Reactive groups such as vinyl units can be introduced as side chains by random copolymerization involving, for example, methylvinylsiloxane trimers or tetramers.[95]

One of the most important uses of end-functionalized polymers is the preparation of block copolymers.[102–103] The reactions are identical to the chain extensions already mentioned, except that the sequences being joined are chemically different. In the case of the—$OSiR_2R'Y$ chain end R' is typically $(CH_2)_{3-5}$ and Y can be NH_2, OH, COOH, CH = CH_2, and so on. The siloxane sequences containing these ends have been joined to other polymeric sequences such as carbonates, ureas, urethanes, amides, and imides. Other functional groups include amines and sulfonic acids,[104–105] ammonium groups,[106] epoxides,[105] and chloroalkyl groups.[107]

It is also possible to prepare polysiloxane ionomers. For example, PDMS with carboxyl side groups has been prepared with a controlled number of

Table 3.1. REACTIVE POLYMERS OF THE TYPE $XSi(CH_3)_2O[-Si(CH_3)_2O-]_nSi(CH_3)]_2X$

X	Reactant
Hydroxyls,—OH	Alkoxysilanes such as [$Si(OC_2H_5)_4$]
Active hydrogens,—H	Unsaturated groups
Unsaturated groups such as—CH = CH_2	Active H atoms
"	Free radicals

repeat units between barium, gallium, zinc, or cobalt cations, and a controlled number of such cations per chain.[108–110]

Groups that are chemically unreactive have also been added to PDMS chains. For example, perylene groups have been placed at the ends of the chains to induce vesicle formation,[111–112] as have been modified bichromoporic perylenes to study ordering effects.[113] Similarly, anthracenyl groups were introduced so that fluorescence spectroscopy could be used to investigate glass transitions and melting of polysiloxanes.[114] Related molecules such as pyrene have been introduced by simple doping techniques.[115] The photochemistry of polysiloxanes has been reviewed recently.[116]

Polysiloxanes have also been pyrolyzed to give ceramics and organic/inorganic hybrid materials.[117–118] The general topic of pre-ceramic polymers is discussed chapter 9.[119]

3.3 DENDRIMERS AND HYPERBRANCHED POLYMERS

The literature[124–127] describes a number of dendrimers and the closely related star-like [120–123] polysiloxanes. The hyperbranched polysiloxanes[128] are the primary example of more random structures. Although the emphasis has been on synthesis and characterization,[129–136] modeling on hyperbranched polymers has also been carried out.[137] Some of the most interesting species involve polysiloxane chains.[124–126, 138–139] Star polymers, some with nanosized silica cores, have also been synthesized.[120, 122–123]

Hyperbranched polysiloxanes have been prepared with controllable molecular weights and polydispersities,[128, 133, 140] with epoxy terminal groups[132]; some are UV-curable[130] and some serve as a source of molecular silica.[129] Hyperbranched polysiloxanes have also been used in the sol-gel preparation of polypropylene/silica nanocomposites.[134]

Heavily branched polysiloxanes can be prepared by a variety of techniques,[131] some of which generate hybrid mesostructures.[141] Branched and linear chains have been compared with regard to their interactions with solvents.[142]

3.4 LIQUID-CRYSTALLINE POLYMERS

3.4.1 Main-Chain Liquid-Crystalline Elastomers

The sequences that give rise to the liquid crystallinity can be in the chain backbone, in the side chains, or in both,[143] as shown schematically in figure 3.1. The polymer in the category that has been the most studied is

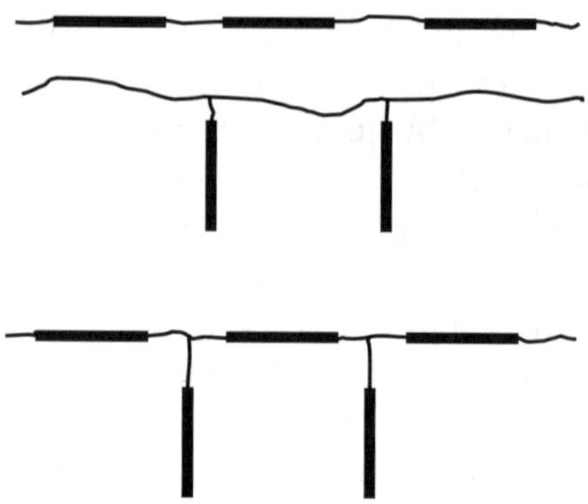

Figure 3.1:
Liquid-crystalline polymers in which the mesomorphic sequences occur in the chain backbone, in the side chains, or in both.

poly(diethylsiloxane) (PDES) [–Si$(OC_2H_5)_2$O–],[99] which forms a nematic mesophase. Relevant studies have include (i) heat capacity,[144] (ii) dielectric relaxation,[145] (iii) nuclear magnetic resonance (NMR) characterization,[146] (iv) structural changes during transitions,[147] (v) effects of stretching on the transitions,[148] (vi) optical properties such as refractive indices and optical rotations,[149] (vii) thermoelastic (force-temperature) properties,[150] (viii) mechanical properties in general,[151] (ix) end linking of the chains into "model networks,"[152–153] (x) effects of molecular weight,[154] (xi) characterization using atomic force microscopy[155] or x-ray diffraction,[156] (xii) thermopolarization effects (cooling a heated sample in an electric field),[157] (xiii) solution properties[26] and swelling behavior,[153] (xiv) segmental orientation,[158] (xv) responses of guest chains in deformed PDES elastomers[159] and the properties of blends.[160]

One item of great interest with regard to these materials is the temperature at which the nematic liquid-crystalline phase becomes isotropic.[99] Table 3.2 lists such *isotropization* ("clearing") temperatures T_i for a variety of symmetric polysiloxanes polymers having repeat units [Si$((CH_2)_mCH_3)_2$–O–]. Although the focus is on PDES (m = 1), PDMS with m = 0 is included for comparison. The table includes polymer molecular weights (M) where important, and some relevant melting points, T_m. PDMS is very different from PDES in that it does not show a liquid-crystalline phase. This situation is explained in connection with table 3.2. PDES shows values of T_i that decrease with decreasing in molecular weight. In fact, the liquid-crystalline phase *does not form at all* if M is below approximately 25,000 g mol^{-1}.[99] This situation is in sharp contrast to the behavior of low molecular weight liquid-crystalline molecules; many cholesterol molecules that show mesophases have molecular weights down in the hundreds.[161]

Table 3.2. TRANSITION TEMPERATURES FOR SOME SYMMETRIC POLYSILOXANE ELASTOMERS HAVING REPEAT UNITS $[Si((C_2)_mCH_3)_2-O-]$

Polymer	m	$10^{-3}M$ (g/mol)	T_m(° C)	T_i(° C)
Poly(dimethylsiloxane)[a]	0	High	−43	None
Poly(diethylsiloxane)[b]	1	765	—	53
"	"	425	—	52
"	"	172	—	46
"	"	100	—	34
"	"	58	—	23
"	"	~25	—	None
Poly(di-*n*-propyl-siloxane)[c]	2	87	—	207
"	"	68	—	177
"	"	51	—	172
"	"	43	—	145
"	"	~10	—	None
Poly(di-*n*-butyl-siloxane)[c]	3	128	—	299
"	3	28	—	216
Poly(di-*n*-pentyl-siloxane)[c]	4	High	—	330
Poly(di-*n*-hexyl-siloxane)[c]	5	"	—	330
Poly(di-*n*-heptyl-siloxane)[d]	6	"	375[e]	None
Poly(di-*n*-octyl-siloxane)[d]	7	"	28[e]	None
Poly(di-*n*-nonyl-siloxane)[d]	8	"	31[e]	None
Poly(di-*n*-decyl-siloxane)[d]	9	"	47[e]	None

[a]See note 345.
[b]See note 154.
[c]See note 346.
[d]See note 167.
[e]Crystallization of the side chains, rather than the backbones.

Since stretching a PDES aligns the chains in the direction of the liquid-crystalline structures, the values of T_i should also increase with elongation,[99] as has been found experimentally.[150] As expected, the values of T_i level off at higher elongations.

There are a number of other acyclic polysiloxanes. For example, symmetric polysiloxanes having $m = 2–9$ have been investigated, with some properties listed in several handbook articles.[162] These specific polymers and some recent relevant studies on them are (i) poly(di-*n*-propyl-siloxane),[163–165] (ii) poly(di-*n*-butyl-siloxane),[163, 166] (iii) poly(di-*n*-pentyl-siloxane,[152, 163] (iv) poly(di-*n*-hexyl-siloxane),[152, 163](v)poly(di-*n*-heptyl-siloxane),[167](vi)poly(di-*n*-octyl-siloxane),[167] (vii) poly(di-*n*-nonyl-siloxane),[167] and (viii) poly(di-*n*-decyl-siloxane).[152, 167, 99]

The isotropization temperatures show an interesting increase with increasing *m*, the number of methylene groups in the side chains.[168] The melting points of the same polymers also increase with *m*, but with a smaller slope. The two lines cross in such a way that the predicted value of T_i for PDMS lies *below* the melting temperature, T_m, meaning that PDMS does not show a liquid-crystalline phase because it crystallizes before it gets to T_i.[99]

Another interesting feature of the table pertains to the melting points for polymers having $m = 6–9$. The side chains are now long enough to crystallize. Such side-chain crystallization has also been seen in polysilane homopolymers[169] and chemical copolymers of polyethylene,[170–171] (so-called linear low-density polyethylene). In any case, such crystallization presumably could interfere with the mesophases that might otherwise form.[99]

Some polysiloxanes have cyclic groups in the backbone, typically cyclics of $–Si(CH_3)_2–O–$ units of various sizes, or such siloxane units mixed with some carbosiloxanes (with additional $–CH_2–$ sequences).[164, 172–176] The cyclic portions can add considerable stiffness, resulting in isotropization temperatures above the decomposition temperatures.

One aspect of the mechanical properties of these polymers is the increase in the isotropization temperatures with increasing elongation.[177–179] Interesting examples are the elongation results reported for PDES as a function of mesomorphic structure.[148] The tensile curves differ greatly depending on the amount of mesophase present.[99] Specifically, for low mesophase content, the isotherms are similar to those of natural rubber, with upturns in force at high elongations. In the case of larger amounts, the curves show yield points akin to those shown by partially crystalline polymers. Crystallinity may be either present initially or induced by the deformation. Elongation-retraction curves demonstrate that formation of a second phase leads to irreversibility in the stress-strain isotherms.[148] The larger the elongation during the deformation, the larger the irreversibility (hysteresis) upon retraction.

In some cases, smectic ordering has been observed.[180] There is also interest in phenylpyrimidine liquid-crystalline hosts[181] and ferroelectric organosiloxanes.[182] Results on chain dynamics have also been reported for polysiloxanes containing *p*-phenyleneterephthalate,[183] or polyimide mesogens.[184]

3.4.2 Side-Chain Liquid-Crystalline Elastomers

The units giving rise to the liquid-crystalline behavior can be in the side chains.[99, 185–187] Although most of the polymers studied have been linear, there has been some work on hyperbranched[188] or comb-like structures.[189] The orientation of the mesogenic groups is important.[190–191] Frequently backbones include siloxanes and acrylates,[192–195] but a variety of other structures have also been studied, including amphiphilics.[141, 196–209] Side chains have included chiral groups[210–212] and groups capable of hydrogen bonding.[213]

The side chains in these structures can rearrange either parallel or perpendicular to the deformed chain backbone,[214] as illustrated in figure 3.2. The outcome depends on the nature and length of the flexible spacer connecting the mesogenic groups to the chain backbone. As expected, the physical properties can become strongly anisotropic.[215]

Side-chain liquid-crystalline materials can be oriented by imposing an electric or magnetic field.[216] The chains can also be aligned when deformed (generally in elongation but also in some cases in compression) and cross linked into network structures. The focus in these experiments was how the mechanical deformation affected the nature of the mesophase (in particular its axial direction relative to the direction of the strain) and its isotropization temperature. The studies generally involved measurements of both stress and birefringence as a function of strain, and the ratio of the former to the latter (the "stress-optical coefficient"). The mesogenic behavior of such networks obviously depends strongly on their structures. The effects of degree of cross linking, and composition in the case of copolymers, for example, have been documented.[217]

The phase transitions depend significantly on spacer length, as has been demonstrated for oligo-oxyethylene spacers.[218] Closer coupling between the mesogenic groups and the polymer backbone tends to make the system more sensitive to the mechanical deformation, at least in the case of methylene groups in the spacer.[219]

Relatively high degrees of cross linking can be introduced without destroying the liquid crystallinity.[222] Cross linking can generally be induced

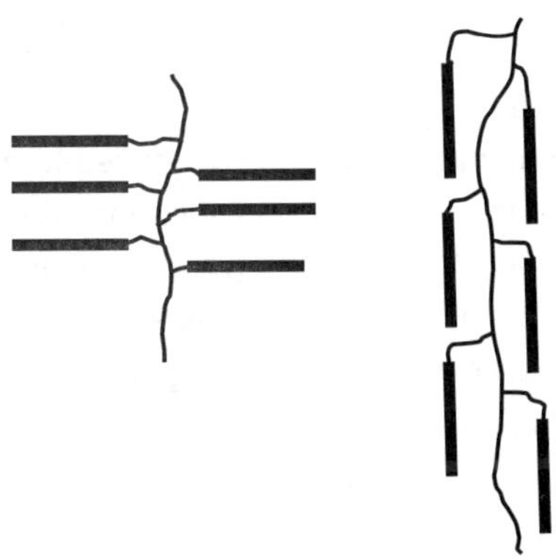

Figure 3.2:
Approximately parallel and perpendicular arrangements of mesogenic side groups on a chain backbone stretched in the vertical direction.

by gamma radiation or chemical means.[220–221] IR (infrared) spectroscopy and stress-strain measurements in extension indicate that relatively low strains can frequently induce significant organization. One relevant experiment involved two cross-linking procedures: the first cross linking produced a network in which the mesogenic units could be oriented[223]; the second subsequently locked in the network anisotropy. Samples were clear and x-ray diffraction patterns were characteristic of a highly ordered nematic material. Transitions from the isotropic phase to the nematic phase caused significant increases in sample length. Using one or two cross-linking agents in the presence of a magnetic field could also be used to prepare "monodomain" nematics in which the director alignment is claimed to be macroscopically uniform.[192]

The thermoelastic behavior of these materials has also been reported.[224] Such experiments resolve the thermodynamics of the nematic to isotropic transition into entropic and enthalpic contributions.[193]

Also in this category are side-chain cholesteric liquid-crystalline polysiloxanes containing groups of varying rigidity,[225] thermotropic liquid-crystalline polyimides with siloxane linkages,[226] and fluorine-containing liquid-crystalline polysiloxanes having cholesteryl cinnamate mesogens and trifluoromethyl-substituted mesogens.[227]

Although experimental studies have been of the greatest interest by far, there has been some molecular modeling of side-chain liquid-crystalline polysiloxanes.[228]

Two of the most recent applications of these materials are in the areas of electrorheology[229] and holography.[230]

3.5 CYCLICS

3.5.1 Introduction

The synthesis of polysiloxanes generally produces significant cyclics as well as linear chains. A number of studies report equilibrium cyclic concentrations under various conditions.[231–241] Cyclic polysiloxanes have also been prepared in dilute solutions by cyclodepolymerization of linear dihydroxy-PDMS,[242] or ring-closing dehydrocoupling of dihydroxy-PDMS and dihydro-PDMS.[243] Diels-Alder reactions have also proved useful for the production of high-purity cyclics,[244] as have been polyfunctional cage oligosilsesquioxanes by thiol-ene additions.[245]

Considerable information on polysiloxane cyclics is available in reviews of cyclic polymers in general.[246–251] Considerable attention has been paid to cyclics of poly(dimethylsiloxane),[252] poly(phenylmethylsiloxane),[253] and poly(vinylmethylsiloxane).[254] Other molecular systems also produce cyclics and general methods have been developed for carrying out cyclizations.[255–266] Organofunctionalized cyclotrisiloxanes have also been reported.[267] Interesting comparisons have been made with results on poly(ethylene oxide),[268] poly(ε-caprolactones),[269] polystyrene,[270] and DNA.[271]

Cyclic structures also provide conceptual frameworks in some molecular dynamics simulations.[272–273]

3.5.2 Miscellaneous Properties of Polysiloxane Cyclics

The conformational dynamics of PDMS cyclics of various sizes have been studied by ultrasound.[274] The magnitudes of the dispersions obtained were used to estimate energy differences between stable and less-stable conformations. Analogous information has been obtained using excimer emission from small probes placed into a cyclic PDMS.[275] Chemical shifts and relaxation times in the ^{29}Si NMR spectra of cyclic PDMS have also been used for this purpose.[276–277] There has been work with regard to the structure and dynamics of cyclics in general,[278] including percolation of linear polymers in melts of cyclic polymers.[279]

PDMS cyclics were combined with γ-cyclodextrins to form "slide ring" gels in which the usual cross links in a network structure are replaced sliding linkages through which the polymer chains are threaded.[280] Similar reactions with γ-cyclodextrins occur even in the case of linear chains of PDMS.[281] Related "disk-necklace" structures were obtained by connecting disk-shaped entities (the cyclics) by flexible linking agents.[282] Also, catenate structures have been studied,[283] and "Olympic" networks of interlinked PDMS cyclics may have been formed in some reactions.[284–285] "Knotted" ring polymers also exist.[286] Finally, PDMS cyclics have been studied as "pseudo crown ethers" for the binding of metal cations.[287]

In some cases, cyclic and linear PDMS have been combined to form "conetworks,"[288] and unsaturated cyclic side-chain fragments have been placed into polysiloxanes to make them thermoreactive.[289] Interactions between ring polymers have also been analyzed and related to loops in chromatin.[290] Also, some cyclic oligosiloxanes having polar end groups show liquid-crystalline behavior, specifically smectic A and E phases.[291] Finally, several PDMS cyclics have been exposed to vacuum pyrolysis and the products analyzed by matrix-isolation spectroscopy.[292] The pyrolysis products obtained under a variety of conditions identified the radical reactions that were involved.

With regard to simulations, improved Monte Carlo methods have been developed for characterizing cyclic polymers in the melt.[293] A simple lattice is employed, and kink translocations are introduced to speed up the computations. Similarly, Monte Carlo simulations have been used to calculate dimensions of ring polymers[294–296] and excluded volume effects in semiflexible ring polymers.[297] Monte Carlo simulations have also been used to characterize stiffening transitions in semiflexible cyclics.[298] Molecular dynamics simulations have given interesting information on both static[299–300] and dynamic properties [299–301] of cyclic polymers. A self-consistent field theory for Gaussian ring polymers is also available.[302] Dimensional analyses using a scaling model have been used to describe cyclic molecules in general, including those of PDMS.[303] Finally, simulations have been carried out on the cyclization of α,ω-telechelic chains.[304]

3.5.3 Comparisons Between Polysiloxane Cyclics and Polysiloxane Linear Chains

There have been many comparisons of cyclic polysiloxanes with the corresponding linear chains of the same molecular weight. Table 3.3 provides

Table 3.3. COMPARISONS BETWEEN CYCLIC POLYSILOXANES AND LINEAR POLYSILOXANES OF THE SAME MOLECULAR WEIGHT

Property	Polymer	Value of property for cyclics relative to linears	Reference
Intrinsic viscosities	Poly(dimethylsiloxane)	Lower	347–348
Gel permeation chromatography	"	Both gave sharp fractions	349–350
Chain dimensions	"	Lower	351
Bulk viscosities	"	Higher at low M, but lower at high M	352
Diffusion coefficients	"	Higher	353–356
Effects of heat	"	Both show bond interchanges	357–358
Second virial coefficients	"	Lower	359
Radii of gyration (Monte Carlo)	"	Lower	360
Ellipsoidal shapes (Monte Carlo)	"	Lower asymmetry	361

(*continued*)

Table 3.3 (CONTINUED)

Property	Polymer	Value of property for cyclics relative to linears	Reference
Dipole moments	"	Lower	362
Particle scattering functions (Monte Carlo)	"	Cyclics show maxima	363
Surface pressures of monolayers	"	Higher plateau levels	364
Glass transition temperatures	"	Higher	305
Crystallization	"	Only low M cyclics and linear chains resist crystallization	365
Lower critical solution temperatures	"	Higher	366
Upper critical solution temperatures	"	Lower	367
Critical masses for entangling	"	No significant difference	368–369
Densities	"	Higher	370
Adsorption onto silica	"	Higher at low M, but lower at high M	371
Chain length dependence of α transition temperatures	"	Increase for cyclics but decrease for linear chains	372
Gel permeation chromatography	Poly(phenylmethylsiloxane)	Both gave sharp fractions	373
Radii of gyration from neutron scattering	"	Lower	374–376
Glass transition temperatures	"	Higher	377
Dielectric relaxations	"	Cyclics have higher dipolar cancellations	378
Intrinsic viscosities	Poly(vinylmethylsiloxane)	Lower	379
Bulk viscosities	"	Higher at low M	379
Densities	"	Higher	379
Glass transition temperatures	"	Higher	379

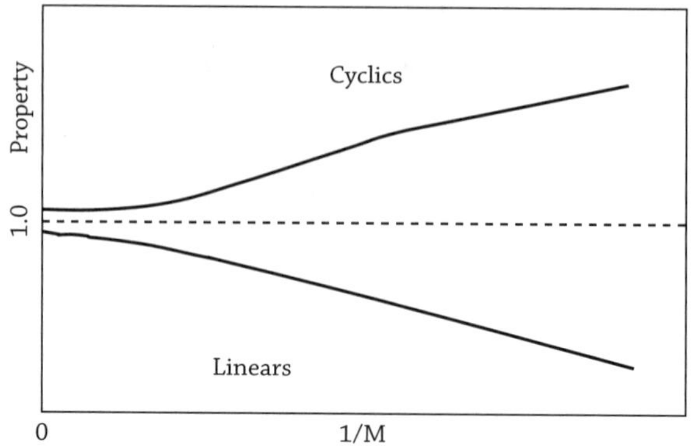

Figure 3.3:
The asymptotic effect of molecular weight on some property such as the glass transition temperature for cyclic and linear polysiloxanes.[305]

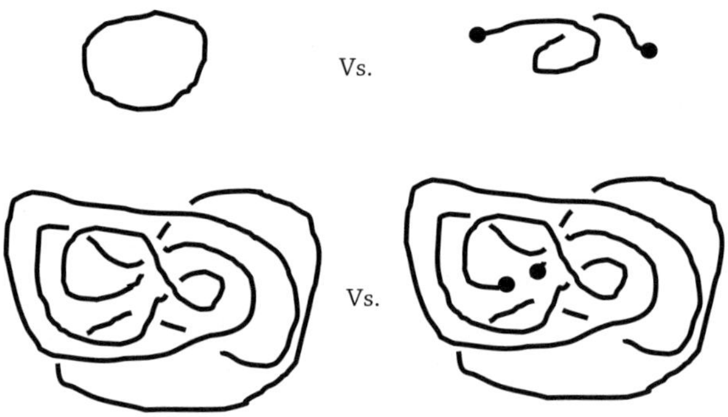

Figure 3.4:
Cutting a cyclic to form two chains (solid circles) can have less of an effect in the case of high molecular weight.

examples. In general, differences that appear at low molecular weight can decrease significantly at higher molecular weight, as occurs in the case of the glass transition temperature.[305] Figure 3.3 illustrates this effect. Cutting a skeletal bond in a cyclic of high molecular weight can have much less effect on some properties than cutting a skeletal bond in a smaller cyclic, as illustrated schematically in figure 3.4.

3.6 OTHER NOVEL MATERIALS

3.6.1 Blends

There have been numerous studies of blends based on the polysiloxanes.[306] Blends with polystyrene are of particular interest, [307–312] including some with the syndiotactic polymer.[313] Other vinyl polymers have included acrylics,[314–315] poly(vinyl chloride),[316] polypropylene,[317] ethylene-propylene-diene monomer elastomers,[318] and ethylene-octene copolymers.[319]

Other studies have focused on poly(dimethylsiloxane) blended with polycarbonates,[320–322] polyisobutylene,[323] poly(ethylene oxide),[324] polyurethanes,[325] epoxies,[326–327] benzoxazines,[328–329] and poly(hexylthiophene).[330] In some cases, a polysiloxane oil was blended into polypropylene to facilitate its processing.[331] A variety of other siloxane materials have been employed—for example, poly(diethylsiloxane),[332] polyurethanes,[333] fluorinated siloxane copoymers[334] and fluororubbers in general,[335] and trimethylsiloxy silicates.[336]

3.6.2 Ceramic Phases and Coatings

Pyrolyses of siloxane materials leads to ceramic-like phases. For example, ceramic fibers based on silicon carbide have been prepared from polycarbosilane/polymethylphenylsiloxane polymer blends.[337]

PDMS coatings have been made less oleophobic by modifying them using copolyacrylate side chains some of which are semifluorinated or PDMS itself.[338] Also relevant here are UV-curable bismaleimides containing PDMS as hydrophobic agents.[339]

3.6.3 Micropatterned Materials

PDMS emerged as the polymer of choice for micropatterned surfaces and microfluidic devices. Fabrication is particularly straightforward since PDMS can be cast against a suitable mold with high fidelity. The optical, thermal, interfacial, permeability, and reactivity properties of PDMS make possible numerous functionalities including optical detection, reversible deformation, reversible wetting, and management of cell proliferation.[340]

3.6.4 Nanofilaments and Molecular Wires

Metallic nanowires with vertical alignment have been prepared on silicone substrates using a nanoporous alumina template.[341] The fact that the substrate is flexible should be an advantage in a number of applications.

3.6.5 Thermosets

Siloxanes that are extremely heavily cross linked can be thought of as thermosets, which have some interesting glass-like properties.[342] Copolymeric analogues have also been prepared, by combining siloxanes with epoxy resins,[343] or with polybenzoxazine prepolymers.[344]

REFERENCES

1. Falender, J. R.; Yeh, G. S. Y.; Chiu, D. S.; Mark, J. E., Thermodynamics of Fusion in Poly(diphenylsiloxane). *J. Polym. Sci., Part B: Polym. Phys.* **1980**, *18*, 389–391.
2. Babu, G. N.; Christopher, S. S.; Newmark, R. A., Poly(dimethylsiloxane-co-diphenylsiloxanes): Synthesis, Characterization, and Sequence Analysis. *Macromolecules* **1987**, *20*, 2654–2659.
3. Dubchak, I. L.; Babchinitser, T. M.; Kazaryan, L. G.; Tartakovskaya, L. M.; Vasilenko, N. G.; Zhdanov, A. A.; Korshak, V. V., Structural Study of Polydiphenylsiloxane. *Polym. Sci. U.S.S.R.* **1989**, *31*, 70–78.
4. Wang, S.; Mark, J. E., Reinforcement of Elastomeric Poly(dimethylsiloxane) by Glassy Poly(diphenylsiloxane). *J. Mater. Sci.* **1990**, *25*, 65–68.
5. Ibemesi, J.; Gvozdic, N.; Keumin, M.; Tarshiani, Y.; Meier, D. J., Synthesis and Properties of Copolymers of Diphenylsiloxane with Other Organosiloxanes. In *Polymer-Based Molecular Composites*, Schaefer, D. W.; Mark, J. E., Eds. Materials Research Society: Pittsburgh, PA, 1990; Vol. 171, pp 105–115.
6. Grigoras, S.; Qian, C.; Crowder, C.; Harkness, B.; Mita, I., Structure Elucidation of Crystalline Poly(diphenylsiloxane). *Macromolecules* **1995**, *28*, 7370–7375.
7. Harkness, B. R.; Tachikawa, M.; Mita, I., Mesomorphic Behavior in Poly(diphenylsiloxane). 1. A Preliminary Examination of Diphenylsiloxane Oligomers. *Macromolecules* **1995**, *28*, 1323–1325.
8. Papkov, V. S.; Gerasimov, M. V.; Buzin, M. I.; Il'ina, M. N.; Karzaryan, L. G., The Structure of Ordered Phases in Poly(diphenylsiloxane). *Polym. Sci. U.S.S.R.* **1996**, *38*, 1097–1102.
9. Buzin, M. I.; Gerasimov, M. V.; Obolonkova, E. S.; Papkov, V. S., Solid-State Polymerization of Hexaphenylcyclotrisiloxane. *J. Polym. Sci. Part A: Polym. Chem.* **1997**, *35*, 1973–1984.
10. Ovchinnikov, Y. E.; Struchkov, Y. T.; Buzin, M. I.; Papkov, V. S., The Crystal Structure and Solid-State Polymerization of Hexaphenylcyclotrisiloxane. *Polym. Sci., U.S.S.R.* **1997**, *39*, 273–279.
11. Harkness, B. R.; Tachikawa, M.; Yue, H.; Mita, I., Mesomorphic Behavior in Poly(diphenylsiloxane). 3. An Examination of High Molecular Weight Polydiphenylsiloxane. *Chem. Mater.* **1998**, *10*, 1700–1705.
12. Li, L.-J.; Yang, M.-H., Phase Behaviours of Diphenylsiloxane Oligomers. *Polymer* **1998**, *39*, 689–695.
13. Rossignol, F.; Nakata, Y.; Nagai, H.; Okutani, T.; Suzuki, M.; Kushibiki, N.; Murakami, M., Fabrication of Poly(diphenylsilylenemethylene) and Poly(diphenylsiloxane) Thin Films Using Fine Metal Particles. *Chem. Mater.* **1999**, *11*, 358–366.

14. Lebedev, B. V.; Smirnova, N. N.; Papkov, V. S.; Buzin, M. I.; Dubovik, I. I., Thermodynamics of Poly(diphenylsiloxane) in the Temperature Range 0–620 K. *Polym. Sci. U.S.S.R.* **2000**, *42*, 1111–1119.
15. Papkov, V. S.; Buzin, M. I.; Gerasimov, M. V.; Obolonkova, E. S., Thermoreversible Gelation in Poly(diphenylsiloxane)/Diphenyl Ether Systems Exhibiting Mesophase Behavior. *Macromolecules* **2002**, *35*, 1079–1090.
16. Evmenenko, G.; S, K.; Dutta, P., Influence of Molecular Rigidity on Interfacial Ordering in Diphenyl-Based Polysiloxane Films. *Polymer* **2007**, *48*, 7163–7168.
17. Meier, D. J., Poly(diphenylsiloxane). In *Polymer Data Handbook, 2nd ed.*; Mark, J. E., Ed. Oxford University Press: New York, 2009; pp 573–575.
18. Meier, D. J., Poly(phenyl/tolylsiloxane). In *Polymer Data Handbook*, 2nd ed.; Mark, J. E., Ed. Oxford University Press: New York, 2009; pp 930–933.
19. Buch, R. R.; Klimish, H. M.; Johannson, O. K., Dilute Solution Parameters of Polymethylphenylsiloxanes. *J. Polym. Sci., Part A-2: Polym. Chem.* **1970**, *8*, 541–554.
20. Mark, J. E.; Ko, J. H., Random-Coil Dimensions of Poly(methylphenylsiloxane) and Their Dependence on Stereochemical Structure. *J. Polym. Sci., Part B: Polym. Phys.* **1975**, *13*, 2221–2235.
21. Freire, J. J.; Pierola, I. F.; Horta, A., Conformational Analysis of Methyl-Phenyl-Siloxane Chains. *Macromolecules* **1996**, *29*, 5143–5148.
22. Van Dyke, M. E.; Clarson, S. J., Reaction Kinetics for the Anionic Ring-Opening Polymerization of Tetraphenyltetramethylcyclosiloxane Using a Fast Initiator System. *J. Inorg. Organomet. Polym.* **1998**, *8*, 111–121.
23. Kuo, A. C. M., Polymethylphenylsiloxane. In *Polymer Data Handbook, 2nd ed.*; Mark, J. E., Ed. Oxford University Press: New York, 2009; pp 838–845.
24. Ahn, H. W.; Clarson, S. J., Synthesis of *Isotactic* Rich Poly(methylphenylsiloxane) by Living Anionic Ring-Opening Polymerization. *Silicon* **2011**, *3*, 157–161.
25. Lee, C. L.; Haberland, G. G., Polymethylsiloxamer. *Polym. Lett.* **1965**, *3*, 883–886.
26. Villegas, J. A.; Olayo, R.; Cervantes, J., Effect of Side Groups on the Conformation of a Series of Polysiloxanes in Solution. *J. Inorg. Organomet. Polym.* **2003**, *13*, 205–222.
27. Mukbaniani, O.; Titvinidze, G.; Tatrishvili, T.; Mukbaniani, N.; Brostow, W.; Pietkiewicz, D., Formation of Polymethylsiloxanes with Alkyl Side Groups. *J. Appl. Polym. Sci.* **2007**, *104*, 1176–1183.
28. Mukbaniani, O.; Titrvinidze, G.; Dundua, A.; Doroshenko, M.; Tatrishvili, T., Synthesis of Polysiloxanes with Pendant Methoxy-Substituted Aromatic Fragments. *J. Appl. Polym. Sci.* **2008**, *107*, 2567–2571.
29. Billmeyer, F. W., *Textbook of Polymer Science*, 3rd ed. Wiley: New York, 1984.
30. Zamora, M.; Bruña, S.; Alonso, B.; Cuadrado, I., Polysiloxanes Bearing Pendant Redox-Active Dendritic Wedges Containing Ferrocenyl and (η6-Aryl)tricarbonylchromium Moieties. *Macromolecules* **2011**, *44* (20), 7994–8007.
31. Ganjicz, T.; Kowalewska, A.; Stanczyk, W. A.; Butts, M.; Nye, S. A.; Rubinsztajn, S., A Novel Organometallic Route to Phenylethenyl-Modified Polysiloxanes. *J. Mater. Chem.* **2005**, *15*, 611–619.
32. Sacarescu, L., Non-Shrinking Polycyclocarbosiloxane Networks. *Polym. Bull.* **2008**, *59*, 873–882.
33. Hui, C.; Xiaohui, T.; Yizhong, Y.; Jiaping, L., Preparation of Polysiloxane with Pendant Cr-Bound Carbazole Chromophore for Non-Linear Optics. *J. Appl. Polym. Sci.* **2008**, *107*, 4076–4081.

34. Zhang, W.; Shiotsuki, M.; Masuda, T., Synthesis and Properties of Polymer Brush Consisting of Poly(phenylacetylene) Main Chain and Poly(dimethylsiloxane) Side Chains. *Polymer* **2007**, *48*, 2548–2553.
35. Mei, J.; Kim, D. H.; Ayzner, A. L.; Toney, M. F.; Bao, Z., Silicone-Terminated Solubilizing Side Chains: Bringing Conjugated Polymer Backbones Closer and Boosting Hole Mobilities in Thin Film Transistors. *J. Am. Chem. Soc.* **2011**, *133*, 20130–20133.
36. Levin, G.; Carmichael, J. B., Molecular Weight Distribution and Thermal Characterization of Polydimethylsilmethylene. *J. Polym. Sci., Part A-1: Polym. Chem.* **1968**, *6*, 1–12.
37. Ko, J. H.; Mark, J. E., Configuration-Dependent Properties of Poly(dimethylsilmethylene) Chains. I. Experimental Results. *Macromolecules* **1975**, *8*, 869–874.
38. Mark, J. E.; Ko, J. H., Configuration-Dependent Properties of Poly(dimethylsilmethylene) Chains. II. Correlation of Theory and Experiment. *Macromolecules* **1975**, *8*, 874–878.
39. Llorente, M. A.; Mark, J. E.; Saiz, E., Stress-Optical Behavior of Polydimethylsilmethylene [–Si$(CH_3)_2CH_2$–], a Hydrocarbon Analog of Polydimethylsiloxane, and a Silicon Analogue of Polyisobutylene. *J. Polym. Sci., Polym. Phys. Ed.* **1983**, *21*, 1173–1185.
40. Galiatsatos, V.; Mark, J. E., Effects of Dimethylsiloxane Diluents on the Optical Properties of Poly(dimethylsilmethylene) Networks. *Polym. Bull.* **1987**, *17*, 197–203.
41. Koopmann, F.; Frey, H., Synthesis of Poly(silylenemethylene)s Symmetrically Substituted with Alkyl Side Groups Containing 4–6 Carbon Atoms. *Macromolecules* **1996**, *29*, 3701–3706.
42. Benouargha, A.; Boutevin, B.; Caporiccio, G.; Essassi, E.; Guida-Pietrasanta, F.; Ratsimihety, A., Hybrid Silalkylene Polysiloxanes: Synthesis and Thermal Properties. *Eur. Polym. J.* **1997**, *33*, 1117–1124.
43. Park, S.-Y.; Interrante, L. V.; Farmer, B. L., The Structure of Poly(di-n-propylsilylenemethylene). *Polymer* **2001**, *42*, 4253–4260.
44. Park, S.-Y.; Interrante, L. V.; Farmer, B. L., The Structure of Poly(di-n-alkylsilylenemethylene). *Polymer* **2001**, *42*, 4261–4269.
45. Guida-Pietrasanta, F.; Boutevin, B., Polyalkylene or Silarylene Siloxanes Said Hybrid Silicones. *Adv. Polym. Sci.* **2005**, *179*, 1–27.
46. Makrodimitri, Z. A.; Raptis, V. E.; Economou, J. G., Molecular Dynamics Simulation of Structure, Thermodynamic, and Dynamic Properties of Poly(dimethylsilamethylene), Poly(dimethylsilatrimethylene) and Their Alternating Copolymer. *J. Phys. Chem. B* **2006**, *110*, 16047–16058.
47. Koopmann, F.; Frey, H., Synthesis and Properties of Poly(dimethyldiphenylsil ylenemethylene). *Macromol. Chem. Phys.* **1998**, *199*, 2119–2127.
48. Rossignol, F.; Konno, S.; Nakata, Y.; Nagai, H.; Okutani, T.; Suzuki, M., Properties of Poly(diphenylsilylmethylene) Thin Films Fabricated with Fine Metal Particles. *Chem. Mater.* **1999**, *11*, 367–373.
49. Xue, L.; Kawakami, Y., Precise Synthesis of Poly(Silphenylenesiloxane)s with Epoxy Side Functional Groups by Tris(pentafluorophenyl)borane as a Catalyst. *Polym. J.* **2007**, *39*, 379–388.
50. Ito, H.; Akiyama, E.; Nagase, Y.; Fukui, S., Syntheses of Poly(Tetraphenyl-p-silphenylenesiloxane)-co-(Tetramethyl-p-silphenylenesiloxane)s and the Physical Properties of Their Films. *Polym. J.* **2007**, *39*, 471–477.

51. Sato, I.; Takeda, S.; Arai, Y.; Miwa, H.; Nagase, Y.; Nemoto, N., Synthesis and Thermal Characterization of Novel Poly(tetramethylsilanthrylsiloxane) and Poly(tetramethylsilphenanthrylsiloxane) Derivatives. *Polym. Bull.* **2007**, *59*, 707–617.
52. Zhang, Z.-X.; Hao, J.; Xie, P.; Zhang, X.; Han, C. C.; Zhang, R., A Well-Defined Ladder Polyphenylsilsesquioxane (Ph-LPSQ) Synthesized via a New Three-Step Approach: Monomer Self-Organization-Lyophilization-Surface-Confined Polycondensation. *Chem. Mater.* **2008**, *20*, 1322–1330.
53. Cela, J.; Rubinzstajn, S., Preparation of Polyaryloxysilanes and Polyaryloxysiloxanes by B(C6F5)3 Catalyzed Polyetherification of Dihydrosilanes and Bis-Phenols. *Macromolecules* **2008**, *41*, 6965–6971.
54. Qu, L.; Huang, G.; Wang, Q.; Xie, Z., Effect of Diphenylsiloxane Unit Content on Aggregation Structure of Poly(dimethylsiloxane-co-diphenylsiloxane). *J. Polym. Sci., Part B: Polym. Phys.* **2008**, *46*, 73–79.
55. Henderson, E. J.; Veinot, J. G. C., From Phenylsiloxane Polymer Composition to Size-Controlled Silicon Carbide Nanocrystals. *J. Am. Chem. Soc.* **2009**, *131*, 809–815.
56. Ogawa, T.; Tachikawa, M.; Kushibiki, N.; Murakami, M., Poly(diarylsilmethylene)s. I. Synthesis and Characterization. *J. Polym. Sci., Part A: Polym. Chem.* **1995**, *33*, 2821–2831.
57. Ogawa, T.; Murakami, M., Poly(diarylsilmethylene)s. II. Thermal and Mechanical Properties. *J. Polym. Sci., Part B: Polym. Phys.* **1996**, *34*, 1317–1326.
58. Beck, H. N.; Chaffee, R. G., Arylene Silicon Compounds. *J. Chem. Eng. Data* **1963**, *8*, 602–603.
59. Dunnavant, W. R., Silarylene and Silarylene-Siloxane Polymers. *Inorg. Macromol. Rev.* **1971**, *1*, 165–189.
60. Dvornic, P. R.; Lenz, R. W., Exactly Alternating Silarylene-Siloxane Polymers. 9. Relationships between Polymer Structure and Glass Transition Temperature. *Macromolecules* **1992**, *25*, 3769–3778.
61. Dvornic, P. R.; Owen, M. J., Poly(silphenylene-siloxanes). In *Polymer Data Handbook, 2nd ed.*; Mark, J. E., Ed. Oxford University Press: New York, 2009; pp 1022–1027.
62. Merker, R. L.; Scott, M. J., Preparation and Properties of Poly(tetramethyl-p-Silphenylene-Siloxane). *J. Polym. Sci., Part A: Polym. Chem.* **1964**, *2*, 15–29.
63. Merker, R. L.; Scott, M. J.; Haberland, G. G., Random and Block Copolymers of Poly(tetramethyl-p-Silphenylene-Siloxane) and Polydimethylsiloxane. *J. Polym. Sci., Part A: Polym. Chem.* **1964**, *2*, 31–44.
64. Gardner, K. H.; Magill, J. H.; Atkins, E. D. T., Structural Studies of Polysiloxanes: Crystal Structure and Fold Conformation of Poly(tetramethyl-p-silphenylene siloxane). *Polymer* **1978**, *19*, 370–378.
65. Ballistreri, A.; Montaudo, G.; Lenz, R. W., Mass Spectral Characterization and Thermal Decomposition Mechanism of Alternating Silarylene-Siloxane Polymers. *Macromolecules* **1984**, *17*, 1848–1854.
66. Wang, S.; Mark, J. E., Computational Analysis of Some Conformational Properties of Poly-(tetramethyl-*p*-silphenylene-siloxane). *Comput. Theor. Polym. Sci.* **1993**, *3*, 33–37.
67. Wang, S.; Mark, J. E., An Experimental Study of the Unperturbed Chain Dimensions of Poly(Tetramethyl-*p*-silphenylene-siloxane). *Polym. Bull.* **1993**, *31*, 205–211.

68. Zhang, R.; Pinhas, A. R.; Mark, J. E., Synthesis of Poly(Tetramethylene-*m*-silphenylene-siloxane), an Elastomer of Enhanced High-Temperature Stability. *Macromolecules* **1997**, *30*, 2513–2515.
69. Dorset, D. L.; McCourt, M. P., Direct Phase Determination for Polymer Fibre X-ray Data—The Structure of Poly(tetramethyl-*p*-silphenylene siloxane). *Polymer* **1997**, *38*, 1985–1988.
70. Yanai, Y.; Sasaki, T.; Otomo, Y.; Ota, S.; Nagase, Y.; Nemoto, N., Functionalized Poly(tetramethyl-1,4-silphenylenesiloxane) with Pendant Hydroxyl Groups. *Polym. J.* **2008**, *40*, 875–881.
71. Jia, M.; Wu, C.; Li, W.; Gao, D., Synthesis and Characterization of a Silicone Resin with Silphenylene Units in Si-O-Si Backbones. *J. Appl. Polym. Sci.* **2009**, *114*, 971–977.
72. Nemoto, N.; Yamaguchi, T.; Ozawa, Y.; Nigorikawa, M.; Yanai, Y.; Nagase, V., Synthesis and Thermal Characterization of Novel Fluorene-Based Polysiloxane Derivatives. *Polym. Bull.* **2008**, *61*, 165–175.
73. Owen, M. J., Fluorosiloxane Polymers. In *Polymer Data Handbook, 2nd ed.*; Mark, J. E., Ed. Oxford University Press: New York, 2009; pp 171–173.
74. Hou, A.; Yu, J.; Shi, Y., Preparation and Surface Properties of the Polysiloxane Material Modified with Fluorocarbon Side Chains. *Eur. Polym. J.* **2008**, *44*, 1696–1700.
75. Madkour, T. M.; Mark, J. E., Monte Carlo Simulations on the Stereoregularity, Crystallinity, and Related Physical Properties of Poly[methyl(3,3,3-trifluoropropyl)siloxane}. *Macromolecules* **1995**, *28*, 6865–6870.
76. Cornelius, D. J.; Monroe, C. M., The Unique Properties of Silicone and Fluorosilicone Elastomers. *Polym. Eng. Sci.* **1985**, *25*, 467–473.
77. Kobayshi, H.; Owen, M. J., Surface Tension of Liquid Polysiloxanes Having Fluorinated Alkyl Side-Chains. *Makromol. Chem.* **1993**, *194*, 1785–1792.
78. Donskoi, A. A.; Baritko, N. V., Fluorosiloxane Sealants for Aviation Industry. *Polym. Sci. U.S.S.R., Ser. C* **2007**, *49*, 182–187.
79. Kilaru, M. K.; Heikenfeld, J.; Lin, G.; Mark, J. E., Strong Charge Trapping and Bistable Electrowetting on Nanocomposite Fluoropolymer: $BaTiO_3$ Dielectrics. *Appl. Phys. Lett.* **2007**, *90*, 212906.
80. Smith, J. A. S.; Brisdon, B. J.; Brewer, S. A.; Willis, C. R., Synthesis of Model Organosiloxanes Containing Perfluoroether Side-Chains. *J. Mater. Chem.* **2000**, *10*, 1765–1769.
81. Kobayashi, H., Surface Tension of Poly(fluoroalkylsilsesquioxane)s. *Makromol. Chem.* **1993**, *194*, 2569–2577.
82. Patwardhan, D. V.; Zimmer, H.; Mark, J. E., Synthesis of Some Fluorinated Phenylmethylsiloxane Polymers and Characterization of Their Surface Properties. *J. Inorg. Organomet. Polym.* **1997**, *7*, 93–109.
83. Zhao, Q.; Mark, J. E., Fluoro-Substituted Polysiloxanes. I. Synthesis of Poly(*p*-Fluorophenylmethylsiloxane). *Macromol. Sci., Macromol. Rep.* **1992**, *A(29)*, 21–26.
84. Zhao, Q.; Mark, J. E., Fluoro-Substituted Polysiloxanes. I. Synthesis of Poly(Pentafluorophenylmethylsiloxane). *Macromol. Sci., Macromol. Rep.* **1992**, *A(29)*, 221–227.
85. Conrad, M. P. C.; Shoichet, M. S., Synthesis and Thermal Stability of Hybrid Fluorosilicone Polymers. *Polymer* **2007**, *48*, 5233–5240.
86. Wei, D.; Wang, L.; Ma, J.; Jiang, H., Synthesis and Evaluation of Hexafluoroisopropanol-Functionalized Polysiloxane as a New Coating Material for Sensors. *J. Appl. Polym. Sci.* **2012**, *124*, 4136–4140.

87. Patwardhan, D. V.; Zimmer, H.; Mark, J. E., Synthesis and Surface Properties of Polybutadiene Chains End-Capped with Two Types of Fluorosilicon Groups. *J. Macromol. Sci., Chem.* **1998**, *A35*, 1941–1955.
88. Liu, C.; Liu, Y.; Shen, Z.; Xie, P.; Zhang, R.; Yang, J.; Bai, F., Study of the Steric Tacticity of Novel Soluble Ladderlike Poly(phenylsilsesquioxne) Prepared by Stepwise Coupling Polymerizatioin. *Macromol. Chem. Phys.* **2001**, *202*, 1581–1585.
89. Tang, H.; Sun, J.; Zhou, X.; Fu, P.; Xie, P.; Zhang, R., Hydrogen-Bonding-Directed Template Synthesis of Novel Stereo-Regular Organo-Bridged Ladder-Like Polymethylsiloxane. *Macromol. Chem. Phys.* **2003**, *204*, 155–163.
90. Hillson, S. D.; Smith, E.; Zeldin, M.; Parish, C. A., Cages, Baskets, Ladders, and Tubes: Conformational Studies of Polyhedral Oligomeric Silsesquioxanes. *J. Phys. Chem. A* **2005**, *109*, 8371–8378.
91. An, Y.-C.; Jang, N.-Y.; Lee, S. W.; Lim, J.-H.; Kim, K.-M., Organic-Inorganic Polymer Hybrids of Polystyrene with Ladder-Like Polyphenylsilsesquioxane. *Polym. J.* **2008**, *40*, 538–542.
92. Zhou, Q.; Yan, S.; Han, C. C.; Xie, P.; Zhang, R., Promising Functional Materials Based on Ladder Polysiloxanes. *Adv. Mater.* **2008**, *20*, 2970–2976.
93. Ren, Z.; Xie, P.; Jiang, S.; Yan, S.; Zhang, R., Study of the Supramolecular Architecture-Directed Synthesis of a Well-Defined Triple-Chain Ladder Polyphenylsiloxane. *Macromolecules* **2010**, *43*, 2130–2136.
94. Furgal, J. C.; Jung, J. H.; Clark, S.; Goodson, T.; Laine, R. M., Beads on a Chain (BoC) Phenylsilsesquioxane (SQ) Polymers via F- Catalyzed Rearrangements and ADMET or Reverse Heck Cross-Coupling Reactions: Through Chain, Extended Conjugation in 3-D with Potential for Dendronization. *Macromolecules* **2013**, *46*(19), 7591–7604.
95. Warrick, E. L.; Pierce, O. R.; Polmanteer, K. E.; Saam, J. C., Silicone Elastomer Developments 1967–1977. *Rubber Chem. Technol.* **1979**, *52*, 437–525.
96. McGrath, J. E.; Riffle, J. S.; Banthia, A. K.; Yilgor, I.; Wilkes, G. L., An Overview of the Polymerization of Cyclosiloxanes. In *Initiation of Polymerization*, Bailey, F. E., Jr., Ed. American Chemical Society: Washington, DC, 1983; pp 145–172.
97. Mark, J. E., Silicon-Containing Polymers. In *Silicon-Based Polymer Science. A Comprehensive Resource*, Zeigler, J. M.; Fearon, F. W. G., Eds. American Chemical Society: Washington, DC, 1990; Vol. 224, pp 47–68.
98. Mark, J. E., Molecular Aspects of Rubberlike Elasticity. *Acc. Chem. Res.* **1985**, *18*, 202–206.
99. Mark, J. E.; Erman, B., *Rubberlike Elasticity: A Molecular Primer*, 2nd ed. Cambridge University Press: Cambridge, 2007.
100. Thami, T.; Nasr, G.; Bestal, H.; van der Lee, A.; Bresson, B., Functionalization of Surface-Grafted Polymethylhydrosiloxane Thin Films with Alkyl Side Chains. *J. Polym. Sci., Part A: Polym. Chem.* **2008**, *46*, 3546–3562.
101. Cazacu, M., Possibilities to Develop Functional Materials on Silicone/Silica Backbones. In *Recent Developments in Silicone-Based Materials*, Cazacu, M., Ed. Nova Science: New York, 2010; pp 1–34.
102. Yilgor, I.; Riffle, J. S.; McGrath, J. E., Reactive Siloxane Oligomers. In *Reactive Oligomers*, Harris, F. W.; Spinelli, H. J., Eds. American Chemical Society: Washington, DC, 1985; Vol. 282, pp 161–174.
103. McGrath, J. E.; Sormani, P. M.; Elsbernd, C. S.; Kilic, S., Kinetics, Mechanisms, and Synthesis Studies of Difunctional Aminopropyl Terminated Polydimethylsiloxane Oligomers. *Makromol. Chem., Macromol. Symp.* **1986**, *6*, 67–80.

104. Wang, B.; Oleschuk, R. D.; Horton, J. H., Chemical Force Titrations of Amine- and Sulfonic Acid-Modified Poly(dimethylsiloxane). *Langmuir* **2005**, *21*, 1290–1298.
105. Chakraborty, R.; Soucek, M. D., Synthesis of Amine and Epoxide Telechelic Siloxanes. *Macromol. Chem. Phys.* **2008**, *209*, 604–614.
106. Novi, C.; Mourran, A.; Keul, H.; Moller, M., Ammonium-Functionalized Polydimethylsiloxanes: Synthesis and Properties. *Macromol. Chem. Phys.* **2006**, *207*, 273–286.
107. Dragan, E. S.; Cazacu, M.; Nistor, A., Ionic Organic/Inorganic Materials. III. Stimuli Responsive Hybrid Hydrogels Based on Oligo(*N,N*-dimethylamino-ethylmethacrylate) and Chloroalkyl-Functionalized Siloxanes. *J. Polym. Sci., Part A: Polym. Chem.* **2009**, *47*, 6801–6813.
108. Batra, A.; Cohen, C., Reversible Modulus Reinforcement of End-Linked Poly(dimethylsiloxane) Ionomer Networks. *Polymer* **2005**, *46*, 12416–12421.
109. Batra, A.; Cohen, C.; Duncan, T. M., Scaling Behavior of the Viscosity of Poly(dimethylsiloxane)/Ionomer Solutions. *Macromolecules* **2006**, *39*, 2398–2404.
110. Batra, A.; Cohen, C.; Kim, H.; Winey, K. I.; Ando, N.; Gruner, S. M., Counterion Effect on the Rheology and Morphology of Tailored Poly(dimethylsiloxane) Ionomers. *Macromolecules* **2006**, *39*, 1630–1638.
111. Yao, D.; Bender, T. P.; Gerroir, P. J.; Sundararajan, P. R., Self Assembled Vesicular Nanostructures of Perylene End-Capped Poly(dimethylsiloxane). *Macromolecules* **2005**, *38*, 6972–6978.
112. Yao, D.; Tuteja, B.; Sundararajan, P. R., Pigment-Mediated Nanoweb Morphology of Poly(dimethylsiloxane)-Substituted Perylene Bisimides. *Macromolecules* **2006**, *39*, 7786–7788.
113. Langhals, H.; Krotz, O., Chiral, Bichromophoric Silicones: Ordering Principles of Structural Units in Complex Molecules. *Angew. Chem. Int. Ed. Engl.* **2006**, *45*, 4444–4447.
114. Domingues, R. A.; Yoshida, I. V. P.; Atvars, T. D. Z., Fluorescence Labeling and Thermal Relaxation Processes of Anthracenyl-Labeled Polysiloxanes. *J. Polym. Sci., Part B: Polym. Phys.* **2010**, *48*, 74–81.
115. Szadkowska-Nicze, M.; Mayer, J., Ionic and Excited Species in Irradiated Poly(dimethylsiloxane) Doped with Pyrene. *J. Polym. Sci., Polym. Chem. Ed.* **2004**, *42*, 6125–6133.
116. Cazaux, F.; Coqueret, X., Photochemistry of Polysiloxanes. In *Silicon-Based Inorganic Polymers*, De Jaeger, R.; Gleria, M., Eds. Nova Science: New York, 2008; pp 191–215.
117. Wild, M. J.; Buhler, P., On the Phase Composition of Polydimethylsiloxane Derived Ceramics. *J. Mater. Sci.* **1998**, *33*, 5441–5444.
118. Stevens, N. S. M.; Rezac, M. E., Nanoporous Organic/Inorganic Hybrid Materials Produced from Poly(dimethylsiloxane). *Polymer* **1999**, *40*, 4289–4298.
119. Mark, J. E.; Allcock, H. R.; West, R., *Inorganic Polymers*, 2nd ed. Oxford University Press: New York, 2005.
120. Liu, Y.-L.; Li, S.-H., Poly(dimethylsiloxane) Star Polymers Having Nanosized Silica Cores. *Macromol. Chem., Rapid Commun.* **2004**, *25*, 1392–1395.
121. Cai, G.; Weber, W. P., Synthesis of Terminal Si-H Irregular Tetra-Branched Star Polysiloxanes. Pt-Catalyzed Hydrosilation with Unsaturated Epoxides. Polysiloxane Films by Photo-Acid Catalyzed Crosslinking. *Polymer* **2004**, *45*, 2941–2948.

122. Fragouli, P.; Iatrou, H.; Hajichristidis, N.; Sakurai, T.; Matsunaga, Y.; Hirao, A., Synthesis of Well-Defined Miktoarm Star Polymers of Poly(dimethylsiloxane) by the Combination of Chlorosilane and Benzyl Chloride Linking Chemistry. *J. Polym. Sci., Part A: Polym. Chem.* **2006**, *44*, 6587–6599.
123. Thompson, D. B.; Brook, M. A., Rapid Assembly of Complex 3D Siloxane Architectures. *J. Am. Chem. Soc.* **2008**, *130*, 32–33.
124. Dvornic, P. R.; Owen, M. J., Poly(amidoamine organosilicon) Dendrimers and Their Derivatives of Higher Degree of Structural Complexity. In *Synthesis and Properties of Silicones and Silicone-Modified Materials*, Clarson, S. J.; Fitzgerald, J. J.; Owen, M. J.; Smith, S. D.; Van Dyke, M. E., Eds. American Chemical Society: Washington, DC, 2003; Vol. 838, pp 236–259.
125. Dvornic, P. R., PAMAMOS: The First Commercial Silicon-Containing Dendrimers and Their Applications. *J. Polym. Sci., Part A: Polym. Chem.* **2006**, *44*, 2755–2773.
126. Chojnowski, J.; Fortuniak, W.; Scibiorek, M.; Rozga-Wijas, K.; Grzelka, A.; Mizerska, U., 3-Chloropropyl Functionalized Dendrigraft Polysiloxanes and Dendritic Polyelectrolytes. *Macromolecules* **2007**, *40*, 9339–9347.
127. Muzafarov, A. M., Siloxane Dendrimers. In *Polymer Data Handbook, 2nd ed.*; Mark, J. E., Ed. Oxford University Press: New York, 2009; pp 1194–1195.
128. Yan, D.; Frey, H., *Hyperbranched Polymers: Synthesis, Properties, and Applications*. Wiley: Hoboken, NJ, 2011.
129. Kazakova, V. V.; Rebrov, E. A.; Myakushev, V. B.; Strelkova, T. V.; Ozerin, A. N.; Ozerina, L. A.; Chenskaya, T. B.; Sheiko, S. S.; Sharipov, E. Y.; Muzafarov, A. M., From a Hyperbranched Polyethoxysiloxane Toward Molecular Forms of Silica: A Polymer-Based Approach to the Monitoring of Silica Properties. In *Silicones and Silicone-Modified Materials*, Clarson, S. J.; Fitzgerald, J. J.; Owen, M. J.; Smith, S. D., Eds. American Chemical Society: Washington, DC, 2000; Vol. 729, pp 503–515.
130. Si, Q.-F.; Wang, X.; Fan, X.-D.; Wang, S.-J., Synthesis and Characterization of Ultraviolet-Curable Hyperbranched Poly(siloxysilanes). *J. Polym. Sci., Part A: Polym. Chem.* **2005**, *43*, 1883–1894.
131. Chojnowski, J.; Rubinzstajn, S.; Fortuniak, W.; Kurjata, J., Synthesis of Highly Branched Alkoxysilane-Dimethylsiloxane Copolymers by Nonhydrolytic Dehydrocarbon Polycondensation Catalyzed by Tris(pentafluorophenyl)borane. *Macromolecules* **2008**, *41*, 7352–7358.
132. Yokomachi, K.; Seino, M.; Grunzinger, S. J.; Hayakawa, T.; Kakimoto, M., Synthesis and Degree of Branching of Epoxy-Terminated Hyperbranched Polysiloxane. *Polym. J.* **2008**, *40*, 198–204.
133. Wang, S.-J.; Fan, X.-D.; Kong, J.; Wang, X.; Liu, Y.-Y.; Zhang, G.-B., A New Controllable Approach to Synthesize Hyperbranched Poly(siloxanes). *J. Polym. Sci., Part A: Polym. Chem.* **2008**, *46*, 2708–2720.
134. Dou, Q.; Zhu, X.; Peter, K.; Demco, D. E.; Muller, M., Preparation of Polypropylene/Silica Composites by In-Situ Sol-Gel Processing Using Hyperbranched Polyethoxysiloxane. *J. Sol-Gel Sci. Technol.* **2008**, *48*, 51–60.
135. Korolev, G. V.; Mogilevich, M. M., *Three-Dimensional Free-Radical Polymerization: Cross-Linked and Hyper-Branched Polymers*. Springer Verlag: Heidelberg, 2009.
136. Ji, L.; Gu, A.; Liang, G.; Yuan, L., Novel Modification of Bismaleimide-Triazine Resin by Reactive Hyperbranched Polysiloxane. *J. Mater. Sci.* **2010**, *45*, 1859–1865.
137. Konkolewicz, D.; Thorn-Seshold, O.; Gray-Weale, A., Models for Randomly Hyperbranched Polymers: Theory and Simulations. *J. Chem. Phys.* **2008**, *129*, 054901.

138. Tomalia, D. A.; Frechet, J. M. J., Discovery of Dendrimers and Dendritic Polymers: A Brief Historical Perspective. *J. Polym. Sci., Part A: Polym. Chem.* **2002**, *40*, 2719–2728.
139. Zheng, Y.; Thurecht, K. J.; Wang, W., Polysiloxanes Polymers with Hyperbranched Structure and Multivinyl Functionality. *J Polym. Sci Pol Chem* **2012**, *50*(4), 629–637.
140. Zheng, Y.; Thurecht, K. J.; Wang, W., Polysiloxane Polymers with Hyperbranched Structure and Multivinyl Functionality. *J. Polym. Sci., Part A: Polym. Chem.* **2012**, *50*, 629–637.
141. Sakamoto, S.; Shimojima, A.; Miyasaka, K.; Ruan, J.; Terasaki, O.; Kurada, K., Formation of Two- and Three-Dimensional Hybrid Mesostructures from Branched Siloxane Molecules. *J. Am. Chem. Soc.* **2009**, *131*, 9634–9635.
142. Samadi, F.; Eckelt, J.; Wolf, B. A.; Schule, H.; Frey, H., Branched Versus Linear Oligo(dimethylsiloxanes): Differences in Their Thermodynamic Interaction with Solvents. *J. Polym. Sci., Part B: Polym. Phys.* **2010**, *48*, 1308–1318.
143. Racles, C., Siloxane-Containing Liquid Crystalline Polymers (LCPs). In *Advances in Functional Heterochain Polymers*, Cazacu, M., Ed. Nova Science: New York, 2008; pp 33–66.
144. Beatty, C. L.; Karasz, F. E., Transitions in Poly(diethylsiloxane). *J. Polym. Sci., Polym. Phys. Ed.* **1975**, *13*, 971–975.
145. Pochan, J. M.; Beatty, C. L.; Hinman, D. D.; Karasz, F. E., Dielectric Studies of Relaxations and Polymorphism in Poly(diethylsiloxane). *J. Polym. Sci., Polym. Phys. Ed.* **1975**, *13*, 977–983.
146. Kanesaka, S.; Kimura, H.; Kuroki, S.; Ando, I.; Fujishige, S., Diffusional Behavior of Poly(diethylsiloxane) in the Liquid-Crystalline and Isotropic Phases as Studied by High Field-Gradient NMR. *Macromolecules* **2004**, *37*, 453–458.
147. Miller, K. J.; Grebowicz, J.; Wesson, J. P.; Wunderlich, B., Conformations of Poly(diethylsiloxane) and Its Mesophase Transitions. *Macromolecules* **1990**, *23*, 849–856.
148. Papkov, V. S.; Turetskii, A. A.; Out, G. J. J.; Moller, M., Uniaxial Extension of Poly(diethylsiloxane) Networks: Mesophase Formation Under Stress. *Int. J. Polym. Mat.* **2002**, *51*, 369–391.
149. Shibanov, Y. D., Optical Properties of Polydiethylsiloxane Mesophase. *Polym. Sci. U.S.S.R.* **1989**, *31*, 2653–2662.
150. Godovsky, Y. K.; Valetskaya, L. A., Mesophase Elastomers: Stress-Induced Mesophase Formation in Poly(diethylsloxane) Networks and Their Thermomechanical Behavior. *Polym. Bull.* **1991**, *27*, 221–226.
151. Godovsky, Y. K., Rubber-Elastic Mesophase Siloxane Systems. *Angew. Makromol. Chem.* **1992**, *202*, 187–212.
152. Out, G. J. J.; Turetskii, A. A.; Moller, M.; Oelfin, D., Poly(di-*n*-alkylsiloxane)s with Long Alkyl Side Groups. *Macromolecules* **1994**, *27*, 3310–3318.
153. Hedden, R. C.; Saxena, H.; Cohen, C., Mechanical Properties and Swelling Behavior of End-Linked Poly(diethylsiloxane) Networks. *Macromolecules* **2000**, *33*, 8676–8684.
154. Godovsky, Y. K.; Papkov, V. S., Poly(diethylsiloxane). In *Polymer Data Handbook, 2nd ed.*; Mark, J. E., Ed. Oxford University Press: New York, 2009; pp 522–527.
155. Godovsky, Y. K.; Papkov, V. S.; Magonov, S. N., Atomic Force Microscopy Visualization of Morphology Changes Resulting from the Phase Transitions in Poly(di-*n*-alkoxysiloxane)s: Poly(diethylsiloxane). *Macromolecules* **2001**, *34*, 976–990.

156. Inomata, K.; Yamamoto, K.; Nose, T., An X-Ray Diffraction Study on Crystalline and Mesomorphic Structures of Poly(diethylsiloxane). *Polym. J.* **2000**, *32*, 1044–1048.
157. Voischchev, V. S.; Popov, I. V.; Sazhin, B. I.; Voishcheva, O. V.; Valetskii, P. M.; Papkov, V. S., Thermopolarization Effects in Poly(diethylsiloxane). *Polym. Sci. U.S.S.R., A* **1999**, *41*, 311–314.
158. Hedden, R. C.; Tachibana, H.; Duncan, T. M.; Cohen, C., Effects of Molecular Structure on Segment Orientation in Siloxane Elastomers. 2. NMR Measurements from Uniaxially Stretched Samples. *Macromolecules* **2001**, *34*, 5540–5546.
159. Batra, A.; Hedden, R. C.; Schofield, P.; Barnes, A.; Cohen, C.; Duncan, T. M., Conformational Behavior of Guest Chains in Uniaxially Stretched Poly(diethylsiloxane) Elastomers: ^{2}H NMR and SANS. *Macromolecules* **2003**, *36*, 9458–9466.
160. Chaika, E. M.; Chalykh, A. E.; Gerasimov, V. K.; Dubovik, I. I.; Rabkina, A. Y.; Zavin, B. G.; Papkov, V. S., Phase State in Poly(dimethylsiloxane) and Poly(diethylsiloxane) Blends. *Polym. Sci. U.S.S.R., Ser. A* **2005**, *47*, 139–150.
161. Collings, P. J., *Liquid Crystals. Nature's Delicate Phase of Matter.* Princeton University Press: Princeton, 1990.
162. Mark, J. E., *Polymer Data Handbook, 2nd ed.* Oxford University Press: New York, 2009.
163. Molenberg, A.; Moller, M.; Sautter, E., Columnar Mesophase Behavior of Poly(di-*n*-alkylsiloxane)s. *Prog. Polym. Sci.* **1997**, *22*, 1133–1144.
164. Belousov, S. I.; Buzin, A. I.; Godovsky, Y. K., Self-Organization of Polysiloxanes into Mono- and Multilayer Structures at Interphase Boundaries. *Polym. Sci. U.S.S.R.* **1999**, *41*, 303–324.
165. Gearba, R. I.; Anokhin, D. V.; Bondar, A. I.; Godovsky, Y. K.; Papkov, V. S.; Makarova, N. N.; Maganov, S. N.; Bras, W.; Koch, M. H. J.; Masin, F.; Goderis, B.; Ivanov, D. A., Mesomorphism, Polymorphism, and Semicrystalline Morphology of Poly(di-*n*-propylsiloxane). *Macromolecules* **2006**, *39*, 988–999.
166. Out, G. J. J.; Turetskii, A. A.; Moller, M., Structural Correlation Between the Columnar Mesophase and the Melt in Poly(di-*n*-alkylsiloxane)s. *Makromol. Chem., Rapid Commun.* **1995**, *16*, 107–112.
167. Turetskii, A. A.; Out, G. J. J.; Klok, H.-A.; Moller, M., Structural Transformations in Poly(di-*n*-alkylsiloxane)s with Alkyl Side Groups Containing 7 to 10 Carbon Atoms. *Polymer* **1995**, *36*, 1303–1308.
168. Godovsky, Y. K.; Makarova, N. N.; Papkov, V. S.; Kuzmin, N. N., Crystalline Phases and a Mesophase in Poly(dipropylsiloxane). *Makromol. Chem.* **1985**, *6*, 443–450.
169. Winokur, M. J.; West, R., X-ray Diffraction and Molecular Modeling Studies of Poly(di-n-alkylsilanes): The Near Planar Type Phases of Poly(di-n-butylsilane) and Poly(di-n-hexylsilane). *Macromolecules* **2003**, *36*(19), 7338–7347.
170. Wagner, J.; Phillips, P. J., The Mechanism of Crystallization of Linear Polyethylene, and Its Copolymers with Octene, Over a Wide Range of Supercoolings. *Polymer* **2001**, *42*, 8999–9013.
171. Zhang, X.-B.; Li, Z.-S.; Lu, Z.-Y.; Sun, C.-C., Roles of Branch Content and Branch Length in Copolyethylene Crystallization: Molecular Dynamics Simulations. *Macromolecules* **2002**, *35*, 106–111.
172. Bergmann, G. H. F.; Finkelmann, H.; Percec, V.; Zhao, M., Liquid-Crystalline Main-Chain Elastomers. *Makromol. Rapid Commun.* **1997**, *18*, 353–360.

173. Godovsky, Y. K.; Makarova, N. N.; Matukhina, E. V., Mesophase Behavior and Structure of Mesophases in Cyclolinear Polyorganosiloxanes. In *Silicones and Silicone-Modified Materials*, Clarson, S. J.; Fitzgerald, J. J.; Owen, M. J.; Smith, S. D., Eds. American Chemical Society: Washington, DC, 2000; Vol. 729, pp 98–114.
174. Ganicz, T.; Stanczyk, W. A., Organosilicon Mesomorphic Polymer Systems. *Prog. Polym. Sci.* **2003**, *28*, 303–329.
175. Godovsky, Y. K.; Belousov, S. I.; Makarova, N. N.; Brezesinski, G.; Mohwald, H.; Jensen, T. R.; Kjaer, K., "Smart" Polysiloxane Monolayers and Self-Organized Ultrathin Films. *Nonlinear Opt., Quantum Opt.* **2004**, *31*, 31–44.
176. Kaneko, K.; Miwa, Y.; Nakamura, N., Phase Transition Behavior and Electro-Rheological Effect of Liquid Crystalline Cyclic-Siloxanes with Fluorine Atoms. *J. Appl. Polym. Sci.* **2007**, *105*, 2474–2481.
177. Disch, S.; Schmidt, C.; Finkelmann, H., Nematic Elastomers Beyond the Critical Point. *Makromol. Rapid Commun.* **1994**, *15*, 303–310.
178. Kupfer, J.; Finkelmann, H., Liquid Crystal Elastomers: Influence of the Orientational Distribution of the Crosslinks on the Phase Behavior and Reorientation Processes. *Makromol. Chem. Phys.* **1994**, *195*, 1353–1367.
179. Kundler, I.; Finkelmann, H., Strain-Induced Director Reorientation in Nematic Liquid Single Crystal Elastomers. *Makromol. Rapid Commun.* **1995**, *16*, 679–686.
180. Patil, H. P.; Liao, J.; Hedden, R. C., Smectic Ordering in Main-Chain Siloxane Polymers and Elastomers Containing p-Phenylene Terephthalate Mesogens. *Macromolecules* **2007**, *40*, 6206–6216.
181. Li, L.; Jones, C. D.; Magolan, J.; Lemieux, R. P., Siloxane-Terminated Phenyl-pyrimidine Liquid Crystal Hosts. *J. Mater. Chem.* **2007**, *17*, 2313–2318.
182. Roberts, J. C.; Kapernaum, N.; Giesselmann, F.; Wand, M. D.; Lemieux, R. F., Fast Switching Organosiloxane Ferroelectric Crystals. *J. Mater. Chem.* **2008**, *18*, 5301–5306.
183. Masser, K. A.; Patil, H. P.; Hedden, R. C.; Runt, J., Dynamics of Main-Chain Liquid Crystalline Polysiloxanes Containing *p*-Phenyleneterephthalate Mesogens. *J. Non-Cryst. Solids* **2010**, *356*, 578–581.
184. Shoji, Y.; Ishige, R.; Higashihara, T.; Kawauchi, S.; Watanabe, J.; Ueda, M., Synthesis and Liquid Crystalline Behavior of Laterally Substituted Polyimides with Siloxane Linkages. *Macromolecules* **2010**, *43*, 8950–8956.
185. Zentel, R.; Reckert, G., Liquid Crystalline Elastomers Based on Liquid Crystalline Side Group, Main Chain and Combined Polymers. *Makromol. Chem.* **1986**, *187*, 1915–1926.
186. Pandey, S.; Kolli, B.; Mishra, S. P.; Samui, A. B., Siloxane Polymers Containing Azo Moieties Synthesized by Click Chemistry for Photo Responsive and Liquid Crystalline Applications. *J. Polym. Sci., Part A: Polym. Chem.* **2012**, *50*, 1205–1215.
187. Kaneko, K.; Oto, K.; Kawai, T.; Choi, H.; Kikuchi, H.; Nakamura, N., Electrorheological Effect and Electro-Optical Properties of Side-on Liquid Crystalline Polysiloxane in a Nematic Solvent. *ChemPhysChem* **2013**, *14*(12), 2704–2710.
188. Ganicz, T.; Pakula, T.; Fortuniak, W.; Bialecka-Florjanczyk, E., Linear and Hyperbranched Liquid Crystalline Polysiloxanes. *Polymer* **2005**, *46*, 11380–11388.
189. Meng, F.-B.; Cui, Y.; Chen, H.-B.; Zhang, B.-Y.; Jia, C., Phase Behaviors of Comb-Like Liquid-Crystalline Polysiloxanes Containing Fluorinated Mesogenic Units. *Polymer* **2009**, *50*, 1187–1196.

190. Anglaret, E.; Brunet, M.; Desbat, B.; Keller, P.; Buffeteau, T., Molecular Orientation of Liquid Crystalline Elastomers. *Macromolecules* **2005**, *38*, 4799–4810.
191. Tammer, M.; Li, J.; Komp, A.; Finkelmann, H.; Kremer, F., FTIR-Spectroscopy on Segmental Reorientation of a Nematic Elastomer Under External Mechanical Fields. *Makromol. Chem. Phys.* **2005**, *206*, 709–714.
192. Lacey, D.; Beattie, H. N.; Mitchell, G. R.; Pople, J. A., Orientation Effects in Monodomain Nematic Liquid Crystalline Polysiloxane Elastomers. *J. Mater. Chem.* **1998**, *8*, 53–60.
193. Shenoy, D. K.; Thomsen, D. L.; Keller, P.; Ratna, B. R., Nonideal Elasticity in Liquid Crystal Elastomers. *J. Phys. Chem.* **2003**, *107*, 13755–13757.
194. Gilles, P.-P.; Milano, J.-C.; Vernet, J.-L., Synthesis and Dielectric Study of New Liquid-crystalline Polysiloxanes Presenting a Thioether Spacer. *Macromol. Chem. Phys.* **2003**, *204*, 2222–2232.
195. Zhang, L.; Chen, S.; Zhao, H.; Shen, Z.; Chen, X.; Fan, X.; Zhou, Q., Synthesis and Properties of a Series of Mesogen-Jacketed Liquid Crystalline Polymers with Polysiloxane Backbones. *Macromolecules* **2010**, *43*, 6024–6032.
196. Loffler, R.; Finkelmann, H., Amphiphilic Liquid-Crystalline Elastomers. *Makromol. Rapid Commun.* **1990**, *11*, 321–328.
197. Fischer, P.; Schmidt, C.; Finkelmann, H., Amphiphilic Liquid-Crystalline Networks—Phase Behavior and Alignment by Mechanical Fields. *Makromol. Rapid Commun.* **1995**, *16*, 435–447.
198. Da Rocha, S. R. P.; Dickson, J.; Cho, D.; Rossky, P. J.; Johnston, K. P., Stubby Surfactants for Stabilization of Water and CO_2 Emulsions: Trisiloxanes. *Langmuir* **2003**, *19*, 3114–3120.
199. Dong, J.; Mao, G.; Hill, R. M., Nanoscale Aggregate Structures of Trisiloxane Surfactants at the Solid-Liquid Interface. *Langmuir* **2004**, *20*, 2695–2700.
200. Kaneko, M.; Matsuzawa, K.; Uddin, M. H.; Lopez-Quintela, M. A.; Kunieda, H., Effect of Hydrophilic Chain Length of Amphiphilic Silicone Oil (Copolymer) on the Nonionic Surfactant-Layer Curvature. *J. Phys. Chem. B* **2004**, *108*, 12736–12743.
201. Liang, W.-J.; Wu, C.-P.; Hsu, C.-Y.; Kuo, P.-L., Synthesis, Characterization, and Proton-Conducting Properties of Organic-Inorganic Hybrid Membranes Based on Polysiloxane Zwitterionomer. *J. Polym. Sci., Part A: Polym. Chem.* **2006**, *44*, 3444–3453.
202. Erdodi, G.; Kennedy, J. P., Third-Generation Amphiphilic Conetworks. I. Synthesis and Swelling Behavior of Poly(N,N-dimethyl acrylamide) / Poly(dimethylsiloxane) Conetworks. *J. Polym. Sci., Part A: Polym. Chem.* **2007**, *45*, 295–307.
203. Srividhya, M.; Chandrasekar, K.; Baskar, G.; Reddy, B. S. R., Physico-Chemical Properties of Siloxane Surfactants in Water and Their Surface Energy Characteristics. *Polymer* **2007**, *48*, 1261–1268.
204. Beppu, K.; Kaneko, Y.; Kadokawa, J.-I.; Mori, H.; Nishiksawa, T., Synthesis of Sugar-Polysiloxane Hybrids Having Rigid Main-Chains and Formation of Their Nano Aggregates. *Polym. J.* **2007**, *39*, 1065–1070.
205. Pavlovic, D.; Linhardt, J. G.; Kunzier, J. F.; Shipp, D. A., Synthesis of Amphiphilic Multiblock and Triblock Copolymers of Polydimethylsiloxane and Poly(*N,N*-dimethylacrylamide). *J. Polym. Sci., Part A: Polym. Chem.* **2008**, *46*, 7033–7048.
206. Iwakiri, N.; Nishikawa, T.; Kaneko, Y.; Kadokawa, J.-I., Synthesis of Amphiphilic Polysiloxanes and Their Properties for Formation of Nano-Aggregates. *Coll. Polym. Sci.* **2009**, *287*, 577–582.

207. Kang, J.; Erdodi, G.; Kennedy, J. P., Thermoplastic Amphiphilic Conetworks. *J. Polym. Sci., Part A: Polym. Chem.* **2009**, *47*, 682–691.
208. Mecham, S.; Sentman, A.; Sambasivam, M., Amphiphilic Silicone Copolymers for Pressure Sensitive Adhesive Applications. *J. Appl. Polym. Sci.* **2010**, *116*, 3265–3270.
209. Di, C.; Jiang, Y.; Wang, R.; Yin, J., Multi-Responsive Nanoparticles from the Amphiphilic Poly(dimethylsiloxane) (PDMS)-Containing Poly(ether amine) (PDMS-gPEA) and Its Potential Applications for Smart Separation. *J. Mater. Chem.* **2011**, *21*, 4416–4423.
210. Meng, F.; Lian, J.; Zhang, B.; Sun, Y., Chiral Side-Chain Liquid-Crystalline Polysiloxanes Bearing Fluorinated Mesogens and Cholesteryl Groups. *Eur. Polym. J.* **2008**, *44*, 504–513.
211. Zhang, B.-Y.; Hu, J.-S.; Ren, S.-C.; Liu, C., Synthesis, Structure, and Phase Behavior of New Chiral Liquid-Crystalline Polysiloxanes Based on Mesogenic Menthyl Monomers. *J. Appl. Polym. Sci.* **2008**, *111*, 3016–3025.
212. Chen, W. H.; Chuang, W.-T.; Jeng, U.-S.; Sheu, H.-S.; Lin, H.-C., New SmCG Phases in a Hydrogen-Bonded Bent-Core Liquid Crystal Featuring a Branched Siloxane Terminal Group. *J. Am. Chem. Soc.* **2011**, *133*, 15674–15685.
213. Meng, F.-B.; Gao, Y.-M.; Zhang, B.-Y.; Xu, X.-X., Synthesis and Characterization of H-Bonded Side-Chain Liquid-Crystalline Polysiloxanes. *J. Macromol. Sci., Pure Appl. Chem.* **2008**, *45*, 186–194.
214. Zentel, R.; Benalia, M., Stress-Induced Orientation in Lightly Crosslinked Liquid-Crystalline Side-Group Polymers. *Makromol. Chem.* **1987**, *188*, 665–674.
215. Hammerschmidt, K.; Finkelmann, H., Stress-Optical and Thermomechanical Measurements on Liquid Crystalline Elastomers. *Makromol. Chem.* **1989**, *190*, 1089–1101.
216. Finkelmann, H.; Kock, H.-J.; Gleim, W.; Rehage, G., Investigations on Liquid Crystalline Polysiloxanes, 5. Orientation of L-C-Elastomers by Mechanical Forces. *Makromol. Rapid Commun.* **1984**, *5*, 287–293.
217. Mitchell, G. R.; Davis, F. J.; Ashman, A., Structural Studies of Side-Chain Liquid Crystal Polymers and Elastomers. *Polymer* **1987**, *28*, 639–647.
218. Percec, V.; Hsu, C.-S., Synthesis and Mesomorphic Behavior of Poly(methylsiloxane)s and Poly(methylsiloxane-co-dimethylsiloxane)s Containing Oligooxyethylene Spacers and Mesogenic Side Groups. *Polym. Bull.* **1990**, *23*, 463–470.
219. Kaufhold, W.; Finkelmann, H.; Brand, H. R., Effect of the Spacer Length on the Mechanical Coupling Between Network Anisotropy and Nematic Order. *Makromol. Chem.* **1991**, *192*, 2555–2579.
220. Zubarev, E. R.; Talroze, R. V.; Yuranova, T. I.; Vasilets, V. N.; Plate, N. A., Influence of Crosslinking Conditions on the Phase Behavior of a Polyacrylate-Based Liquid-Crystalline Elastomer. *Makromol. Rapid Commun.* **1996**, *17*, 43–49.
221. Jana, R. N.; Cho, J. W., Silicone-Based Cholesteric Liquid Crystalline Polymers: Effect of Crosslinking Agent on Phase Transition Behavior. *J. Appl. Polym. Sci.* **2009**, *114*, 3566–3573.
222. Davis, F. J.; Gilbert, A.; Mann, J.; Mitchell, G. R., Liquid Crystal Elastomers: Synthesis and Characterization. *J. Polym. Sci., Part A: Polym. Chem.* **1990**, *28*, 1455–1472.
223. Kupfer, J.; Finkelmann, H., Nematic Liquid Single Crystal Elastomers. *Makromol. Rapid Commun.* **1991**, *12*, 717–726.

224. Talroze, R. V.; Gubina, T. I.; Shibaev, V. P.; Plate, N. A., Peculiarities of the Thermoelastic Behavior of Liquid-Crystalline Elastomers. *Makromol. Rapid Commun.* **1990**, *11*, 67–71.
225. Wang, J.-W.; Meng, F.-B.; Li, Y.-H.; Zhang, B.-Y., Synthesis and Characterization of Side-Chain Cholesteric Liquid-Crystalline Polysiloxanes Containing Different Space Groups. *J. Appl. Polym. Sci.* **2009**, *111*, 2078–2084.
226. Shoji, Y.; Ishige, R.; Higashihara, T.; Watanabe, J.; Ueda, M., Thermotropic Liquid Crystalline Polyimides with Siloxane Linkages: Synthesis, Characterization, and Liquid Crystalline Behavior. *Macromolecules* **2010**, *43*, 805–819.
227. Meng, F.-B.; Wang, Z.-Y.; Chai, G.-W.; Wang, H.-G.; Chen, Y. C.; Zhang, B.-Y., Synthesis and Characterization of Fluorine-Containing Liquid Crystalline Polysiloxanes Bearing Cholesteryl Cinnamate Mesogens and Trifluoromethyl-Substituted Mesogens. *J. Appl. Polym. Sci.* **2010**, *116*, 2384–2395.
228. Lee, K.-J.; Hsiue, G.-H.; Wu, J.-L.; Sha, Y.-A., Molecular Structure of Side-Chain Liquid Crystalline Polysiloxane in the Smectic C Phase from X-Ray Diffraction and Molecular Modeling *Polymer* **2007**, *48*, 5161–5173.
229. Kaneko, K.; Kawai, T.; Nakamura, N., Electrorheological Effect of "Side-On" Liquid Crystalline Polysiloxane. *ChemPhysChem* **2008**, *9*, 2457–2460.
230. Woo, J. Y.; Kim, E. H.; Kim, B. Y., Transmission Holographic Polymer Dispersed Liquid Crystals Based on a Siloxane Polymer. *ChemPhysChem* **2008**, *9*, 141–146.
231. Carmichael, J. B.; Winger, R., Cyclic Distributions in Dimethylsiloxane. *J. Polym. Sci., Part A: Polym. Chem.* **1965**, *3*, 971–984.
232. Carmichael, J. B.; Gordon, D. J.; Isackkson, F. J., Dilution Effects on Dimethylsiloxane Ring-Chain Equilibria. *J. Phys. Chem.* **1967**, *71*, 2011–2015.
233. Semlyen, J. A.; Wright, P. V., Equilibrium Ring Concentrations and the Statistical Conformations of Polymer Chains. I—Oligomeric Dimethylsiloxanes. *Polymer* **1969**, *10*, 543–553.
234. Wright, P. V.; Semlyen, J. A., Equilibrium Ring Concentrations and the Statistical Conformations of Polymer Chains: Part 3. Substituent Effects in Polysiloxane Systems. *Polymer* **1970**, *11*, 462–471.
235. Beevers, M. S.; Semlyen, J. A., Equilibrium Ring Concentrations and the Statistical Conformations of Polymer Chains: Part 5. Stereoisomeric Cyclics in Poly(phenylmethylsiloxane) Equilibrates. *Polymer* **1971**, *12*, 373–382.
236. Beevers, M. S.; Semlyen, J. A., Equilibrium Ring Concentrations and the Statistical Conformations of Polymer Chains: Part 8. Calculation of Small Ring Concentrations in Polydihydrogensiloxne and Polydimethylsiloxane Equilibrates. *Polymer* **1972**, *13*, 385–390.
237. Beevers, M. S.; Semlyen, J. A., Equilibrium Ring Concentrations and the Statistical Conformations of Polymer Chains: Part 10. Cyclics in a Polymeric Paraffin-Siloxane. *Polymer* **1972**, *13*, 523–526.
238. Wright, P. V., Cyclization Equilibrium Constants and the Conformations of Polydimethylsiloxane Chains. *J. Polym. Sci., Polym. Phys. Ed.* **1973**, *11*, 51–64.
239. Scales, L. E.; Semlyen, J. A., Equilibrium Ring Concentrations and the Statistical Conformations of Polymer Chains: Part 14. Calculation of the Concentrations of Medium Sized Cyclics in Poly(dimethylsiloxane) Equilibrates. *Polymer* **1976**, *17*, 601–604.
240. Dagger, A. C.; Semlyen, J. A., Hydrogenated and Deuterated Cyclic Poly(dimethylsiloxanes). In *Silicones and Silicone-Modified Materials*, Clarson, S.

J.; Fitzgerald, J. J.; Owen, M. J.; Smith, S. D., Eds. American Chemical Society: Washington, DC, 2000; Vol. 729, pp 38–80.

241. White, B. M.; Watson, W. P.; Barthelme, E. E.; Beckham, H. W., Synthesis and Efficient Purification of Cyclic Poly(dimethylsiloxane). *Macromolecules* **2002**, *35*, 5345–5348.

242. Foston, M.; Beckham, H. W., Cyclic Poly(dimethylsiloxane) from Kinetically Controlled Cyclodepolymerization of Linear Precursors in Dilute Solution. *Polymer* **2010**, *51*, 2112–2118.

243. Foston, M.; Beckham, H. W., Cyclic Poly(dimethylsiloxane) via Ring-Closing Dehydrocoupling ±,ω-Dihydroxy-PDMS with ±,ω-dihydro-PDMS in Dilute Solution. *Polymer* **2010**, *51*, 2515–2519.

244. Glassner, M.; Blinco, J. P.; Barner-Kowollik, C., Diels-Alder Reactions as an Efficient Route to High Purity Cyclic Polymers. *Macromol. Rapid Commun.* **2011**, *32*, 724–728.

245. Lewicki, J. P.; Mayer, B. P.; Alviso, C. T.; Maxwell, R. S., Thermal Degradation Behavior and Product Speciation in Model Poly(dimethylsiloxane) Networks. *J. Inorg. Organomet. Polym. Mater.* **2012**, *22*, 636–645.

246. Semlyen, J. A.; Wood, B. R.; Hodge, P., Cyclic Polymers: Past, Present and Future. *Polym. Adv. Technol.* **1994**, *5*, 473–478.

247. Semlyen, J. A., Cyclic Polysiloxanes. *Makromol. Chem., Macromol. Symp.* **1985**, *6*, 155–163.

248. Semlyen, J. A., Cyclic Polymers. *Pure Appl. Chem* **1981**, *53*(9), 1797–1804.

249. Semlyen, J. A.; Wood, B. R., Advances in Cyclic Polymer Chemistry. *Polym. News* **1996**, *21*, 335–339.

250. Semlyen, J. A., *Cyclic Polymers*, 2nd ed. Kluwer Academic Publishers: Dordrecht, 2000.

251. Chivers, T.; Manners, I., *Inorganic Rings and Polymers of the p-Block Elements: From Fundamentals to Applications.* Royal Society of Chemistry: Cambridge, 2009.

252. Clarson, S. J., Cyclic Poly(dimethylsiloxane). In *Polymer Data Handbook, 2nd ed.*; Mark, J. E., Ed. Oxford University Press: New York, 2009; pp 113–115.

253. Clarson, S. J., Cyclic Poly(phenylmethylsiloxanes). In *Polymer Data Handbook*, 2nd ed.; Mark, J. E., Ed. Oxford University Press: New York, 2009; pp 116–118.

254. Clarson, S. J., Cyclic Poly(vinylmethylsiloxanes). In *Polymer Data Handbook*, 2nd ed.; Mark, J. E., Ed. Oxford University Press: New York, 2009; pp 119–120.

255. Endo, K., Synthesis and Properties of Cyclic Polymers. *Adv. Polym. Sci.* **2008**, *217*, 121–183.

256. Schappacher, M.; Deffieux, A., Synthesis and Macrocyclic Copolymer Brushes and Their Self-Assembly into Supramolecular Tubes. *Science* **2008**, *319*, 1512–1515.

257. Feng, W.; Yamato, K.; Yang, L.; Ferguson, J. S.; Zhang, L.; Zou, S.; Yuan, L.; Zeng, X. C.; Gong, B., Efficient Kinetic Macrocyclization. *J. Am. Chem. Soc.* **2009**, *131*, 2629–2637.

258. Laurent, B. A.; Grayson, S. M., Synthetic Approaches for the Preparation of Cyclic Polymers. *Chem. Soc. Rev.* **2009**, *38*, 2202–2213.

259. Kricheldorf, H. R., Cyclic and Multicyclic Polymers by Three-Dimensional Polycondensation. *Acct. Chem. Res.* **2009**, *42*, 981–992.

260. Monfette, S.; Fogg, D. E., Equilibrium Ring-Closing Methathesis. *Chem. Rev.* **2009**, *109*, 3783–3816.
261. Yashiro, T.; Kricheldorf, H. R.; Schwarz, G., Polymerization of Cyclosiloxanes by Means of Triflic Acid and Metal Triflates. *Macromol. Chem. Phys.* **2010**, *211*, 1311–1321.
262. Xiao, Y.; Hu, A., Bergman Cyclization in Polymer Chemistry and Material Science. *Macromol. Rapid Commun.* **2011**, *32*, 1688–1698.
263. Quirk, R. D.; Wang, S.-F.; Foster, M. D.; Wesdemiotis, C.; Yol, A. M., Synthesis of Cyclic Polystyrenes Using Living Anionic Polymerization and Methathesis Ring Closure. *Macromolecules* **2011**, *44*, 7538–7545.
264. Davis, F.; Hifson, S., *Macrocycles: Construction, Chemistry and Nanotechnology Applications*. Wiley & Sons: Hoboken, NJ, 2011.
265. Harada, A., *Supramolecular Polymer Chemistry*. Wiley-VCH: Weinheim, 2012.
266. Chen, F.; Liu, G.; Zhang, G., Synthesis of Cyclic Polyelectrolyte via Direct Copper(I)-Catalyzed Click Chemistry. *J. Polym. Sci., Part A: Polym. Chem.* **2012**, *50*, 831–838.
267. Nataro, C.; Cleaver, W. M.; Allen, C. W., 1-Methyl-1-Vinyl-3,3,5,5,-Tetraphenylcyclotrisiloxane: An Organofunctional Cyclotrisiloxane. *J. Inorg. Organomet. Polym.* **2009**, *19*, 566–569.
268. Chen, S.; Duhamel, J.; Winnik, M. A., Probing End-to-End Cyclization beyond Willemski and Fixman. *J. Phys. Chem. B* **2011**, *115*, 3289–3302.
269. Cordova, M. E.; Lorenzo, A. T.; Miller, A. J.; Hoskins, J. N.; Grayson, S. M., A Comparative Study on the Crystallization Behavior of Analogous Linear and Cyclic Poly(ε-caprolactones). *Macromolecules* **2011**, *44*, 1742–1746.
270. Takano, A.; Ohta, Y.; Matsuoka, K.; Matsubara, K.; Nakano, T.; Hieno, A.; Itakura, M.; Takahashi, K.; Kinugasa, S.; Kawaguchi, D.; Takahashii, Y.; Matsuhita, Y., Radii of Gyration of Ring-Shaped Polystyrenes with High Purity in Dilute Solutions. *Macromolecules* **2012**, *45*, 369–371.
271. Cherstvy, A. G., DNA Cyclization: Suppression or Enhancement by Electrostatic Repulsions? *J. Phys. Chem. B* **2011**, *115*, 4286–4294.
272. Collepardo-Guevara, R.; Suleimanov, Y. V.; Manolopoulos, D. E., Bimolecular Reaction Rates from Ring Polymer Molecular Dynamics. *J. Chem. Phys.* **2009**, *130*, 174713.
273. Perez, A.; Tuckerman, M. E.; Muser, M. H., A Comparative Study of the Centroid and Ring-Polymer Molecular Dynamics Methods for Approximating Quantum Time Correlation Functions from Path Integrals. *J. Chem. Phys.* **2009**, *130*, 184105.
274. Orrah, D. J.; Semlyen, J. A.; Dodgson, K.; Pethrick, R. A., Studies of Cyclic and Linear Poly(dimethylsiloxanes): 29. An Investigation of the Conformational Dynamics of Cyclic Poly(dimethylsiloxane). *Polymer* **1990**, *29*, 25–29.
275. Pham-Van-Cang, C.; Bokobza, L.; Monnerie, L.; Clarson, S. J.; Semlyen, J. A.; Vandendriessche, J.; Schryver, F. C. D., Studies of Cyclic and Linear Poly(dimethylsiloxanes): 26. An Investigation of the Mobility of Cyclic Pol(dimethylsiloxane) in the Melt Through the Excimer Emission of Small Probes. *Polymer* **1987**, *28*, 1561–1566.
276. Burton, D. J.; Harris, R. K.; Dodgson, K.; Pellow, C. J.; Semlyen, J. A., Silicon-29 N.M.R, Chemical Shifts of Cyclic Dimethyl Siloxane Oligomers. *Polym. Commun.* **1983**, *24*, 278–281.
277. Kosfeld, R.; Kreuzburg, C.; Krause, R., Mobility and ^{29}Si NMR Relaxation in Some Cyclic Siloxanes. *Makromol. Chem.* **1988**, *189*, 2077–2084.

278. Rosa, A.; Orlandini, E.; Tubiana, L.; Micheletti, C., Structure and Dynamics of Ring Polymers: Entanglement Effects Because of Solution Density and Ring Topology. *Macromolecules* **2011**, *44*, 8668–8680.
279. Vasquez, R.; Shanbhag, S., Percolation of Trace Amounts of Linear Polymers in Melts of Cyclic Polymers. *Macromol. Theory Simul.* **2011**, *20*, 205–211.
280. Kato, K.; Inoue, K.; Kidowaki, M.; Ito, K., Organic-Inorganic Hybrid Slide-Ring Gels: Polyroraxanes Consisting of Poly(dimethylsiloxane) and g-Cyclodextrin and Subsequent Topological Cross-Linking. *Macromolecules* **2009**, *42*, 7129–7136.
281. Okumura, H.; Okada, M.; Kawaguchi, Y.; Harada, A., Complex Formation Between Poly(dimethylsiloxane) and Cyclodextrins: New *Pseudo*-Polyrotaxanes Containing Inorganic Polymers. *Macromolecules* **2000**, *33*, 4297–4298.
282. Anokhin, D. V.; Gearba, R. I.; Godovsky, Y. K.; Magonov, S. N.; Makarova, N. N.; Ivanov, R. I.; Bras, W.; Ivanov, D. A., Structure and Phase Behavior of a Disk-Necklace Polymer: Cyclolinear Polymethylsiloxane. *Polymer* **2007**, *48*, 4837–4848.
283. Bohn, M.; Heermann, D. W.; Lourenco, O.; Cordeiro, C., On the Influence of Topological Catenation and Bonding Constraints on Ring Polymers. *Macromolecules* **2010**, *43*, 2564–2573.
284. Rigbi, Z.; Mark, J. E., Concurrent Chain Extension and Crosslinking of Hydroxyl-Terminated Poly(dimethylsiloxane): Possible Formation of Catenate Structures. *J. Polym. Sci., Polym. Phys. Ed.* **1986**, *24*, 443–449.
285. Fyvie, T. J.; Frisch, H. L.; Semlyen, J. A.; Clarson, S. J.; Mark, J. E., Polymeric Catenanes from Cross-Linked Poly(2,6-dimethyl-1,4-phenylene oxide) and Cyclic Poly(dimethylsiloxane). *J. Polym. Sci., Polym. Chem. Ed.* **1987**, *25*(9), 2503–2509.
286. Ohta, T.; Nakamura, M.; Matsushita, Y.; Takano, A., Synthesis, Separation and Characterization of Knotted Ring Polymers. *Polymer* **2012**, *53*, 466–470.
287. Decken, A.; Passmore, J.; Wang, X., Cyclic Dimethylsiloxanes as Pseudo Crown Ethers: Syntheses and Characterization of $\mathbf{Li(Me_2SiO)_5[Al\{OC(CF_3)_3\}_4]}$, $\mathbf{Li(Me_2SiO)_6[Al\{OC(CF_3)_3\}_4]}$, **and** $\mathbf{Li(Me_2SiO)_6[Al\{OC(CF_3)_2Ph\}_4]}$. *Angew. Chem. Int. Ed. Engl.* **2006**, *44*, 2773–2777.
288. Nugay, N.; Erdodi, G.; Kennedy, J. P., Cyclosiloxane-Based Networks: Synthesis, Thermal Characterization, and Microstructure. *J. Polym. Sci., Part A: Polym. Chem.* **2005**, *43*, 630–637.
289. Mukbaniani, O.; Tatrishvili, T.; Titvinidze, G.; Patwsatsia, S., Synthesis of Thermoreactive Polysiloxanes with Cyclic Fragments in the Side Chain. *J. Appl. Polym. Sci.* **2009**, *114*, 892–900.
290. Bohn, M.; Heermann, D. W., Topological Interactions Between Ring Polymers: Implications for Chromatin Loops. *J. Chem. Phys.* **2010**, *132*, 044904.
291. Soltysiak, J. T.; Bialecka-Florjanczyk, E.; Przedmojski, J., Liquid Crystalline Cyclic Oligosiloxanes with Polar End Groups. *Eur. Polym. J.* **2006**, *42*, 1662–1669.
292. Almond, M. J.; Becerra, R.; Bowes, S. J.; Cannady, J. P.; Ogden, J. S.; Walsh, R., A Mechanistic Study of Cyclic Siloxane Pyrolyses at Low Pressures. *Phys. Chem. Chem. Phys.* **2008**, *10*, 6856–6861.
293. Vettorel, T.; Reigh, S. Y.; Yoon, D. Y.; Kremer, K., Monte Carlo Method for Simulations of Ring Polymers in the Melt. *Macromol. Rapid Commun.* **2009**, *30*, 345–352.

294. Suzuki, J.; Takano, A.; Deguchi, T.; Matsushita, Y., Dimension of Ring Polymers in Bulk Studied by Monte-Carlo Simulations and Self-Consistent Theory. *J. Chem. Phys.* **2009**, *131*, 144902.
295. Fuereder, I.; Zifferer, G., Monte Carlo Simulation Studies of Ring Polymers at Athermal and Theta Conditions. *J. Chem. Phys.* **2011**, *135*, 184906.
296. Suzuki, J.; Takano, A.; Matsushita, Y., The Theta-Temperature Depression Caused by Topological Effect in Ring Polymers Studied by Monte Carlo Simulation. *J. Chem. Phys.* **2011**, *135*, 204903.
297. Drube, F.; Alim, K.; Witz, G.; Dietler, G.; Frey, E., Excluded Volume Effects on Semiflexible Ring Polymers. *Nano Lett.* **2010**, *10*, 1445–1449.
298. Benkova, Z.; Cifra, P., Stiffening Transition in Semiflexible Cyclic Macromolecules. *Macromol. Theory Simul.* **2011**, *20*, 65–74.
299. Halverson, J. D.; Lee, W. B.; Grest, G. S.; Grosberg, A. Y., Molecular Dynamics Simulation Study on Nonconcatenated Ring Polymers in a Melt. I. Statics. *J. Chem. Phys.* **2011**, *134*, 204904.
300. Hegde, G. A.; Chang, J.-F.; Chen, Y.-L.; Khare, R., Conformation and Diffusion Behavior of Ring Polymers in Solution: A Comparison Between Molecular Dynamics, Multiparticle Collision Dynamics, and Lattice Boltzmann Simulations. *J. Chem. Phys.* **2011**, *135*, 184901.
301. Hur, K.; Jeong, C.; Winkler, R. G.; Lacevic, N.; Gee, R. H.; Yoon, D. Y., Chain Dynamics of Ring and Linear Polyethylene Melts from Molecular Dynamics Simulations. *Macromolecules* **2011**, *44*, 2311–2315.
302. Kim, J. U.; Yang, Y.-B.; Lee, W. B., Self-Consistent Field Theory of Gaussian Ring Polymers. *Macromolecules* **2012**, *45*, 3261–3269.
303. Beaucage, G.; Kulkarni, A. S., Dimensional Description of Cyclic Macromolecules. *Macromolecules* **2010**, *43*, 532–537.
304. Lonsdale, D. A.; Monteiro, M. J., Kinetic Simulations for Cyclization of α,ω-Telechelic Polymers. *J. Polym. Sci., Part A: Polym. Chem.* **2010**, *48*, 4496–4503.
305. Clarson, S. J.; Dodgson, K.; Semlyen, J. A., Studies of Cyclic and Linear Poly(dimethylsiloxanes): 19. Glass Transition Temperatures and Crystallization Behaviour. *Polymer* **1985**, *26*, 930–934.
306. Lucas, P.; Robin, J.-J., Silicone-Based Polymer Blends: An Overview of the Materials and Processes. *Adv. Polym. Sci.* **2007**, *209*, 111–147.
307. Dong, J.; Liu, Z.; Cao, X.; Zhang, C., Elastomers Based on α,ω-Dihydroxy-Polydimethylsiloxane/Polystyrene Blends: Morphology and Mechanical Properties. *J. Appl. Polym. Sci.* **2006**, *101*, 2565–2572.
308. Wang, Y.; Hu, Y.; Gong, X.; Jiang, W.; Zhang, P.; Chen, Z., Preparation and Properties of Magnetorheological Elastomers Based on Silicone Rubber/Polystyrene Blend Matrix. *J. Appl. Polym. Sci.* **2007**, *103*, 3143–3149.
309. Daimatsu, K.; Anno, Y.; Sugimoto, H.; Nakanishi, E.; Inomata, K., Preparation, Morphology, and Physical Properties of Transparent PSt Hybrid Materials Containing Silicone Macromonomer. *J. Appl. Polym. Sci.* **2008**, *108*, 362–369.
310. Dong, J.; Zhang, N.; Liu, Z., Preparation, Morphology, and Mechanical Properties of Elastomers Based on Polydimethylsiloxane/Polystyrene Blends. *J. Appl. Polym. Sci.* **2009**, *112*, 985–990.
311. Busche, B. J.; Tonelli, A. E.; Balik, C. M., Morphology of Polystyrene/Poly(dimethylsiloxane) Blends Compatibilized with Star Polymers Containing a Gamma-Cyclodextrin Core and Polystyrene Arms. *Polymer* **2010**, *51*, 1465–1471.

312. Busche, B. J.; Tonelli, A. E.; Balik, C. M., Properties of Polystyrene/Poly(dimethylsiloxane) Blends Partially Compatibilized with Star Polymers Containing a γ-Cyclodextrin Core and Polystyrene Arms. *Polymer* **2010**, *51*, 6013–6020.
313. Bose, S.; Mukherjee, M.; Das, C. K., Silicone Rubber Compatibilized Syndiotactic Polystyrene and Thermotropic Liquid-Crystalline Polymer (Vectra A950) Blend. *Polym. Plast. Technol. Eng.* **2009**, *48*, 158–163.
314. Rodriguez, R.; De Las Heras Alarcon, C.; Ekanayake, P.; McDonald, P. J.; Keddie, J. L.; Barandiaran, M. J.; Asua, J. M., Correlation of Silicone Incorporation into Acrylic Coatings with the Resulting Hydrophobic and Thermal Properties. *Macromolecules* **2008**, *41*, 8537–8546.
315. Dos Anjos, D. S. C.; Revoredo, E. C. V.; Galembeck, A., Silicone-Polyacrylate Chemical Compatibilization with Organosilanes. *Polym. Eng. Sci.* **2010**, *50*, 606–612.
316. Gordin, C.; Delaite, C.; Bistac, S.; Rusu, D.; Rusu, M., DSC Studies of Poly(vinyl chloride)/Poly(ε-caprolactone)/Poly(ε-caprolactone)-b-Poly(dimethylsiloxane) Blends. *Polym. Bull.* **2009**, *63*, 517–529.
317. Munoz, M. P.; Werlang, M. M.; Nachtigall, S. M. B.; Cardoza, N. S. M.; Mauler, R. S., Blends of Polypropylene with Solid Silicone Additive. *J. Appl. Polym. Sci.* **2007**, *104*, 226–233.
318. Zhang, C.; Pal, K.; Byeon, J.-U.; Han, S.-M.; Kim, J. K., A Study on Mechanical and Thermal Properties of Silicone Rubber/EPDM Damping Materials. *J. Appl. Polym. Sci.* **2011**, *119*, 2737–2741.
319. Basuli, U.; Chaki, T. K.; Naskar, K., Mechanical Properties of Thermoplastic Elastomers Based on Silicone Rubber and an Ethylene-Octene Copolymer by Dynamic Vulcanization. *J. Appl. Polym. Sci.* **2008**, *108*, 1079–1085.
320. Kumar, S.; Maiti, S. N., Studies on Polycarbonate and Polydimethylsiloxane Rubber Blends. *Polym. Plastics Technol. Eng.* **2007**, *46*, 427–433.
321. Kim, S.-R.; Kim, D.-J., Effect of Silicone Oil on the Morphology and Properties of Polycarbonate. *J. Appl. Polym. Sci.* **2008**, *109*, 3439–3446.
322. Zhou, W.; Osby, J., Siloxane Modification of Polycarbonate for Superior Flow and Impact Toughness. *Polymer* **2010**, *51*, 1990–1999.
323. Tong, W.; Huang, Y.; Liu, C.; Chen, X.; Yang, Q.; Li, G., The Morphology of Immiscible PDMS/PIB Blends Filled with Silica Nanoparticles under Shear Flow. *Coll. Polym. Sci.* **2010**, *288*, 753–760.
324. Lee, J. Y.; Bhattacharya, B.; Kim, D.-W.; Park, J.-K., Poly(ethylene oxide/) Poly(dimethylsiloxane) Blend Solid Polymer Electrolyte and Its Dye-Sensitive Solar Cell Applications. *J. Phys. Chem. C* **2007**, *112*, 12576–12582.
325. Yen, M.-S.; Tsai, P.-Y., Effects of Soft Segments on the Surface Properties of Polydimethylsiloxane Waterborne Polyurethane Prepolymer Blends and Treated Nylon Fabrics. *J. Appl. Polym. Sci.* **2010**, *115*, 3550–3558.
326. Pant, R. R.; Buckley, J. L.; Fulmer, P. A.; Wynne, J. H.; McCluskey, D. M.; Phillips, J. P., Hybrid Siloxane Epoxy Coatings Containing Quaternary Ammonium Moieties. *J. Appl. Polym. Sci.* **2008**, *110*, 3080–3086.
327. Sobhari, S.; Jannesari, A.; Bashini, S., Effect of Molecular Weight and Content of PDMS on Morphology and Properties of Silicone-Modified Epoxy Resin. *J. Appl. Polym. Sci.* **2012**, *123*, 162–178.
328. Ardhyananta, H.; Wahid, M. H.; Sasaki, M.; Agag, T.; Kawauchi, T.; Ismail, H.; Takeichi, T., Performance Enhancement of Polybenzoxazine by Hybridization with Polysiloxane. *Polymer* **2008**, *49*, 4585–4591.

329. Ardhyananta, H.; Kawauchi, T.; Ismail, H.; Takeichi, T., Effect of Pendant Group of Polysiloxanes on the Thermal and Mechanical Properties of Polybenzoxazine Hybrida. *Polymer* **2009**, *50*, 5959–5969.
330. Carpi, F.; Gallone, G.; Galantini, F.; De Rossi, D., Silicone-Poly(hexylthiophene) Blends as Elastomers with Enhanced Electromechanical Transduction Properties. *Adv. Funct. Mater.* **2008**, *18*, 235–241.
331. Zhang, W.; Deodhar, S.; Yao, D., Processing Properties of Polypropylene with a Minor Addition of Silicone Oil. *Polym. Eng. Sci.* **2010**, *50*, 1340–1349.
332. Chaika, E. M.; Vasil'ev, V. G.; Papkov, V. S., Rheological Properties of Poly(dimethylsiloxane)-Poly(diethylsiloxane) Blends. *Polym. Sci. U.S.S.R., A* **2007**, *49*, 337–348.
333. Fenouillot, F.; Mechin, F.; Boisson, F.; Alcouffe, P.; Pokropski, T.; Kalllei, T.; Mnif, M., Coarsening of Nanodomains by Reorganization of Polysiloxane Segments at High Temperature in Polyurethane/α,ω-Aminopropyl Polydimethylsiloxane Blends. *Eur. Polym. J.* **2012**, *48*, 284–295.
334. Mielczarski, J. A.; Mielczarski, E.; Galli, G.; Morelli, A.; Martinelli, E.; Chiellini, E., The Surface-Segregated Nanostructure of Fluorinated Copolymer-Poly(dimethylsiloxane) Blend Films. *Langmuir* **2010**, *26*, 2871–2876.
335. Kim, D. H.; Hwang, S. H.; Kim, B. S., The Effects of Technological Compatibility for Silicone Rubber/Fluororubber Blends. *J. Appl. Polym. Sci.* **2012**, *125*, 1625–1635.
336. Schmidt, R. G.; Badour, L. R.; Gordon, G. V., Modulus Reduction Mechanism of Trimethylsiloxy Silicates in a Polyorganosiloxane. In *Synthesis and Properties of Silicones and Silicone-Modified Materials*, Clarson, S. J.; Fitzgerald, J. J.; Owen, M. J.; Smith, S. D.; Van Dyke, M. E., Eds. American Chemical Society: Washington, DC, 2003; Vol. 838, pp 170–180.
337. Kita, K.; Narisawa, M.; Nakahira, A.; Mabuchi, H.; Sugimoto, M.; Yoshikawa, M., Synthesis and Properties of Ceramic Fibers from Polycarbosilane/Polymethylphenylsiloxane Polymer Blends. *J. Mater. Sci.* **2010**, *45*, 3397–3404.
338. Zhang, W.; Zheng, Y.; Orsini, L.; Morelli, A.; Galli, G.; Chiellini, E.; Carpenter, E. E.; Wynne, K. J., More Fluorous Surface Modifier Makes It Less Oleophobic: Fluorinated Siloxane Copolymer/PDMS Coatings. *Langmuir* **2010**, *26*, 5848–5855.
339. Vasquez, C. P.; Tayouo, R.; Joly-Duhamel, C.; Boutevin, B., UV-Curable Bismaleimides Containing Poly(dimethylsiloxane): Use as Hydrophobic Agent. *J. Polym. Sci., Part A: Polym. Chem.* **2010**, *48*, 2123–2134.
340. McDonald, J. C.; Whitesides, G. M., Poly(dimethylsiloxane) as a Material for Fabricating Microfluidic Devices. *Acct. Chem. Res.* **2002**, *35* (7), 491–499.
341. Matefi-Tempfli, S.; Matefi-Tempfli, M., Vertically Aligned Nanowires on Flexible Silicone Using a Supported Alumina Template Prepared by Pulsed Anodization. *Adv. Mater.* **2009**, *21*, 4005–4010.
342. Zheng, P.; McCarthy, T. J., Rediscovering Silicones: Molecularly Smooth, Low Surface Energy, Unfilled, UV/Vis-Transparent, Extremely Cross-Linked, Thermally Stable, Hard, Elastic PDMS. *Langmuir* **2010**, *26*, 18585–18590.
343. Yang, X.; Huang, W.; Yu, Y., Synthesis, Characterization, and Properties of Silicone-Epoxy Resins. *J. Appl. Polym. Sci.* **2011**, *120*, 1216–1224.
344. Takeichi, T.; Kano, T.; Agag, T.; Kawauchi, T.; Furukawa, N., Preparation of High Molecular Weight Polybenzoxazine Prepolymers Containing Siloxane Units and Properties of Their Thermosets. *J. Polym. Sci. Part A: Polym. Chem.* **2010**, *48*, 5945–5952.
345. Kuo, A. C. M., Poly(dimethylsiloxane). In *Polymer Data Handbook*, 2nd ed.; Mark, J. E., Ed. Oxford University Press: New York, 2009; pp 539–561.

346. Godovsky, Y. K.; Papkov, V. S., Poly(di-*n*-propylsiloxane). In *Polymer Data Handbook, 2nd ed.*; Mark, J. E., Ed. Oxford University Press: New York, 2009; pp 576–578.
347. Dodgson, K.; Semlyen, J. A., Studies of Cyclic and Linear Poly(dimethylsiloxanes): 1. Limiting Viscosity Number-Molecular Weight Relationships. *Polymer* **1977**, *18*, 1265–1268.
348. Clarson, S. J.; Semlyen, J. A.; Horska, J.; Stepto, R. F. T., Studies of Cyclic and Linear Poly(dimethylsiloxanes): 20. Dilute Solution Viscosities in Bromocyclohexane Under Reported Theta-Point Conditions. *Polymer* **1986**, *27*, 31–32.
349. Dodgson, K.; Sympson, D.; Semlyen, J. A., Studies of Cyclic and Linear Poly(dimethylsiloxanes): 2. Preparative Gel Permeation Chromatography. *Polymer* **1978**, *19*, 1285–1288.
350. Dagger, A. C.; Semlyen, J. A., Studies of Cyclic and Linear Poly(dimethylsiloxanes): 34. Preparation, Fractionation and Characterization of the First Per-Deuterated Macrocyclic Poly(dimethylsiloxane). *Polym. Commun.* **1999**, *40*, 3243–3245.
351. Higgins, J. S.; Dodgson, K.; Semlyen, J. A., Studies of Cyclic and Linear Poly(dimethylsiloxanes): 3.Neutron Scattering Measurements of the Dimensions of Ring and Chain Polymers. *Polymer* **1979**, *20*, 553–558.
352. Dodgson, K.; Bannister, D. J.; Semlyen, J. A., Studies of Cyclic and Linear Poly(dimethylsiloxanes): 4. Bulk Viscosities. *Polymer* **1980**, *21*, 663–667.
353. Edwards, C. J. C.; Stepto, R. F. T.; Semlyen, J. A., Studies of Cyclic and Linear Poly(dimethylsiloxanes): 5. Diffusion Behaviour in Dilute Solution. *Polymer* **1980**, *21*, 781–786.
354. Edwards, C. J. C.; Stepto, R. F. T.; Semlyen, J. A., Studies of Cyclic and Linear Poly(dimethylsiloxanes): 7. Diffusion Behaviour in a Poor Solvent. *Polymer* **1982**, *23*, 865–868.
355. Higgins, J. S.; Ma, K.; Nicholson, L. K.; Hayter, J. B.; Dodgson, K.; Semlyen, J. A., Studies of Cyclic and Linear Poly(dimethylsiloxanes): 12. Observation of Diffusion Behaviour by Quasielastic Neutron Scattering *Polymer* **1983**, *24*, 793–799.
356. Dodgson, K.; Edwards, C. J. C.; Stepto, R. F. T., Calculations of Diffusion Coefficients of Cyclic Poly(dimethylsiloxanes). *Brit. Polym. J.* **1985**, *17*, 14–18.
357. Bannister, D. J.; Semlyen, J. A., Studies of Cyclic and Linear Poly(dimethylsiloxanes): 6. Effect of Heat. *Polymer* **1981**, *22*, 377–381.
358. Clarson, S. J.; Semlyen, J. A., Studies of Cyclic and Linear Poly(dimethylsiloxanes): 21. High Temperature Thermal Behaviour. *Polymer* **1986**, *27*, 91–95.
359. Edwards, C. J. C.; Stepto, R. F. T.; Semlyen, J. A., Studies of Cyclic and Linear Poly(dimethylsiloxanes): 8. Light Scattering Measurements in Good and Poor Solvents. *Polymer* **1983**, *23*, 869–872.
360. Edwards, C. J. C.; Rigby, D.; Stepto, R. F. T.; Dodgson, K.; Semlyen, J. A., Studies of Cyclic and Linear Poly(dimethylsiloxanes): 10. Calculations of Radii of Gyration. *Polymer* **1983**, *24*, 391–394.
361. Edwards, C. J. C.; Rigby, D.; Stepto, R. F. T.; Semlyen, J. A., Studies of Cyclic and Linear Poly(dimethylsiloxanes): 10. Shapes of Ring and Chain Molecules. *Polymer* **1983**, *24*, 395–399.
362. Beevers, M. S.; Mumby, S. J.; Clarson, S. J.; Semlyen, J. A., Studies of Cyclic and Linear Poly(dimethylsiloxanes): 13. Static Dielectric Measurements and Dipole Moments. *Polymer* **1983**, *24*, 1565–1570.

363. Edwards, C. J. C.; Richards, R. W.; Stepto, R. F. T.; Dodgson, K.; Higgins, J. S.; Semlyen, J. A., Studies of Cyclic and Linear Poly(dimethylsiloxanes): 14. Particle Scattering Functions *Polymer* **1984**, *25*, 365–368.
364. Granick, S.; Clarson, S. J.; Formoy, T. R.; Semlyen, J. A., Studies of Cyclic and Linear Poly(dimethylsiloxanes): 18. Surface Pressures of the Monolayers in the Plateau Region. *Polymer* **1985**, *26*, 925–929.
365. Orrah, D. J.; Semlyen, J. A.; Dodgson, K.; Ross-Murphy, S. B., Studies of Cyclic and Linear Poly(dimethylsiloxanes): 23. Low Temperature Behaviour as Studied by Dynamic Oscillatory Shear Viscometry and Differential Scanning Calorimetry. *Polymer* **1986**, *28*, 985–988.
366. Barbarin-Castillo, J.-M.; McLure, I. A.; Clarson, S. J.; Semlyen, J. A., Studies of Cyclic and Linear Poly(dimethylsiloxanes): 25. Lower Critical Threshold Temperatures in Tetramethyl Solvents. *Polymer* **1986**, *28*, 212–215.
367. Kuo, C. M.; Clarson, S. J.; Semlyen, J. A., Studies of Cyclic and Linear Poly(dimethylsiloxanes): 31. Effect of Molecular Architecture/Topology on Blends of Poly(methylphenylsiloxane) and Poly(dimethylsiloxane). *Polymer* **1994**, *35*, 4623–4626.
368. Orrah, D. J.; Semlyen, J. A.; Ross-Murphy, S. B., Studies of Cyclic and Linear Poly(dimethylsiloxanes): 27. Bulk Viscosities Above the Critical Molar Mass for Entanglement. *Polymer* **1988**, *29*, 1452–1454.
369. Stepto, R. F. T., Entanglements in Linear and Cyclic Poly(dimethylsiloxane) and Other Polymers. *Eur. Polym. J.* **1993**, *29*, 415–418.
370. Orrah, D. J.; Semlyen, J. A.; Ross-Murphy, S. B., Studies of Cyclic and Linear Poly(dimethylsiloxanes): 28. Viscosities and Densities of Ring and Chain Poly(dimethylsloxane) Blends. *Polymer* **1988**, *29*, 1455–1458.
371. Patel, A.; Cosgrove, T.; Semlyen, J. A., Studies of Cyclic and Linear Poly(dimethylsiloxanes): 30. Adsorption Studies on Silica in Solution. *Polymer* **1991**, *32*, 1313–1317.
372. Goodwin, A. A.; Beevers, M. S.; Clarson, S. J.; Semlyen, J. A., Studies of Cyclic and Linear Poly(dimethylsiloxanes): 32.Dielectric Relaxation Investigations. *Polymer* **1994**, *37*, 2603–2607.
373. Clarson, S. J.; Semlyen, J. A., Cyclic Polysiloxanes: 1. Preparation and Characterization of Poly(phenylmethylsiloxane). *Polymer* **1986**, *27*, 1633–1636.
374. Clarson, S. J.; Dodgson, K.; Semlyen, J. A., Cyclic Polysiloxanes: 2. Neutron Scattering from Poly(phenylmethylsiloxane). *Polymer* **1986**, *28*, 189–192.
375. Arrighi, V.; Gagliardi, S.; Dagger, A. C.; Semlyen, J. A.; Higgins, J. S.; Shenton, M. J., Conformations of Cyclics and Linear Chain Polymers in Bulk by SANS. *Macromolecules* **2004**, *37*, 8057–8065.
376. Gagliardi, S.; Arrighi, V.; Ferguson, R.; Dagger, A. C.; Semlyen, J. A.; Higgins, J. S., On the Difference in Scattering Behavior of Cyclic and Linear Polymers in Bulk. *J. Chem. Phys.* **2003**, *122*, 064904.
377. Clarson, S. J.; Semlyen, J. A.; Dodgson, K., Cyclic Polysiloxanes: 4. Glass Transition Temperatures of Poly(phenylmethylsiloxane). *Polymer* **1991**, *32*, 2823–2827.
378. Goodwin, A. A.; Beevers, M. S.; Clarson, S. J.; Semlyen, J. A., Cyclic Polysiloxanes: 6. Dielectric Relaxation of Polymethylphenylsiloxane. *Polymer* **1996**, *37*, 2597–2602.
379. Formoy, T. R.; Semlyen, J. A., Cyclic Polysiloxanes: 3. Preparation, Characterization and Properties of Poly(vinylmethylsiloxane). *Polymer* **1989**, *30*, 86–89.

CHAPTER 4

Some Characterization Techniques Useful for Polysiloxanes

4.1 GENERAL COMMENTS

The general approach used in choosing a polymer suitable for a particular application is:

$$\text{Polymerization} \leftrightarrow \text{Structure} \leftrightarrow \text{Properties} \leftrightarrow \text{Application}$$

For example, if one wants a polymer for fire-resistant fabrics, then a polymer with good high-temperature properties is required, which implies aromatic structures, which suggest condensation polymerizations. More relevant here, however, would be that a polymer remains elastomeric at low temperatures. This requirement evokes a polymer with high flexibility (low glass transition temperature), which indicates use of the polymerization techniques used with the polysiloxanes.

4.2 OPTICAL AND SPECTROSCOPIC TECHNIQUES

An example of a relevant optical property is the birefringence of a deformed polymer network.[1] This strain-induced birefringence can be used to characterize segmental orientation, and both Gaussian and non-Gaussian elasticity.[2–9] Infrared dichroism has also been helpful in this regard.[10, 11] In the case of the crystallizable polysiloxane elastomers, orientation is of critical importance with regard to strain-induced crystallization and the tremendous reinforcement it provides.[11] Segmental orientation has also been characterized by fluorescence polarization, deuterium nuclear magnetic resonance (NMR), and polarized infrared spectroscopy.[1, 3, 12]

Infrared spectroscopy has been used to characterize the structures of silica-filled polydimethylsiloxane (PDMS).[9, 13–15]

Other optical and spectroscopic techniques are also important, including positron annihilation lifetime spectroscopy,[16, 17] spectroscopic ellipsometry,[18] confocal Raman spectroscopy,[19] and photoluminescence spectroscopy.[20] Surface-enhanced Raman spectroscopy has been made tunable using gold nanorods and strain control on elastomeric PDMS substrates.[21]

4.3 MICROSCOPIES

A great deal of information is now being obtained on filler dispersion and other aspects of elastomer structure and morphology through the use of scanning probe microscopy, which consists of several approaches.[22–27] One approach is that of scanning tunneling microscopy (STM), in which an extremely sharp metal tip on a cantilever is passed along the surface while measuring the electric current flowing through quantum mechanical tunneling. Monitoring the current then permits maintaining the probe at a fixed height above the surface. Display of probe height as a function of surface coordinates then gives the desired topographic map. One limitation of this approach is the requirement that the sample be electrically conductive. Atomic force microscopy (AFM), on the other hand, does not require a conducting surface. The probe simply responds to attractions and repulsions from the surface, and its corresponding downward and upward motions are directly recorded to give the relief map of the surface structure. The probe can be either in contact with the surface or adjacent to it, sensing only Coulombic or van der Waals forces.

Both transmission electron microscopy and AFM have been used to characterize the structures of silica-filled PDMS.[7, 28] Another example of an application to polysiloxane elastomers is the characterization of binodal and spinodal phase-separated structures occurring in model PDMS networks.[29–31]

4.4 NUCLEAR MAGNETIC RESONANCE

Although NMR has been used to characterize some aspects of polymerization (e.g., copolymerization kinetics),[32] the applications most relevant here have to do with the structure of the linear polymers or cross-linked networks.[33–35] In this regard, NMR has been much used to study the characteristics of the polysiloxanes, particularly with regard to orientation,[36, 37] molecular motion,[38–46] and the effect on the diffusion and other properties of small molecules.[47–50] Small-molecule diffusion is particularly relevant to the properties of silicone breast implants, and NMR has been extensively used to image such implants.[51–55] Aspects related to the structure of the networks include the degree of cross linking,[56–58] the distribution of cross links,[59, 60] concentration of pendant chains,[61] elastic stress,[62] and topology.[63–65] Another example is the use of NMR to clarify aging and phase separation.[66] Deuterium NMR has also been used to determine segment orientation distributions in polymer networks, including those of PDMS.[67, 68]

Dipolar coupling constant distribution analyses have been carried out using proton multiple-quantum NMR on elastomers, including unimodal and bimodal PDMS materials.[69] Also, a universal polymer analysis technique has been developed using ^{1}H NMR on complementary trimethylsilyl end groups placed on the polymer.[70]

Most elastomers require reinforcing fillers to function effectively, and NMR has been used to characterize the structure of such composites as well. One examples is the adsorption of chains onto filler surfaces,[71, 72] and the strong absorption of these chains into "bound rubber"—for example, PDMS immobilized onto high surface area silica.[41, 46] Another example is the use of NMR to image the filler or polymer itself.[73–80] NMR has also been used to study the phase separation and order of water molecules and silanol groups in polysiloxane networks[81] and the activation of transport and local dynamics in polysiloxane-based salt-in-polymer electrolytes.[82]

4.5 THERMOPOROMETRY

Thermooporometry involves measuring the crystallization temperatures T_c of small molecules constrained within a porous medium. The pores keep the crystallites from growing beyond the size of the pore leading to a large surface area to volume ratio. The interfacial free energy is positive, which destabilizes the crystallites and decreases T_c. The average values can be used to estimate average pore sizes, and the distribution of T_cs can be used to determine the pore size distribution. The technique was originally applied to inorganic materials such as glasses,[83] mesoporous silica,[84] alcogels,[85] or zeolites.[86] Similar decreases in T_c are observed in nanocrystals or metal clusters prepared for use as "quantum dots."[87, 88]

Swollen elastomers have domains of solvent between the network chains that may also be considered "pores." Thermoporometry is useful to characterize the pore size and size distribution in such gels. Examples include natural rubber,[89, 90] styrene-divinylbenzene polymers,[91] poly(ethylene oxide),[92] polyethylene and polypropylene,[93] and ethylene-propylene-diene polymers and *cis*-polycyclo-octene.[94]

Relevant here are the corresponding studies on polysiloxane gels such as PDMS of various degrees of cross linking (giving various pore sizes). The samples are either unfilled[95–97] or filled with silica.[96] Aged PDMS has also been studied.[98] In some cases, PDMS networks were prepared by end-linking chains, to have known values of the molecular weight between cross links. These model elastomers had chain-length distributions that were unimodal,[99, 100] bimodal,[99] or trimodal.[99]

The effects of constraints in general on the freezing temperature have been studied theoretically, specifically in an entanglement/frozen tube model,[101] and by molecular dynamics simulations.[102] Confinement of liquids also affects the glass transition temperature, which typically decreases as the pore size decreases.[103–105]

4.6 SCATTERING OF LIGHT, X-RAYS, AND NEUTRONS

Static and dynamic light scattering have been used to investigate PDMS in both liquid and supercritical carbon dioxide.[106] The solvent quality of the CO_2 was found to be adjustable by independently varying temperature or density. The results give the theta temperature and strength of excluded volume interactions.[107]

Small-angle scattering techniques have been applied to polysiloxane materials. One important example is the characterization of fillers introduced into polysiloxane elastomers, or the reverse, the incorporation of such elastomers into ceramic matrices (in both cases to improve mechanical properties).[3, 108, 109] Another example is characterization of the anisotropy induced by strain in silica-PDMS composites.[110] Chapter 9 describes some of this work. Elastic neutron scattering can be illustrated by the characterization of polysiloxane blends,[111] and quasielastic neutron scattering by studies of the dynamics of PDMS.[112] There have also been Monte Carlo calculations of PDMS particle scattering functions, including how they varied with chain length, chain structure, and temperature.[113]

4.7 BRILLOUIN SCATTERING

The application of Brillouin scattering to the characterization of PDMS networks was found to be particularly useful for looking at glassy-state properties of such elastomers at very high frequencies.[114–116]

4.8 PULSE PROPAGATION

Another example of a relatively new technique for the noninvasive, nondestructive characterization of network structures involves ultrasound pulse-propagation measurements.[117, 118] The goal here is the rapid determination of the spacing between junctions and between entanglements in a network structure. The delay in a pulse passing through the network is used to obtain such information on the network structure.[114] Ultrasonic methods have also been used to study interdiffusion between two samples of PDMS differing only in molecular weight.[119]

This list is essentially an extension of parts of chapter 2, and many of the techniques described here are also useful for characterizing the composites described in chapter 9.

4.9 THEORY AND SIMULATIONS

Some of the earliest studies involved determination of the potential function governing low-frequency bending modes of disiloxane,[120] the importance of (p-d)π bonding,[121] characterization of helical polysiloxane chains,[122] the molecular states of silicon-containing compounds in general,[123] *ab initio* structure calculations[124] and molecular modeling of elastic behavior.[125] More recent examples include molecular dynamics and integral equations,[126] force fields,[127, 128] and finite element simulations of mechanical property evolution during vulcanization.[129] Artificial intelligence approaches have been used in the polysiloxane area, including studies of synthesis, solution properties, metal ion sorption, liquid-crystalline behavior, and fluorescence of composites containing complexed rare earths.[130]

Simulations and theoretical calculations have also been carried out for PDMS networks[131–133] and single chains[134, 135] under deformation. Finally, investigations have clarified important aspects of polysiloxane permeabilities,[136, 137] interfacial structure,[138] and elasticity using an expansion approach.[139]

REFERENCES

1. Erman, B.; Mark, J. E., Rubber-like elasticity. *Ann. Rev. Phys. Chem.* **1989**, *40*, 351–374.
2. Treloar, L. R. G., *The Physics of Rubber Elasticity*, 3rd ed. Clarendon Press: Oxford, 1975.
3. Mark, J. E.; Erman, B., *Rubberlike Elasticity: A Molecular Primer.* Wiley-Interscience: New York, 1988.
4. Bokobza, L.; Clément, F.; Monnerie, L.; Lapersonne, P., IR Dichroism and Birefringence. In *The Wiley Polymer Networks Group Review Series*, te Nijenhuis, K.; Mijs, W. J. Eds. John Wiley & Sons Ltd.: New York, 1998; Vol. 1, pp 321–333.
5. Bokobza, L.; Nugay, N., Orientational Effects of Mica in Fumed Silica Reinforced Composites. *J. Appl. Polym. Sci.* **2001**, *81*, 215–222.
6. Bokobza, L.; Erman, B., A Theoretical and Experimental Study of Filler Effect on Stress-Deformation-Segmental Orientation Relations for Poly(dimethylsiloxane) Networks. *Macromolecules* **2000**, *33*, 8858–8864.
7. Clement, F.; Lapra, A.; Bokobza, L.; Monnerie, L.; Menez, P., Atomic Force Microscopy Investigation of Filled Elastomers and Comparison with Transmission Electron Microscopy—Application to Silica-Filled Silicone Elastomers. *Polymer* **2001**, *42*, 6259–6270.
8. Bokobza, L., Reinforcement of Elastomeric Networks by Fillers. *Macromol. Symp.* **2001**, *169*, 243–260.
9. Bokobza, L.; Rapoport, O., Filled Elastomers: Mid- and Near-IR Spectroscopic Determination of Rubber Dimensions in Composite in Unstretched State and Under Uniaxial Extension. *J. Appl. Polym. Sci.* **2003**, *87*, 1204–1208.
10. Besbes, S.; Cermelli, I.; Bokobza, L.; Monnerie, L.; Bahar, I.; Erman, B.; Herz, J., Segmental Orientation in Model Networks of Poly(dimethylsiloxane): Fourier Transform Infrared Dichroism Measurements and Theoretical Interpretation. *Macromolecules* **1992**, *25*, 1949–1954.
11. Erman, B.; Mark, J. E., *Structures and Properties of Rubberlike Networks.* Oxford University Press: New York, 1997.
12. Noda, I.; Dowrey, A. E.; Marcott, C., Characterization of Polymers Using Polarization-Modulation Infrared Techniques: Dynamic Infrared Linear Dichroism (DIRLD) Spectroscopy. In *Fourier Transform Infrared Characterization of Polymers*, Ishida, H., Ed. Plenum Press: New York, 1987; pp 33–57.
13. Dewimille, L.; Bresson, B.; Bokobza, L., Synthesis, Structure and Morphology of Poly(dimethylsiloxane) Networks Filled with In Situ Generated Silica Particles. *Polymer* **2005**, *46*, 4135–4143.
14. Bokobza, L., Spectroscopic Investigations of Polymer Nanocomposites. *Macromol. Symp.* **2005**, *230*, 87–94.
15. Bokobza, L., Elastomeric Composites. I. Silicone Composites. *J. Appl. Polym. Sci.* **2004**, *93*, 2095–2104.
16. Dlubek, G.; De, U.; Pionteck, J.; Arutyunov, N. Y.; Edelmann, M.; Krause-Rehberg, E., Temperature Dependence of Free Volume in Pure and Silica-Filled Poly(dimethylsiloxane) from Positron Lifetime and PVT Measurements. *Macromol. Chem. Phys.* **2005**, *206*, 827–840.
17. Winberg, P.; Eldrup, M.; Maurer, F. H. J., Nanoscopic Measurements of Silica Filled Poly(dimethylsiloxane) by Means of Positron Annihilation Lifetime Spectroscopy. *Polymer* **2004**, *45*, 8253–8264.

18. Graubner, V.-M.; Clemens, D.; Gutberlet, T.; Kotz, R.; Lippert, T.; Nuyken, O.; Schnyder, B.; Wokaun, A., Neutron Reflectometry and Spectroscopic Ellipsometry Studies of Cross-Linked Poly(dimethylsiloxane) After Irradiation at 172 nm. *Langmuir* **2005**, *21*, 8940–8946.
19. Bae, S. C.; Lee, H.; Lin, Z.; Granick, S., Chemical Imaging in a Surface Forces Apparatus: Confocal Raman Spectroscopy of Confined Poly(dimethylsiloxane). *Langmuir* **2005**, *21*, 5685–5688.
20. Rossi, N. A. A.; Duplock, E. J.; Meegan, J.; Roberts, D. R. T.; Murphy, J. J.; Patel, M.; Holder, S. J., Synthesis and Characterization of Pyrene-Labelled Polydimethylsiloxane Networks: Towards the In Situ Detection of Strain in Silicone Elastomers. *J. Mater. Chem.* **2009**, *19*, 7674–7686.
21. Alexander, K. D.; Skinner, K.; Zhang, S.; Wei, H.; Lopez, R., Tunable SERS in Gold Nanorod Dimers Through Strain Control on an Elastomeric Substrate. *Nano Lett.* **2010**, *10*, 4488–4493.
22. Wiesendanger, R., *Scanning Probe Microscopy and Spectroscopy.* Cambridge University Press: Cambridge, 1994.
23. Wiesendanger, R., *Scanning Probe Microscopy: Analytical Methods.* Springer: Berlin, 1998.
24. Minne, S. C.; Manalis, S. R.; Quate, C. F., *Bringing Scanning Probe Microscopy up to Speed.* Kluwer Academic Publishers: Boston, 1999.
25. Bonnell, D. A., *Scanning Probe Microscopy and Spectroscopy: Theory, Techniques, and Applications*, 2nd ed. Springer: Berlin, 2001.
26. Stefanis, A. D.; Tomlinson, A. A. G., *Scanning Probe Microscopies: From Surface Structure to Nano-Scale Engineering.* Trans Tech Publications Ltd.: Uetikon-Zurich, 2001.
27. Allcock, H. R.; Lampe, F. W.; Mark, J. E., *Contemporary Polymer Chemistry*, 3rd ed. Prentice Hall: Englewood Cliffs, NJ, 2003.
28. Lapra, A.; Clement, F.; Bokobza, L.; Monnerie, L., Straining Effects in Silica-Filled Elastomers Investigated by Atomic Force Microscopy: From Macroscopic Stretching to Nanoscale Strainfield. *Rubber Chem. Technol.* **2003**, *76*, 60–81.
29. Viers, B. D.; Sukumaran, S.; Beaucage, G.; Mark, J. E., On the Observation of Novel Morphologies in "Model" PDMS Elastomers. Evidence of Trapped Spinodal Structures. *Polym. Preprints* **1997**, *38*(2), 333–334.
30. Viers, B. D.; Mark, J. E., Atomic Force Microscopy Investigations of Binodal and Spinodal Phase Separated Structures in Model PDMS Elastomers. *Polym. Preprints* **1998**, *39*(1), 520–521.
31. Viers, B. D. On the Supramolecular Structure in Model PDMS Elastomers. Ph.D. Chemistry, The University of Cincinnati, Cincinnati, 1998.
32. Rankin, S. E.; Macosko, C. W.; McCormick, A. V., Copolymerization Kinetics of a Model Siloxane System. *J. Polym. Sci., Part A: Polym. Chem.* **1997**, *35*, 1293–1302.
33. Eldred, D. V.; Taylor, R. B., Implementation of Low Temperature Physical Changes on Silicone Elastomers to Achieve Sensitivity Gains through CPMAS NMR. *Silicon Chem.* **2005**, *3*, 229–237.
34. Mayer, B. P.; Lewicki, J. P.; Weisgraber, T. H.; Small, W.; Chinn, S. C.; Maxwell, R. S., Linking Network Microstructure to Macroscopic Properties of Siloxane Elastomers Using Combined Nuclear Magnetic Resonance and Mesoscale Computational Modeling. *Macromolecules* **2011**, *44*, 8106–8115.

35. Chasse, W.; Lang, M.; Sommer, J.-U.; Saalwachter, K., Cross-Link Density Estimates of PDMS Networks with Precise Consideration of Network Defects. *Macromolecules* **2012**, *45*, 899–912.
36. Litvinov, V. M., Strain-Induced Phenomena in Amorphous and Semicrystalline Elastomers. Solid State ^{1}H NMR T_2 Relaxation under Uniaxial Compression. *Macromolecules* **2001**, *34*, 8468–8474.
37. Freudenberger, J. C.; Spiteller, P.; Baure, R.; Kessler, H.; Luy, B., Stretched Poly(dimethylsiloxane) Gels as NMR Alignment Media for Apolar and Weakly Polar Organic Solvents: An Ideal Tool for Measuring RDCs at Low Molecular Concentrataions. *J. Am. Chem. Soc.* **2003**, *126*, 14690–14691.
38. Powles, J. G.; Hartland, A.; Kail, J. A. E., Proton Magnetic Resonance Relaxation in a Series of Dimethylsiloxane Polymers. *J. Polym. Sci.* **1961**, *55*, 361–380.
39. Cuniberti, C., Proton Magnetic Relaxation Times and Molecular Motions in Polydimethylsiloxane. *J. Polym. Sci., Part A-2: Polymer Chemistry* **1970**, *8*, 2051–2065.
40. Litvinov, V. M.; Spiess, H. W., ^{2}H NMR Study of Molecular Motions in Polydimethylsiloxane and Its Mixtures with Aerosils. *Macromol. Chem.* **1991**, *192*, 3005–3019.
41. Litvinov, V. M.; Spiess, H. W., Molecular Mobility in the Adsorption Layer and Chain Orientation in Strained Poly(dimethylsiloxane) Networks by ^{2}H NMR. *Macromol. Chem.* **1992**, *193*, 1181–1194.
42. Hedden, R. C.; McCaskey, E.; Cohen, C.; Duncan, T. M., Effects of Molecular Structure on Segment Orientation in Siloxane Elastomers. 1. NMR Measurements from Compressed Samples. *Macromolecules* **2001**, *34*, 3285–3293.
43. Lorthioir, C.; Deloche, B.; Courtieu, J., Experimental Evidence for a Negative Orientational Order Parameter in a Uniaxially Compressed Poly(dimethylsiloxane) Network. *Macromolecules* **2001**, *34*, 8835–8837.
44. Hedden, R. C.; Tachibana, H.; Duncan, T. M.; Cohen, C., Effects of Molecular Structure on Segment Orientation in Siloxane Elastomers. 2. NMR Measurements from Uniaxially Stretched Samples. *Macromolecules* **2001**, *34*, 5540–5546.
45. Garasanin, T.; Cosgrove, T.; Marteaux, L.; Kretschmer, A.; Goodwin, A.; Zick, K., NMR Self-Diffusion Studies on PDMS Oil-in-Water Emulsion. *Langmuir* **2002**, *18*, 10298–10304.
46. Litvinov, V. M.; Barthel, H.; Weis, J., Structure of a PDMS Layer Grafted onto a Silica Surface Studied by Means of DSC and Solid-State NMR. *Macromolecules* **2002**, *35*, 4356–4364.
47. Garrido, L.; Mark, J. E.; Clarson, S. J.; Semlyen, J. A., Studies of Cyclic and Linear Poly(dimethylsiloxanes): 15. Diffusion Coefficients from Network Sorption Measurements. *Polym. Comm.* **1984**, *25*, 218–220.
48. Garrido, L.; Mark, J. E.; Ackerman, J. L.; Kinsey, R. A., Studies of Self-Diffusion of Poly(dimethylsiloxane) Chains in PDMS Model Networks by Pulsed Field Gradient NMR. *J. Polym. Sci., Polym. Phys. Ed.* **1988**, *26*, 2367–2377.
49. Garrido, L.; Ackerman, J. L.; Mark, J. E., Self-Diffusion of Poly(dimethylsiloxane) Chains. In *New Trends in Physics and Physical Chemistry of Polymers*, Lee, L.-H., Ed. Plenum Press: New York, 1989; pp 355–365.
50. Klingenfus, J.; Palmas, P., Weak Molecular Associations Investigated by ^{1}H-NMR Spectroscopy on a Neat Organosiloxane-Solvent Mixture. *Phys. Chem. Chem. Phys.* **2011**, *13*, 10661–10669.

51. Pfleiderer, B.; Ackerman, J. L.; Garrido, L., In Vivo ^{1}H Chemical Shift Imaging of Silicone Implants. *Magn. Reson. Med.* **1993**, *29*, 656–659.
52. Garrido, L.; Kwong, K. K.; Pfleiderer, B.; Crawley, A. P.; Hulka, C. A.; Whitman, G. L.; Kopans, D. B., Echo-Planar Chemical Shift Imaging of Silicone Gel Prostheses. *Magn. Reson. Imaging* **1993**, *11*, 625–634.
53. Pfleiderer, B.; Ackerman, J. L.; Garrido, L., In Vivo Localized Proton NMR Spectroscopy of Silicone. *Magn. Reson. Medicine* 1993, *30*, 149–154.
54. Pfleiderer, B.; Ackerman, J. L.; Garrido, L., Migration and Biodegradation of Free Silicone from Silicone Gel-Filled Implants after Long-Term Implantation. *Magn. Reson. Med.* **1993**, *30*, 534–543.
55. Garrido, L.; Pfleiderer, B.; Papisov, M.; Ackerman, J. L., In Vivo Degradation of Silicones. *Magn. Reson. Med.* **1993**, *29*, 839–843.
56. Simon, G.; Birnstiel, A.; Schimmel, K.-H., Network Characterisation of End-Linked Poly(dimethylsiloxane) by ^{1}H-NMR-Spin-Spin Relaxation. *Polym. Bull.* **1989**, *21*, 235–241.
57. Sandakov, G. I.; Smirnov, L. P.; Sosikov, A. I.; Summanen, K. T.; Volkova, N. N., MMR Analysis of Distribution of Chain Lengths Between Crosslinks of Polymer Networks. *J. Polym. Sci., Polym. Phys. Ed.* **1994**, *32*, 1585–1592.
58. Garbarczyk, M.; Grinberg, F.; Nestle, N.; Kuhn, W., A Novel Approach to the Determination of the Crosslink Density in Rubber Materials with the Dipolar Correlation Effect in Low Magnetic Fields. *J. Polym. Sci., Part B: Polym. Phys.* **2001**, *39*, 2207–2216.
59. Litvinov, V. M.; Vasilev, V. G., Network Structure of Filled Siloxane Elastomers. *Polym. Sci. U.S.S.R.* **1990**, *32*, 2231–2237.
60. Litvinov, V. M.; Dias, A. A., Analysis of Network Structure of UV-Cured Acrylates by ^{1}H NMR Relaxation, ^{13}C NMR Spectroscopy, and Dynamic Mechanical Experiments. *Macromolecules* **2001**, *34*, 4051–4060.
61. Acosta, R. H.; Vegas, D. A.; Villar, M. A.; Monti, G. A.; Valles, E. H., Double Quantum NMR Applied to Polymer Networks with Low Concentration of Pendant Chains. *Macromolecules* **2006**, *39*, 4788–4792.
62. Schneider, M.; Demco, D. E.; Blumich, B., NMR Images of Proton Residual Dipolar Coupling from Strained Elastomers. *Macromolecules* **2001**, *34*, 4019–4026.
63. Beshah, K.; Mark, J. E.; Himstedt, A.; Ackerman, J. L., Characterization of PDMS Model Junctions and Networks by Solution and Solid State Silicon-29 NMR Spectroscopy. *J. Polym. Sci., Polym. Phys. Ed.* **1986**, *24*, 1207–1225.
64. Beshah, K.; Mark, J. E.; Ackerman, J. L., Topology of PDMS Endlinked Elastomeric Networks Studied by Variable Temperature Solid State NMR. *Macromolecules* **1986**, *19*, 2194–2196.
65. Smirnov, L. P., NMR Study of Network Polymer Structures. *Polym. Sci. U.S.S.R.* **2000**, *42*, 1775–1792.
66. Blumler, P.; Blumich, B., Aging and Phase Separation of Elastomers Investigated by NMR Imaging. *Macromolecules* **1991**, *24*, 2183–2188.
67. Genesky, G.; Duncan, T. M.; Cohen, C., Effect of Precursor Molar Mass on the ^{2}H NMR Line Shape of End-Linked PDMS Elastomers. *Macromolecules* **2009**, *42*, 8882–8888.
68. Aguilera-Mercado, B. M.; Cohen, C.; Escobedo, F. A., Extraction of Segment Orientation Distributions in Polymer Networks by Inversion of ^{2}H NMR Spectra through the Maximum-Entropy Method. *Macromolecules* **2009**, *42*, 8889–8898.

69. Chasse, W.; Valentin, J. L.; Genesky, G. D.; Cohen, C.; Saalwachter, K., Precise Dipolar Coupling Constant Distribution Analysis in Proton Multi-Quantum NMR of Elastomers. *J. Chem. Phys.* **2011**, *134*, 044907.
70. Pach, M.; Zehm, D.; Lange, M.; Dambowsky, I.; Weiss, J.; Laschewsky, A., Universal Polymer Analysis by ^{1}H NMR Using Complementary Trimethylsilyl End Groups. *J. Am. Chem. Soc.* **2010**, *132*, 8757–8765.
71. Cohen-Addad, J. P.; Ebengou, R., Silica-Siloxane Mixtures. Investigations into Adsorption Properties of End-Methylated and End-Hydroxylated Chains. *Polymer* **1992**, *33*, 379–383.
72. Cohen-Addad, J. P., NMR Approach to Siloxane Properties: Melts, Gels, Silica Mixtures. In *Siloxane Polymers*, Clarson, S. J.; Semlyen, J. A., Eds. Prentice Hall: Englewood Cliffs, NJ, 1993; pp 511–566.
73. Glaser, R. H.; Wilkes, G. L.; Bronnimann, C. E., Solid-State ^{29}Si NMR of TEOS-Based Multifunctional Sol-Gel Materials. *J. Non-Cryst. Solids* **1989**, *113*, 73–87.
74. Garrido, L.; Ackerman, J. L.; Mark, J. E., NMR Imaging of Silica-Silicone Composites. In *Polymer-Based Molecular Composites*, Schaefer, D. W.; Mark, J. E., Eds. Materials Research Society: Pittsburgh, PA, 1990; Vol. 171, pp 65–70.
75. Garrido, L.; Mark, J. E.; Sun, C. C.; Ackerman, J. L.; Chang, C., NMR Characterization of Elastomers Reinforced with In-Situ Precipitated Silica. *Macromolecules* **1991**, *24*, 4067–4072.
76. Garrido, L.; Mark, J. E.; Wang, S.; Ackerman, J. L.; Vevea, J. M., Polymer Distribution in Silica Aerogels Impregnated with Siloxanes by ^{1}H NMR Imaging. *Polymer* **1992**, *33*, 1826–1830.
77. Cosgrove, T.; Turner, M. J.; Weatherhead, I.; Roberts, C.; Garasanin, T.; Schmidt, R. G.; Gordon, G. V.; Hannington, J. P., NMR Spin-Spin Relaxation Studies of Reinforced Poly(dimethylsiloxane) Melts. In *Silicones and Silicone-Modified Materials*, Clarson, S. J.; Fitzgerald, J. J.; Owen, M. J.; Smith, S. D., Eds. American Chemical Society: Washington, DC, 2000; Vol. 729, pp 204–213.
78. Gussoni, M.; Greco, F.; Mapelli, M.; Vezzoli, A.; Ranucci, E.; Ferruti, P.; Zetta, L., Elastomeric Polymers. 1. Application of Proton NMR Imaging to the Morphological Study of a Silicone Rubber. *Macromolecules* **2002**, *35*, 1714–1721.
79. Cosgrove, T.; Roberts, C.; Garasanin, T.; Schmidt, R. G.; Gordon, G. V., NMR Spin-Spin Relaxation Studies of Silicate-Filled Low Molecular Weight Poly(dimethylsiloxane)s. *Langmuir* **2002**, *18*, 10080–10085.
80. Gussoni, M.; Greco, F.; Mapelli, M.; Vezzoli, A.; Ranucci, E.; Ferruti, P.; Zetta, L., Elastomeric Polymers. 2. NMR and NMR Imaging Characterization of Cross-Linked PDMS. *Macromolecules* **2002**, *35*, 1722–1729.
81. Brus, J., Solid-State NMR Study of Phase Separation and Order of Water Molecules and Silanol Groups in Polysiloxane Networks. *J. Sol-Gel Sci. Technol.* **2002**, *25*, 17–28.
82. Kunze, M.; Karatas, Y.; Wiemhofer, H.-D.; Eckert, H.; Schonhoff, M., Activation of Transport and Local Dynamics in Polysiloxane-Based Salt-In-Polymer Electrolytes: A Multinuclear NMR Study. *Phys. Chem. Chem. Phys.* **2010**, *12*, 6844–6851.
83. Jackson, C. L.; McKenna, G. B., The Melting Behavior of Organic Materials Confined in Porous Media. *J. Chem. Phys.* **1990**, *93*, 9002–9011.
84. Endo, A.; Yamamoto, T.; Inagi, Y.; Iwakabe, K.; Ohmori, T., Characterization of Nonfreezable Pore Water in Mesoporous Silica by Thermoporometry. *J. Phys. Chem. C* **2008**, *112*, 9034–9039.

85. Phalippou, J.; Ayral, A.; Woignier, T.; Quinson, J. F.; Pauthe, M.; Chatelut, M., Fractal Geometry of Silica Alcogels from Thermoporometry Experiments. *Europhys. Lett.* **1991**, *14*, 249–254.
86. Morishige, M.; Nobuoka, K., X-Ray Diffraction Studies of Freezing and Melting of Water Confined in a Mesoporous Adsorbent (MCM-41). *J. Chem. Phys.* **1997**, *107*, 6965–6969.
87. Goldstein, A. N.; Echer, C. M.; Alivisatos, A. P., Melting in Semiconductor Nanocrystals. *Science* **1992**, *256*, 1425–1427.
88. Bertsch, G., Melting in Clusters. *Science* **1997**, *277*, 1619.
89. Jackson, C. L.; McKenna, G. B., On the Anomalous Freezing and Melting of Solvent Crystals in Swollen Gels of Natural Rubber. *Rubber Chem. Technol.* **1991**, *64*, 760–768.
90. Valentin, J. L.; Fernandez-Torrres, A.; Posadas, P.; Marco-Fernandez, A.; Rodriguez, A.; Gonzalez, L., Measurements of Freezing-Point Depression to Evaluate Rubber Network Structure. Crosslinking of Natural Rubber with Dicumyl Peroxide. *J. Polym. Sci., Part B: Polym. Phys.* **2007**, *45*, 544–556.
91. Ogino, K.; Sato, H., NMR Characterization of Styrene-Divinylbenzene Gel Beads Using Freezing Point Depression of Benzene. *J. Polym. Sci., Polym. Phys. Ed.* **1995**, *33*, 445–451.
92. Mayen, M., Measurement of Molecular Mass-Distribution between Cross-Links in Networks. *Eur. Polym. J.* **1986**, *22*, 987–992.
93. Billamboz, N.; Nedelec, J.-M.; Grivet, M.; Baba, M., Cross-Linking of Polyolefins: A Study by Thermoporosimetry with Benzene Derivatives as Swelling Solvents. *ChemPhysChem* **2005**, *4*, 1126–1132.
94. Baba, M.; Nedelec, J.-M.; Billamboz, N.; Lacoste, J., Porous Volume of Inorganic Materials and Degree of Swelling of Elastomers Monitored by DSC Measurements. *J. Phys. Chem. B* **2003**, *107*, 12884–12890.
95. Arndt, K. F.; Zander, P., Characterization of Polymer Networks by Measurements of the Freezing Point Depression. *Coll. Polym. Sci.* **1990**, *268* (9), 806–813.
96. Pouchelon, A.; Soria, M.; Moll, L., Structure of Filled and Unfilled Elastomers: Thermoporometric Measurements. *Macromol. Chem., Macromol. Symp.* **1990**, *40*, 147–152.
97. Pouchelon, A.; George, C.; Menez, P., The Swelling Behavior of Siloxane Elastomers in Relation to Their Macroscopic Structure. *Macromol. Symp.* **2001**, *171*, 233–242.
98. Nedelec, J.-M.; Grolier, J.-P. E.; Baba, M., Thermoporosimetry: A Powerful Tool to Study the Cross-Linking in Gels Networks. *J. Sol-Gel Sci. Technol.* **2006**, *40*, 191–200.
99. Madkour, T.; Mark, J. E., Some Evidence on Pore Sizes in Poly(dimethylsiloxane) Elastomers Having Unimodal, Bimodal, or Trimodal Distributions of Network Chain Lengths. *Polym. Bull.* **1993**, *31*, 615–621.
100. Wu, J.; McKenna, G. B., Anomalous Melting behavior of Cyclohexane and Cyclooctane in Poly(dimethylsiloxane) Precursors and Model Networks. *J. Polym. Sci., Part B: Polym. Phys.* **2008**, *46*, 2779–2791.
101. Hoei, Y.; Ikeda, Y.; Sasaki, M., A Frozen Tube Model of Melting Point Depression of Solvents in Entangled Polymer Systems. *J. Phys. Chem. B* **2003**, *107*, 1483–1490.
102. Ma, W.-J.; Banavar, J. R.; Koplik, J., A Molecular Dynamics Study of Freezing in a Confined Geometry. *J. Chem. Phys.* **1992**, *97*, 485–493.
103. Jackson, C. L.; McKenna, G. B., The Glass Transition of Organic Liquids Confined to Small Pores. *J. Non-Cryst. Solids* **1991**, *131–133*, 221–224.

104. Zhang, J.; Liu, G.; Jonas, J., Effects of Confinement on the Glass Transition Temperature of Molecular Liquids. *J. Phys. Chem.* **1992**, *96*, 3478–3480.
105. Ellison, C. J.; Torkelson, J. M., The Distribution of Glass-Transition Temperatures in Nanoscopically Confined Glass Formers. *Nat. Mater.* **2003**, *2*, 695–700.
106. Andre, P.; Folk, S. L.; Adam, M.; Rubinstein, M.; DeSimone, J. M., Light Scattering Study of Polydimethyl Siloxane in Liquid and Supercritical Carbon Dioxide. *J. Phys. Chem. A* **2004**, *108*, 9901–9907.
107. Flory, P. J., *Principles of Polymer Chemistry*. Cornell University Press: Ithaca NY, 1953.
108. Schaefer, D. W.; Mark, J. E.; McCarthy, D. W.; Jian, L.; Sun, C.-C.; Farago, B., Structure of Microphase-Separated Silica/Siloxane Molecular Composites. In *Polymer-Based Molecular Composites*, Schaefer, D. W.; Mark, J. E., Eds. Materials Research Society: Pittsburgh, PA, 1990; Vol. 171, pp 57–63.
109. Schaefer, D. W.; Chen, C., Structure Optimization in Colloidal Reinforcing Fillers: Precipitated Silica. *Rubber Chem. Technol.* **2002**, *75*, 773–793.
110. Schneider, G. J.; Goritz, D., Strain Induced Anisotropies in Silica Polydimethylsiloxane Composites. *J. Chem. Phys.* **2010**, *133*, 024903.
111. Yun, S. I.; Melnichenko, Y. B.; Wignall, G. D., Small-Angle Scattering from Symmetric Blends of Poly(dimethylsiloxane) and Poly(ethylmethylsiloxane). *Polymer* **2004**, *45*, 7969–7977.
112. Roots, J. E.; Ma, K. T.; Higgins, J. S.; Arrighi, V., Quasielastic Neutron Scattering Study of Poly(dimethylsiloxane) at High Pressure. *Phys. Chem. Chem. Phys.* **1999**, *1*, 137–141.
113. Edwards, C. J. C.; Richards, R. W.; Stepto, R. F. T., Monte Carlo Calculations of Particle Scattering Functions of Polymer Chains. *Macromolecules* **1984**, *17*, 2147–2153.
114. Sinha, M. Probing Polymer Networks Using Pulse Propagation and Brillouin Light Scattering Techniques. Ph.D. Physics, University of Cincinnati, Cincinnati, 2000.
115. Sinha, M.; Mark, J. E.; Jackson, H. E.; Walton, D., A Brillouin Scattering Study of End-Linked Poly(dimethylsiloxane) Networks. *J. Chem. Phys.* **2002**, *117*, 2968–2974.
116. Stevens, L. L.; Orler, E. B.; Dattelbaum, D. M., Brillouin-Scattering Determination of the Acoustic Properties and Their Pressure Dependence for Three Polymeric Elastomers. *J. Chem. Phys.* **2007**, *127*, 104906.
117. Gent, A. N.; Marteny, P., The Effect of Strain Upon the Velocity of Sound and the Velocity of Free Retraction for Natural Rubber. *J. Appl. Phys.* **1982**, *53*, 6069–6075.
118. Sinha, M.; Erman, B.; Mark, J. E.; Ridgway, T. H.; Jackson, H. E., Pulse Propagation in Poly(dimethylsiloxane) Networks. *Macromolecules* **2003**, *36*, 6127–6134.
119. Juliac, E.; Arman, J.; Harran, D., An Ultrasonic Method to Study the Interdiffusion Phenomenon Between Two Polydimethylsiloxanes. *Eur. Polym. J.* **2002**, *38*, 2161–2169.
120. Durig, J. R.; Flanagan, M. J.; Kalasinsky, V. F., The Determination of the Potential Function Governing Low Frequency Bending Mode in Disiloxane. *J. Chem. Phys.* **1977**, *66*, 2775–2785.
121. Oberhammer, H.; Boggs, J. E., Importance of (p-d)π Bonding in the Siloxane Bond. *J. Am. Chem. Soc.* **1980**, *102*, 7241–7244.

122. Dubchak, I. L.; Pertsin, A. I.; Zhdanov, A. A., Calculation of Helical Chains of Polyorganosiloxanes. *Polym. Sci. U.S.S.R.* **1985**, *27*, 2340–2347.
123. Bock, H., Fundamentals of Silicon Chemistry: Molecular States of Silicon-Containing Compounds. *Angew. Chem. Int. Ed. Engl.* **1989**, *28*(12), 1627–1650.
124. Nicholas, J. B.; Winans, R. E.; Harrison, R. J.; Iton, L. E.; Curtiss, L. A.; Hopfinger, A. J., An ab Initio Investigation of Disiloxane Using Extended Basis Sets and Electron Correlation. *J. Phys. Chem.* **1992**, *96*, 7958–7965.
125. Stepto, R. F. T.; Taylor, D. J. R., Molecular Modeling of the Elastic Behavior of Polymer Chains in Networks. *J. Chem. Soc., Faraday Trans.* **1995**, *91*, 2639–2647.
126. Sides, S. W.; Curro, J.; Grest, G. S.; Stevens, M. J.; Soddemann, T.; Habenschuss, A.; Londono, J. D., Structure of Poly(dimethylsiloxane) Melts: Theory, Simulation, and Experiment. *Macromolecules* **2002**, *35*, 6455–6465.
127. Frischknecht, A. L.; Curro, J., Improved United Atom Force Field for Poly(dimethylsiloxane). *Macromolecules* **2003**, *36*(6), 2122–2129.
128. Smith, J. S.; Borodin, O.; Smith, G. D., A Quantum Chemistry Based Force Field for Poly(dimethylsiloxane). *J. Phys. Chem. B* **2004**, *108*, 20340–20350.
129. Jia, Y.; Sun, S.; Liu, L.; Ji, Z., Finite Element Simulation of Mechanical Property Evolvement of Silicone Rubber in the Hot Vulcanizing Process. *Polym. Int.* **2004**, *53*, 41–47.
130. Curteanu, S., Artificial Intelligence Instruments Applied in Siloxane Chemistry. In *Recent Developments in Silicone-Based Materials*, Cazacu, M., Ed. Nova Science: New York, 2010; pp 223–270.
131. Heine, D. R.; Grest, G. S.; Lorenz, C. D.; Tsige, M.; Stevens, M. J., Atomistic Simulations of End-Linked Poly(dimethylsiloxane) Networks: Structure and Relaxations. *Macromolecules* **2004**, *33*, 3857–3864.
132. Lupton, E. M.; Achenbach, F.; Weis, J.; Brauchle, C.; Frank, I., Modified Chemistry of Siloxanes Under Tensile Stress: Interaction with Environment. *J. Phys. Chem. B* **2006**, *110*, 14557–14563.
133. Liu, J.; Cao, D.; Zhang, L., Static and Dynamic Properties of Model Elastomer with Various Cross-Linking Densities: A Molecular Dynamics Study. *J. Chem. Phys.* **2009**, *131*, 034903.
134. Smith, J. S.; Bedrov, D.; Smith, G. D.; Kober, E. M., Thermodynamics and Conformational Changes upon Stretching a Poly(dimethylsiloxane) Chain in the Melt. *Macromolecules* **2005**, *38*, 8101–8107.
135. Hanson, D. E.; Martin, R. L., How Far Can a Rubber Molecule Stretch Before Breaking? Ab Initio Study of Tensile Elasticity and Failure in Single-Molecule Polyisoprene and Polybutadiene. *J. Chem. Phys.* **2009**, *130*, 064903.
136. Makrodimitri, Z. A.; Economou, J. G., Atomistic Simulations of Poly(dimethylsiloxane) Permeability Properties to Gases and n-Alkanes. *Macromolecules* **2008**, *41*, 5899–5907.
137. Abou-Hussein, R.; Wu, S.; Zhang, L.; Mark, J. E., Effects of Some Structural Features of Poly(dimethylsiloxane) on Its Unusually High Gas Permeability. *J. Inorg. Organomet. Polym. Mater.* **2008**, *18*, 100–103.
138. Ismail, A. E.; Grest, G. S.; Heine, D. R.; Stevens, M. J.; Tsige, M., Interfacial Structure and Dynamics of Siloxane Systems: PDMS-Vapor and PDMS-Water. *Macromolecules* **2009**, *42*, 3186–3194.
139. Bienzobaz, P. F.; Malacarne, L. C.; Mendes, R. S.; Lenzi, E. K., An Expansion Approach in Rubber Elasticity. *Polymer* **2008**, *49*, 1968–1971.

CHAPTER 5
General Properties

5.1 SOME GENERAL INFORMATION

Many of the properties of the polysiloxanes have been tabulated in handbooks of polymer science and engineering.[1–3] Recent work has included the stretching of polydimethylsiloxane (PDMS) chains,[4] in some cases to their rupture points.[5]

The nature of the bonding in siloxane molecules has been of longstanding interest.[6–9] Force fields for calculations of PDMS properties have been revised over the years and are now at an advanced state of development.[10–13] Some of the simplest approaches employ the methods of molecular mechanics.[14] Most of the experimental results have been obtained on solutions of polysiloxanes in thermodynamically good solvents.[15]

5.2 CONFORMATIONS AND SPATIAL CONFIGURATIONS

5.2.1 Symmetrically Substituted Polysiloxanes

The first member of this series, poly(dimethylsiloxane) (PDMS), $[-Si(CH_3)_2O-]_x$, has been studied extensively with regard to its configuration-dependent properties.[16–21] PDMS (figure 2.1) is very similar in structure to the polyphosphate chain in that the successive bond angles are not equal. The Si–O bond length in polysiloxanes is 1.64 Å, and bond angles at the Si and O atoms are 110 and 143°, respectively. This inequality of bond angles causes the all-*trans* form of the molecule (with rotational angles $\phi = 0°$) to form a closed structure after approximately eleven repeat units. The torsional barrier for rotations about the skeletal bonds is very low, which accounts for the high dynamic flexibility and low glass transition temperature of the PDMS chain. Not surprisingly, low temperature properties are superb.[22]

Trans states are of lower energy than *gauche* states ($\phi = \pm 120°$) in the PDMS chain.[16, 17, 20, 21] This conformational preference may arise from favorable van der Waals interactions between pairs of CH_3 groups separated by four bonds in *trans* states. This factor is apparently more important than favorable coulombic interactions between oppositely charged Si and O atoms separated by three bonds, which are larger in *gauche* states because of the reduced distance. Comparisons between experimental and theoretical values of various configuration-dependent properties, however, yield a value for this energy difference that is significantly larger than that obtained from the semi-empirical calculations of interactions between nonbonded atoms. Conformations involving the unlike pairs $g^{\pm}g^{\mp}$ about O–Si–O skeletal bond pairs give rise to "pentane-type interferences"[16, 17, 20, 21] between the bulky $Si(CH_3)_2$ groups. Such configurations are therefore completely excluded. The same conformations about Si–O–Si bond pairs cause interferences between the smaller O atoms; these configurations can occur with low probability. Conflicting arrangements between groups separated by four bonds can be visualized by rotations about pairs of consecutive skeletal bonds in figure 2.1.

There is renewed interest in relating conformational descriptions to crystal structures,[23] which has raised questions about the applicability of this simple rotational isomeric state model to wide-angle scattering results.[24] Conformational rearrangements have been reported for PDMS chains at the air/water interface,[25] and near surfaces of silica[26] or mica.[27]

The equilibrium flexibility of PDMS can be characterized by its unperturbed dimensions, as measured by characteristic ratio $\langle r^2 \rangle_0 / nl^2$ of the unperturbed dimensions of the chain relative to the product of the number n of its skeletal bonds and the square of their length, l. Experimental values of this ratio are in the range 6.2–7.6, the precise value depending on the nature of the solvent.[28] The origin of this "specific solvent" effect is obscure but may involve specific interactions between solvent molecules and polymer segments in a way that changes the conformational preferences in the chain. The effect is significant only in the case of *polar* polymers. The unperturbed dimensions appearing in the definition of the characteristic ratio also appear in the equations for the modulus of the chains when cross linked into an elastomeric network. Not surprisingly, therefore, the specific solvent interactions can effect the modulus of swollen PDMS networks as well as the dimensions of isolated PDMS chains in solution.[29]

The characteristic ratio of PDMS is known to increase with increasing temperature. Numerous computations have been carried out to interpret this result.[16, 20, 21, 30] Thermal expansion is expected since the low energy conformation is the closed polygon discussed in section 1.1, figure 1.1,[16, 20, 21] and an increase in temperature provides thermal energy for switching from these compact, low-energy conformations to higher-energy states that are less compact.

Comparisons between the experimental and theoretical values of the characteristic ratio and its temperature coefficient give values of the chain conformational energies, which are then used to predict a number of other configuration-dependent properties. Dipole moments calculated in this way are in excellent agreement with experiment for small chain length; the agreement at longer chain length is less satisfactory, possibly because of the large specific solvent effect[31] already mentioned. Additional information on the dipole moments of siloxanes is given in section 5.5

Stress-optical coefficients have been determined using PDMS networks both unswollen and swollen with a variety of solvents. Only qualitative agreement was obtained, presumably because of the vanishingly small optical anisotropy of the PDMS chain.[17] Similar studies have been carried out on other polysiloxanes[16, 20, 21]—for example, on poly(methylphenylsiloxane)[32] and poly(tetramethyl-*p*-silphenylene-siloxane).[33]

Molecular mechanics and more sophisticated computational techniques are being applied to siloxane conformational problems to study both the chain backbone and the side chains,[34] and polysiloxane molecular motions in general.[35]

It is interesting to note that the PDMS chain and polyphosphate chain have approximately the same characteristic ratio. Isolated *gauche* states, of relatively high spatial extension, are more prevalent in the polyphosphate chain, but *pairs* of *gauche* states ($g^{\pm}g^{\pm}$) of the same sign are less prevalent so the two effects largely offset one another.

Much experimental and theoretical work has been reported on the cyclization of dimethylsiloxane chains, and on the properties of these cyclics.[36–41] Cyclization has been investigated for a wide range of polymer chain lengths, but in this chapter we discuss only the results for the behavior of long chains. The interpretation of the results for shorter chains may be complicated by failure of the Gaussian distribution function employed for the end-to-distances, directional correlations between terminal bonds prior to cyclization, ring-strain contributions to the heat of the reaction, and the need to revise some of the statistical weight factors for the chains. In the limit of large chain length, the agreement between theory and experiment is excellent and thus supports the proposed model for PDMS. In addition, cyclization studies have generated useful information on excluded volume effects (their absence in the undiluted amorphous state and their magnitudes in solutions, particularly at high polymer concentrations), the critical chain length at which the Gaussian distribution becomes inaccurate, the magnitude of specific solvent effects, and the validity of gel permeation chromatography theories pertaining to both linear and branched chain molecules.

Physical properties of cyclics in comparison with linear chains of the same degree of polymerization have been investigated extensively,[37, 38, 42] as described in chapter 3. Example of such comparisons are solution viscosity-molecular-weight relationships, bulk viscosities, densities, refractive indices, glass transition temperatures, ^{29}Si NMR (nuclear magnetic resonance) chemical shifts, chain dimensions from neutron scattering, diffusion coefficients and their concentration dependence, thermal stability, second virial coefficients, radii of gyration, equilibrium shapes (from Monte Carlo simulations), static dielectric permittivity, particle scattering functions, monolayer surface pressures, melting points, theta temperatures (at which the chains are unperturbed by excluded volume effects),[16] critical temperatures for phase separations, melt mobility (by excimer emission), and conformational dynamics (by ultrasonic relaxation measurements).

Other symmetrically substituted polysiloxanes have been investigated less thoroughly.[17] Poly(diethylsiloxane) $[-Si(C_2H_5)_2-O-]_x$ has been reported to have a characteristic ratio of 7.7 ± 0.2, which is essentially the same as that of PDMS; its dipole moment is difficult to measure because of the low polarity of the repeat unit, but it too is approximately the same as that of PDMS. These results suggest that lengthening of the side chains must generate self-compensating effects. Furthermore, poly(di-*n*-propylsiloxane) $[-Si(C_3H_7)_2-O-]_x$ has been reported to have a characteristic ratio of 13.0 ± 1.0.[43] The high spatial extension in this chain could result from the fact that an articulated side chain, such as $-CH_2CH_2CH_3$, can adopt more conformations in the form of *trans-gauche* states along the chain backbone than in the more restrictive *trans-trans* states shown in figure 5.1. Although this "entropic destabilization" of compact *trans* states would increase the chain dimensions, the large increase in the characteristic ratio over that for PDMS would not have been anticipated. This intuitive conclusion is supported by some rotational isomeric state calculations that do take into account the conformational variability of the siloxane side chains. It should be mentioned, however, that results of cyclization studies carried out on some stereochemically

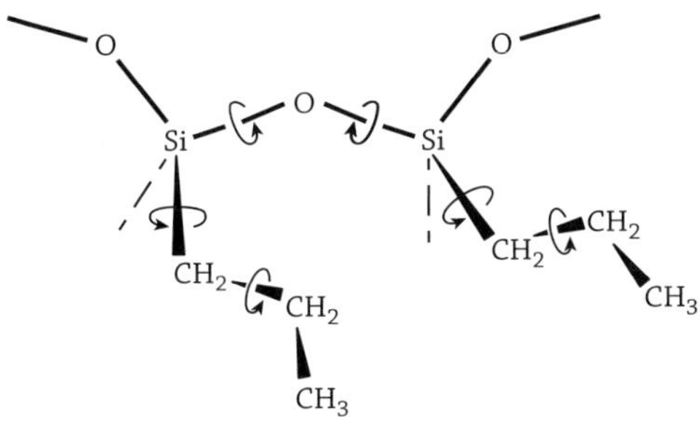

Figure 5.1:
The poly(di-*n*-propylsiloxane) chain, showing the conformational variability of the propyl side chains.[17] Reproduced by permission of the American Chemical Society.

variable polysiloxanes (see section 2.2) suggest that the characteristic ratio increases with increase in length or size of the side groups.

Cyclization studies have also been carried out on the chemical copolymers poly(ethylene, dimethylsiloxane) and poly(styrene, dimethylsiloxane).[36–38] Numerous intramolecular interactions need to be taken into account in a chemical copolymer. Consequently, the results on the copolymers have been given only a preliminary interpretation in terms of rotational isomeric state theory. Cyclization calculations have also been carried out for poly(dihydrogensiloxane) $[-SiH_2O-]_x$, but at present there are no experimental data available for comparison with theory.

Finally, melting point depression measurements have been conducted on several symmetrically substituted polysiloxanes, specifically the dimethyl, diethyl, di-*n*-propyl, and diphenyl polymers. Interpretation of such experimental results yields entropies of fusion. Although it is difficult to extract a reliable configurational entropy from this quantity, such results could help elucidate the configurational characteristics of the chains thus investigated.[17]

5.2.2 Stereochemically Variable Polysiloxanes

In unsymmetrically disubstituted chains, the substituents of one type can be on the same side of the all-*trans* chain, on opposite sides, or on either side in a random arrangement, yielding isotactic, syndiotactic, and atactic forms. Poly(methylphenylsiloxane) was one of the chains chosen to illustrate this stereochemical variability. The relatively large Si–O bond length and Si–O–Si bond angle place apposed side groups at distances of separation (ca. 3.8 Å) at which there is a favorable energy of interaction.

Conformational energy calculations[17, 44] on this polymer indicate that the attractions should be particularly strong in the case of a pair of phenyl groups in adjacent repeat units. Therefore, the chains should have a tendency to adopt conformations in which two phenyl groups are apposed on the same side of the chain.[17] For the syndiotactic polymer this effect favors *gauche* states, which favor relatively high spatial extension, but suppress of such states for the isotactic polymer. As a result, the characteristic ratio is predicted to be very small for the isotactic polymer and to increase linearly with increases in the number of syndiotactic placements in the chain. These results are quite different from those calculated for monosubstituted $[-CHRCH_2-]_x$ vinyl or disubstituted $[-CRR'CH_2-]_x$ vinylidene chains, including the structurally analogous poly(α-methylstyrene) $[-C(CH_3)(C_6H_5)-CH_2-]_x$.[17]

A characteristic ratio of 8.8 was reported for several samples of poly(methylphenylsiloxane),[17, 45] at least some of which were known to be essentially atactic. This experimental result, however, can be reproduced from the model only by assuming a large fraction of syndiotactic placements; the temperature coefficient predicted for this degree of syndiotacticity is then also in good agreement with experiment. The assumption of significant syndiotacticity is in disagreement with NMR results and with the results of cyclization studies, both of which suggest that poly(methylphenylsiloxane) is essentially atactic. The cyclization results, however, yield a prediction for the characteristic ratio that is significantly larger than the experimental value of 8.8. The two tentative conclusions regarding the stereochemical structure might be brought into closer agreement by improving the calculation of the interaction energy of two apposed phenyl groups to take into account the fact that they would be less exposed to favorable interactions with the solvent in such conformations. This effect is apparently quite important in polystyrene $[-CH(C_6H_5)-CH_2-]_x$ but may be less so in poly(methylphenylsiloxane) because of the larger distance of separation between side groups in the siloxane polymers. Such revision could increase the number of isotactic placements without decreasing its predicted value of the characteristic ratio to below its known experimental value. In any case, resolution of this point really requires reliable experimental values of the characteristic ratio, determined on samples of known stereochemical structure.

Cyclization measurements have also been conducted on other stereochemically variable polysiloxanes $[-Si(CH_3)R-O-]_x$, where R is H, CH_2CH_3, $CH_2CH_2CH_3$, and $CH_2CH_2CF_3$. The conclusion from these investigations was that such polymers are generally atactic and that an increase in the length or size of the side chains increases the characteristic ratio.[37, 38, 42]

It would be interesting to find catalysts that parallel the Zeigler-Natta catalysts used to prepare isotactic poly(α-olefin) polymers.[46] Poly(methylphenylsiloxane) samples, for example, having sufficient isotacticity (or syndiotacticity) to crystallize could be quite important. There is promising work based on ring-opening polymerization of cyclics of isotactic stereochemistry using catalysts that prevent scrambling of the cyclic stereochemistry during polymerization. This route has yielded poly[(methyl(3,3,3-trifluoropropyl)siloxane] samples ranging from the usual noncrystallizable (atactic) form to materials that undergo strain-induced crystallization to thermoplastics (which are crystalline at ambient temperatures).[47–50]

Control of stereochemistry, in general, is an interesting and potentially important subject with regard to commercial materials.[51, 52]

5.2.3 Some Unusual Side Groups

Including fluoro groups in the side chains of a polysiloxane can improve solvent resistance and modify surface properties,[53–55] as discussed in chapter 6. Most interest has focused on replacing some hydrogen with fluorine in *n*-alkyl groups—for example, the three fluorines in the trifluoropropyl group $-(CH_2)_2CF_3$. One, two, or three fluorines have also been substituted into the phenyl groups of poly(methylphenylsiloxane).[56, 57] Also of interest is the use of aromatic side groups different from the phenyl groups present in the commercially important poly(methylphenylsiloxane) or the highly intractible poly(diphenylsiloxane).[58–65] Homopolymers and copolymers having *p*-tolylsiloxane groups are of interest with regard to the formation of liquid-crystalline phases.[66, 67]

If the side groups are sufficiently hydrophilic, the polysiloxane can even become water soluble.[68] Attaching groups such as those of ethylene oxide to polysiloxane networks can also be used to make them hydrophilic.[69, 70]

A final example is the use of optically active groups as side chains, the simplest example being the *secondary* butyl group $-CH(CH_3)(C_2H_5)$. Such polymers could be studied using any of the numerous characterization techniques developed to characterize optically active polymers such as the proteins.

Other side chains include unusually long *n*-alkyl groups,[71, 72] cyclics (cyclolinear carbosiloxanes),[73] fluorocarbons,[74] phenylethenyl substituents,[75] Cr-bound carbazole chromophores,[76] and pendant bicyclic fragments.[77]

In contrast to the foregoing examples, the polysiloxane sometimes appears as side chains (e.g., on poly(phenylacetylene) main chains).[78]

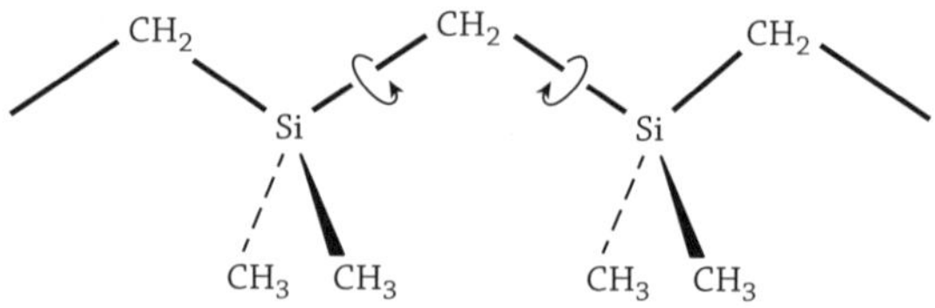

Figure 5.2:
The poly(dimethylsilmethylene) chain.[17] Reproduced by permission of the American Chemical Society.

5.2.4 Poly(dimethylsilmethylene)

Poly(dimethylsilmethylene), $[-Si(CH_3)_2-CH_2-]_x$, can be thought of either as a hydrocarbon analogue of PDMS (in which O atoms are replaced by CH_2 groups) or as a silicon analogue to polyisobutylene $[-C(CH_3)_2-CH_2-]_x$ (in which Si atoms replace one of the two skeletal C atoms in the repeat unit).[17] Figure 5.2 shows the polymer schematically. The Si–C bonds are 1.90 Å and, in contrast to siloxane chains, the two types of skeletal bond angles are essentially identical and tetrahedral. Since CH_2 and CH_3 groups have very similar interactions, this chain molecule should have some characteristics reminiscent of the idealized "freely rotating" chain.[16, 20, 21] This conclusion is supported by experimental evidence, which indicates that the characteristic ratio of the polymer is relatively small and that both its unperturbed dimension and dipole moment are essentially independent of temperature.[17]

5.3 FLEXIBILITY OF THE POLYMER CHAINS

5.3.1 Equilibrium Flexibility

Equilibrium flexibility has a profound effect on the melting point, T_m, of a polymer. Since crystallites in turn have a profound effect on the properties, the crystallization PDMS has been a subject of considerable activity.[23, 79–82] The melting points of small molecules in PDMS networks have also been reported.[83]

High flexibility in the equilibrium sense means high conformational randomness in the amorphous state, and thus high entropy of fusion and low melting point. This entropy can be reduced by stretching, in what is called "strain-induced crystallization." Such strain-induced crystallization has been studied extensively, both experimentally[84–85] and theoretically.[86–87] The crystallites thus generated can be very important since they

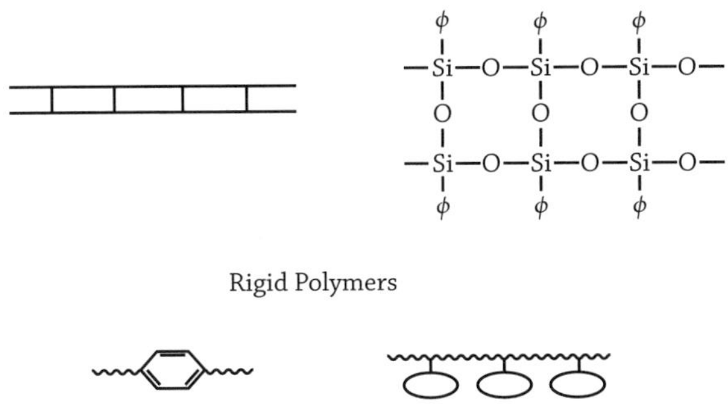

Figure 5.3:
Some ways for making a polymer more rigid.[29]
Reproduced by permission of John Wiley and Sons.

may provide considerable reinforcement for a network. Most polysiloxane elastomers, however, have melting points that are too low to benefit from this effect.

Figure 5.3 shows examples of ways in which a polymer can be made more rigid.[88] Possibilities include combining two chains in a ladder structure (figure 5.4), insertion of rigid units such as *p*-phenylene groups into the chain backbone and the addition of bulky side groups. Attempts have been made to prepare the silsesquioxane ladder polymer shown in the upper portion of figure 5.3 using trifunctional silanes.[89, 90] The basic strategy is to decrease the entropy of fusion and thus increase the melting point T_m (which is inversely proportional to the entropy of fusion). If the chains are combined into a ladder structure they cannot disorder as much as when they are separate, as shown in figure 5.5. The same argument holds for the other two methods for increasing T_m. This decreased equilibrium flexibility is generally paralleled by decreased dynamic flexibility, and thus by an increased glass transition temperature T_g. An advantage of the ladder structure is its resistance to degradative chain scission. The

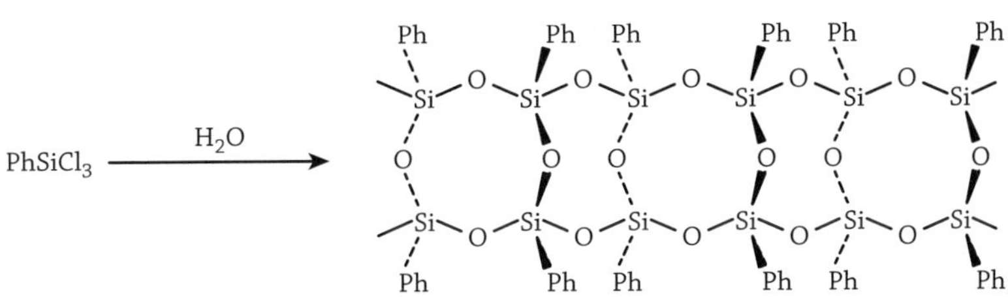

Figure 5.4:
A possible reaction for preparing a sesquisiloxane ladder polymer.[302]
Reproduced by permission of Prentice Hall, Inc.

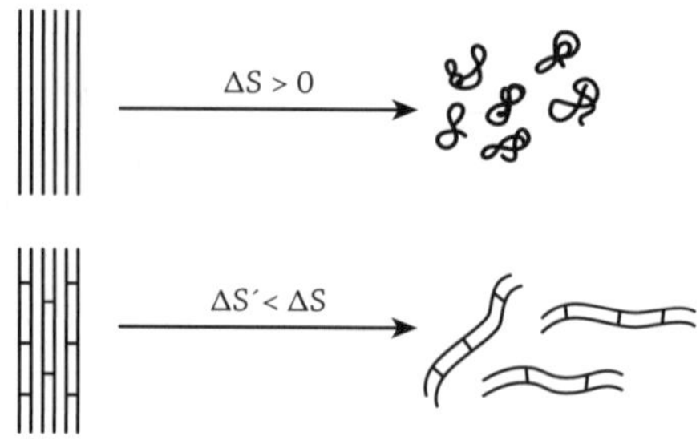

Figure 5.5:
Sketch explaining the increase in melting point with increase in chain rigidity.[29]
Reproduced by permission of John Wiley and Sons.

chain will not be degraded into two shorter ladder structures except in the unlikely event that two single-chain scissions occur directly across from one another.[91]

There is considerable interest in inserting a silphenylene group [–Si$(CH_3)_2C_6H_4$–] into polysiloxane backbones.[29, 92–98] In the case of the PDMS repeat unit, insertion yields the *meta* and *para* silphenylene polymers shown in figure 5.6.[29] The T_g is increased to -48°C compared to -125°C for PDMS, but no crystallinity has been detected. Since the repeat unit is symmetric, it should be possible to induce crystallinity by stretching. The explanation here is the same as that given in figure 5.5 except that the chains are prevented from completely disordering by the stretching force, rather than

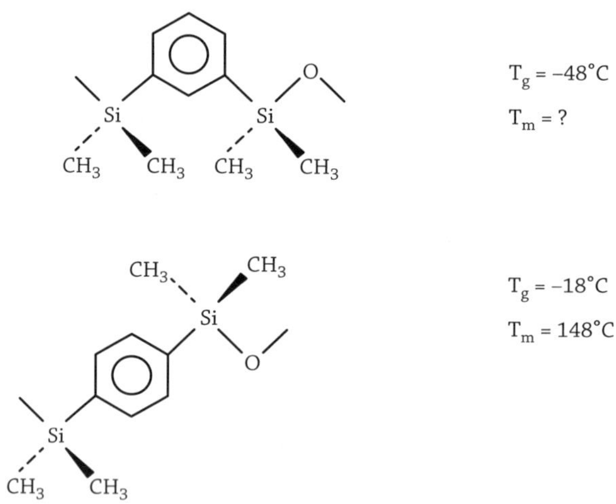

Figure 5.6:
Meta and *para* silphenylene polymers and their transition temperatures.[29]
Reproduced by permission of John Wiley and Sons.

by the structural features of the chains. As expected, the *p*-silphenylene group has a larger stiffening effect, increasing the T_g to –18°C and giving rise to crystallinity with a T_m of 148°C. The resulting polymer is thus a *thermoplastic* siloxane. Apparently, *ortho* (*o*) silphenylene units have not been introduced in this way; they are probably much harder to incorporate because of steric problems. Even if they could be incorporated, such groups would not be expected to have much of a stiffening effect on the chain.

Silarylene polymers that contain more than one phenylene group in the repeat unit could be of considerable interest because of the various *meta*, *para* combinations that could presumably be synthesized. Cyclohexylene (C_6H_{10}) groups can switch between boat and chair forms and should thus be less stiff than phenylene groups. Both *meta*- and *para*-groups are of interest in this regard. Table 5.1 summarizes some of these modifications and their likely effects. Figure 5.3 also has a sketch showing the use of bulky side groups to make a chain stiffer, such as by replacing one of the

Table 5.1. SOME POLYSILOXANE STRUCTURAL CHANGES AND EXPECTED CHANGES IN PROPERTIES

Part of repeat unit	Structural change	Expected changes in properties
Backbone	Replacement of O by CH_2	Decrease in polarity and tendency to cyclize
	Insertion of *p*-phenylene	Large decrease in flexibility
	Insertion of *m*-phenylene	Smaller decrease in flexibility
	Insertion of *p*-cyclo-hexylene	Small decrease in flexibility
	Insertion of *m*-cyclo-hexylene	Small decrease in flexibility
Side group	$(CH_2)_mCH_3$, m > 0	Tendency to form liquid-crystalline phases
	Phenyl (atactic placements)	Suppression of crystallinity, and increase in radiation resistance
	Fluorinated groups	Increase in solvent resistance
	Bulky groups in general	Decrease in flexibility
	H atoms	Capacity for cross linking by vinyl groups
	Vinyl groups	Capacity for cross linking by H atoms
	Hydrophilic groups	Increase in water solubility
	Chiral groups, e.g., $-CH(CH_3)C_2H_5$	Generation of optical activity

methyl groups in the PDMS repeat unit by a phenyl group. The resulting polymer, poly(methylphenylsiloxane), has a glass transition temperature of –28°C, which is significantly higher than the value, –125°C, shown by PDMS.[99]

There has been success in simulating the crystallinity of some polysiloxanes. This work was based on the model of Windle and coworkers[100] for simulating chain ordering in copolymers composed of two comonomers, at least one of which is crystallizable. Monte Carlo techniques are first used to generate chains with sequences that are representative of those occurring in the polymers of interest. In the case of chemical copolymers the generation is based on standard copolymerization equations containing the usual reactivity ratios. In the case of stereochemical copolymers, generation is based on replication probabilities (*meso* rather than *racemic* placements). Typically, the generated chains are placed in parallel, two-dimensional arrangements. Neighboring chains are then searched for sequence matches that could lead to the formation of crystallites.[101]

An example of modeling chemical copolymer crystallization is random versus semi-blocky poly(diphenylsiloxane-*co*-dimethylsiloxane) copolymers. The chains were placed alongside one another in a two-dimensional array, with black squares representing dimethylsiloxane (DMS) units and white squares representing diphenylsiloxane (DPS) units.[102] "Like" squares neighboring each other in the same row were then viewed as coalescing into blocks the lengths of which were compared with the known minimum length required to form a crystallite. Crystallizable DPS regions were identified as distinct from noncrystallizable DMS component, or units of the crystallizable DPS component that were not long enough to crystallize.[102] A value of the degree of crystallinity of a simulated sample was then determined by counting the units involved in the matching sequences relative to the total number of units of all the chains. The crystallites thus identified presumably act as physical cross-linking sites and reinforcing domains, providing the additional toughness over random comonomer distribution.

An example of modeling of a stereochemically variable polysiloxane focused on poly[methyl(3,3,3-trifluoropropyl)siloxane],[103] a polymer of interest because of new synthetic techniques for controlling stereoregularity and thus crystallizability.[47–50]

It is intriguing that some flexible siloxane polymers form mesomorphic (liquid-crystalline) phases.[104–132] Table 5.2 provides some illustrative data. Both poly(diethylsiloxane) and poly(di-*n*-propylsiloxane) show two crystalline modifications as well as a mesomorphic phase. The other major

Table 5.2. EXAMPLES OF LIQUID-CRYSTALLINE POLYSILOXANES[a]

Polymer Repeat Unit	~T, °C	Transition
$[-Si(CH_3)_2O-]$ (Reference polymer)	−40	Cryst →Isotropic
$[-Si(C_2H_5)_2O-]$	−60	Cryst →Cryst'
"	0	Cryst' → Mesomorphic
"	40	Mesomorphic → Isotropic
$[-Si(n\text{-}C_3H_7)O-]$	−55	Cryst →Cryst'
"	60	Cryst' →Mesomorphic
"	205	Mesomorphic → Isotropic

[a] Reproduced by permission of the American Chemical Society.

class of semi-inorganic polymers, the polyphosphazenes, are also relatively flexible, and show similarly interesting behavior.[17, 133] Some polysiloxanes form liquid-crystalline phases because of the presence of relatively stiff side chains.[134–141] This group has been studied with regard to the effect of deformation of the elastomeric polysiloxane phase on the mesomorphic behavior exhibited by the side chains.

Although the polysiloxanes are much more flexible than their organic counterparts, the polysilanes seem to display intermediate properties. It is instructive to compare a polysilane with its hydrocarbon analogue in terms of chain flexibility. For example, relevant conformational energy calculations have been carried out on polysilane itself $[-SiH_2-]_x$.[142–143] Energy maps[143] for two consecutive results skeletal rotation angles ϕ suggest that the lowest energy conformation should be a sequence of *gauche* states ($\phi = \pm 120°$) of the same sign.[142, 143] This conclusion is in contrast to polyethylene $[-CH_2-]_x$, which has a preference for *trans* states. Such preferences generally dictate the regular conformation chosen by a polymer chain when it crystallizes. Polyethylene crystallizes in the all-*trans* planar zig-zag conformation[16, 20, 21] It would be interesting to determine whether polysilane crystallizes in the predicted helical form generated by placing all of its skeletal bonds in *gauche* states of the same sign. The calculations also predict that polysilane should have a higher equilibrium flexibility than polyethylene.[143] Solution characterization techniques could be used to test this expectation. Dynamic flexibility, can also be estimated from such energy maps by determining the barriers between energy minima. Relevant experimental results could be obtained by a variety of dynamic techniques.[144]

Also relevant here is the fact that PDMS in Langmuir films has been shown to undergo an interesting layering transition, which has been highlighted using poly(methyl methacrylayte).[145]

5.3.2 Dynamic Flexibility

Dynamic flexibility refers to a molecule's ability to *change* spatial arrangements by rotations around its skeletal bonds.[35] The more flexible a chain is in this sense, the more it can be cooled before the chains lose their flexibility and mobility and the bulk material becomes glassy. Chains with high dynamic flexibility thus have low glass transition temperatures, T_g.[146] Since exposing a polymer to a temperature below its T_g generally causes it to become brittle, low values of T_g can be advantageous, particularly in the case of fluids and elastomers. Structural changes that increase a chain's equilibrium stiffness generally also increase its dynamic stiffness and thus increase T_g. Conversely, the very high flexibility of PDMS is the origin of its low T_m (–40°C)[1, 99, 147] as well as its very low T_g (–125°C).[1, 99, 147] The general effect of increased rigidity is thus to increase a polymer's "softening temperature," which is approximately T_m if the polymer is crystalline and approximately T_g (typically ~ 2/3 T_m in °K) if it is not.[91]

5.3.3 Viscoelasticity

Dynamic aspects of PDMS have long been of interest. Examples of such studies in the undiluted state include measurements of the onset of non-Newtonian flow,[148] effects of molecular weight[149] and strand-length polydispersity,[150] effects of frequency,[151] methyl group rotations by quasi-elastic neutron scattering,[152] and the dynamics in grafted layers.[153] There have also been studies of the onset of shear thinning,[154] entanglement,[155] transiently trapped entanglements,[156] and electromechanical response of PDMS actuators.[157, 158]

Solution studies have also been reported, for example, on un-cross-linked PDMS in toluene.[159, 160] Dynamic properties have even been carried out for the chains in emulsions,[161] and in diblock lamellar mesophases.[162]

Cross-linked elastomers have been studied with regard to their moduli (particularly plateau values),[163] effects of peroxide cross linking,[164] adhesive interactions with atomic force microscopy tips,[165] and the effects of phenyl-group modifications.[166] Investigations on networks containing fillers include the effects of silica or polysilicate nanoparticles,[167, 168] zero mean-strain measurements,[169] microcapsule dispersions,[170] nanoclay-modified PDMS copolymers,[171] and the effects of strain history.[172]

5.4 PERMEABILITY

Siloxane polymers have much higher permeability to gases than most other elastomeric materials. For this reason, they have long been of interest for gas separation membranes, the goal being to vary the basic siloxane structure to improve selectivity without decreasing permeability. Polysiloxanes for pervaporation separations have also been of considerable interest.[173–178] The techniques employed in this area include NMR and fluorescence correlation spectroscopy[179] and inverse gas chromatography.[180]

The repeat units of some of the polymers which have been investigated[181–185] include [$-Si(CH_3)$ –RO–], [$-Si(CH_3)$ –XO–], [$-Si(C_6H_5)$ –RO–], [$-Si(CH_3)_2-$ $(CH_2)_m-$], [$-Si(CH_3)_2-$ $(CH_2)_m-Si(CH_3)_2-O-$], and [$-Si(CH_3)_2-$ $(C_6H_4)_m-Si(CH_3)_2-O-$], where R is typically an *n*-alkyl group and X is an *n*-propyl group made polar by substitution of atoms such as Cl or N. Polymers with oligo-organosiloxane side chains[186] and co-networks of PDMS with poly(ethylene glycol)[187] have also been studied for possible use as oxygen-permeable membranes. Unfortunately, structural changes that increase the gas-diffusion selectivity are generally found to decrease the permeability, and vice versa. Interest has not focused entirely on gases; for example, liquids such as polychlorinated biphenyls and polycyclic aromatic hydrocarbons have also been studied.[188]

In some cases, the polysiloxane was in the form of a composite—for example, with sulfonated cross-linked polystyrene particles,[189] carbon black,[190] acrylate latexes,[191] or sodium dodecyl sulfate.[192] Counterintuitively, the addition of impenetrable nanofillers can actually *increase* the permeability of a membrane.[193] Also, siloxane-imide copolymers have shown some interesting properties in membrane separations,[194] as have polysiloxanes containing poly(ether amine) groups.[195]

The interdiffusion of two samples of PDMS of different molecular weights has also been of interest.[196]

In addition to numerous experiments, there have been attempts at modeling[197] using solubility parameters[198] and Flory-Huggins interaction parameters[199] to predict separation characteristics and permeability of polysiloxane membranes.[200–202] Simulations indicate that at least the asymmetrically substituted polysilanes [–SiRR'–] have gas permeabilities comparable to that of PDMS.[203] The permeability, *P*, is the product of the solubility, *S*, of the gas in the polymer and its diffusivity, *D*.[204] Values of *P* for the

Table 5.3. SOME GAS PERMEABILITY INFORMATION[a]

Polymer Repeat Unit	Gas	$10^8 P^b$	(P_{O_2} / P_{N_2})
$[-Si(CH_3)_2O-]$	O_2	6.0	1.9
"	N_2	3.1	—
$[-C(Si(CH_3)_3) = C(CH_3)-]$	O_2	72.	1.7
"	N_2	42	—

[a] Reproduced by permission of the American Chemical Society.
[b] Units of cm^3(STP)cm/(cm^2 s cm Hg).

polysilanes studied seem to be high because of very high gas solubility in the polysilanes. It is interesting to note that these simulations indicate a much lower permeability for poly(dimethylsilmethylene) $[-Si(CH_3)_2-CH_2-]$. The presence of a trimethylsilyl group $[-Si(CH_3)_3]$ as a side chain in an acetylene repeat unit increases the permeability of the polymer to a value above that of PDMS. The specific polymer is poly[1-(trimethylsilyl)-1-propyne][205–213]; Table 5.3 gives some comparisons between it and PDMS.[113] Remarkably, its permeability, P, is about an order of magnitude higher than that of PDMS without much decrease in selectivity (as measured by the ratio of the P values for oxygen and nitrogen). The greatly increased values of P are apparently due to the unusually high solubility of gases in this polymer.[206] Studies of the effects of substituting the trimethylsilyl group onto other polymer backbones are in progress. Membranes containing a liquid-crystalline PDMS have also been studied.[214]

Another type of membrane designed as an artificial skin coating for burns also exploits the high permeability of siloxane polymers.[215, 216] The inner layer of the membrane consists primarily of protein and serves as a template for the regenerative growth of new tissue. The outer layer is a sheet of silicone polymer that not only provides mechanical support but also permits outward escape of excess moisture while preventing ingress of harmful bacteria.

Soft contact lens prepared from PDMS provide a final example, as shown in figure 5.7.[113] The oxygen required by the eye for its metabolic processes must be obtained by inward diffusion from the air rather than through blood vessels.[217, 218] PDMS is ideal for such lenses [215] because of its high oxygen permeability, but it is too hydrophobic to be adequately wetted by the fluids covering the eye. Poor wetting means the lens does not "feel right" and can also cause adhesion of the lens to the eye itself. One way to remedy this problem is to graft a thin layer of a hydrophilic polymer to the inner surface of the lens. Because of the thinness of the coating the high permeability of the PDMS is essentially unaffected.

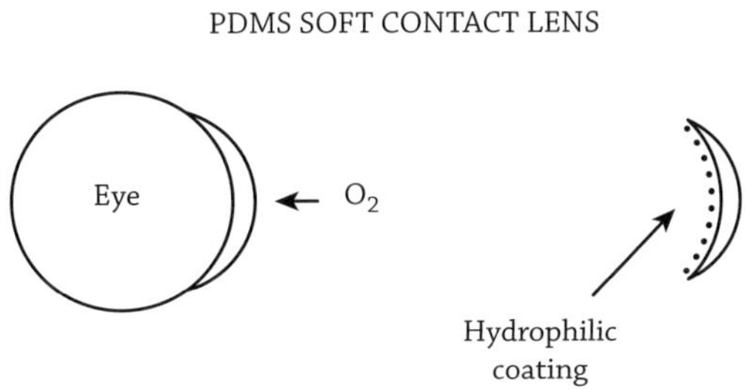

Figure 5.7:
An example of the use of grafting to change only the surface properties of a polymeric material.[303]
Reproduced by permission of the American Chemical Society.

The contact lens application just cited illustrates the use of one of the most striking properties of PDMS, its superb transparency.[219, 220] PDMS can retain this transparency even when filled (reinforced) with rather large amounts of silica. The transparencies[219, 221] and indices of refraction[222] of polysiloxane have been of considerable interest in general.

5.5 DIELECTRIC CONSTANTS AND DIPOLE MOMENTS

Some of the earliest relevant work in this area involved experimental studies of the dielectric relaxation times, polarizations, and dipole moments of the small molecules hexamethyldisiloxane, hexaethyldisiloxane, hexamethylcyclotrisiloxane, and octamethylcyclotetrasiloxane.[223]

Research on the corresponding polymers has focused almost exclusively on PDMS, but there has been some work on poly(diethylsiloxane).[224] In the case of the PDMS, most studies have involved the linear chains,[31, 225–235] some the cyclics,[233–235] and at least one cross-linked elastomer.[236] Improvements in dielectric properties continue to be important goals in the case of the polysiloxanes.[237, 238]

5.6 STABILITY, SAFETY ASPECTS, AND ENVIRONMENTAL IMPACTS

Siloxane polymers possess a number of properties that seem almost contradictory. One example is the tremendous stability and durability shown by polysiloxanes in a wide variety of applications.[239–241] One reason for this stability is the fact that the chain is already in a high oxidation state, and reduction takes place only at very high temperatures. Most scientists are familiar with the now-commonplace constant-temperature bath containing "silicone oil," which operates at high temperatures for years without any evidence for thermal degradation. Body implants made of polysiloxanes show little evidence of degradation, hydrolytic or otherwise, after decades of useful service, due to their resistance to hydrolysis and oxidation. The inertness of the siloxanes should not be much of a surprise, if one thinks of them as simple hydrocarbon modifications of the silicates we commonly refer to as "glass." In spite of this robustness, polysiloxanes do *not* present severe environmental problems. For example, in the case of a spill, or rupture of an electrical device such as a transformer, the polymers released degrade completely and relatively rapidly under normal environmental conditions.[239–240]

Examples of polysiloxane degradation studies include the role of surfactants in suppressing aging of silica-PDMS gels,[242] the effects of pigments on the stability of montmorillonite-PDMS composites,[243] and the use of NMR and mass spectrometry to characterize degradation processes.[244]

Degradation can occur in water, in air, and particularly in the soil[245] when polymers come into contact with one or more reactive species, such as nitrate ion present in natural waterways. Nitrate is a source of atomic oxygen and, from it, hydroxyl radicals, which initiate the degradation process. Another reagent is ozone, split by UV (ultraviolet) light into oxygen atoms, followed again by the production of hydroxyl radicals.

It is interesting to note that UV light itself has very little effect on the siloxane structure. Only the very shortest wavelengths present in sunlight have any influence and, in this case, generate methyl radicals from the side groups. Polysiloxanes are generally resistant to all types of radiation,[246–250] particularly if they contain aromatic groups (e.g., the phenyl groups in poly(methylphenylsiloxane)).

Even when methyl radicals are replaced by silanol units, the surface of the material does not remain hydrophilic (water-wettable) very long. Either the silanol groups condense with other silanol units to restore the siloxane structure or unmodified chain segments migrate to the surface. In any case, a "self-repair" mechanisms underlies the "recoverability" of siloxane surfaces.

Clay minerals present in many soils have high interfacial areas with strongly acidic groups on their surfaces. These materials can react with siloxane chains and reorganize them into much smaller molecules. In fact, water readily reacts with the Si–O bond in the presence of catalytic amounts of either acids or bases. Some of these small molecules are volatile enough to evaporate into the atmosphere. Others become capped with silanol (–SiOH) groups that frequently makes them water-soluble, and thus environmentally degradable. At later stages in the process, even the hydrocarbon groups are affected. Although the organosiloxane structure is completely unknown in nature, the introduction of these small molecules into the biosphere is thought to be entirely harmless. Furthermore, some of the degradation processes lead ultimately to silica, water, and carbon dioxide or inorganic carbonates.[239, 240]

The degradative reactions experienced by siloxane chains usually generate silanol groups, often at the chain ends, and can produce various cyclics.[251] Pairs of such silanol groups can condense with one another, forming new siloxane linkages. This condensation reaction provides an interesting "healing" mechanism for the siloxane backbone.

Resistance to degradation or cross linking by radiation is also important.[252, 253] Resistance is one advantage that poly(methylphenylsiloxane) has over poly(dimethylsiloxane); the phenyl groups are efficient energy sinks that suppress the usual effects of radiation on a polymer.

There is interest in synthesizing polysiloxane in environmentally friendly supercritical carbon dioxide.[254] Studies of the interactions between PDMS and supercritical carbon dioxide have been reported.[255] This medium has also been used in the preparation of composites between PDMS and polystyrene[256] and nanoclays.[257] Supercritical fluids have been used to study various properties[258] including surface behavior and carrying out depth analyses in PE/PDMS composites.[259]

Recycling of polymers has become an important environmental issue. Not surprisingly given its relatively high cost, such studies are underway on silicone elastomers as part of a general effort addressed to various types of elastomeric materials, including ones that are heavily filled with silica.[260–263]

5.7 THERMODYNAMICS

Typical thermodynamic properties of the polysiloxanes, such as heat capacities, solubility parameters, thermodynamic interactions parameters, and so on, have been extensively tabulated.[1, 99, 264] Of considerable interest have been their low-temperature properties, particularly stiffening effects.[22] Static and dynamic light scattering measurements have even been carried out on PDMS in supercritical carbon dioxide.[265] In some cases, dual techniques have been employed, for example, the joint use of differential scanning calorimetry[266] and Raman spectroscopy.[267] Equation-of-state parameters for PDMS itself have been reported,[268–270] as have excess volumes and chemical potentials for PDMS in several solvents.[271]

The strength of a siloxane bond has been estimated,[272] using both density functional theory and single-molecule measurements.[273] *Ab initio* methods have been applied to gain a better understanding of siloxane-functionalized ionic liquids, with the specific focus being on ion pairs in 1-methyl-3-pentamethyldisiloxymethylimidazolium chloride.[274]

Phase diagrams show miscibility and phase separation among various polysiloxanes, including cyclics.[275–277] Related to phase separation is the segregation of siloxanes to surfaces in multiphase systems because of the low surface energies of these polymers.[55, 278–280] As a result, a blend or block copolymer having only a few percent of siloxane can have surfaces consisting almost entirely of this minor constituent.

There is also much interest in applying various types of liquid-state theory, to PDMS in particular.[10, 11]

5.8 CRYSTALLINITY

The crystallization of PDMS in the undeformed state has long been of interest, including unfilled elastomers[23, 80, 82] and block copolymers reinforced with silica.[81] Of greater interest, however, is PDMS, generally in the filled state, in elongation[79, 84, 85, 281] where strain-induced crystallization provides considerable reinforcement of the elastomer. Strain-induced crystallization is of practical as well as fundamental importance.[282, 283] Theoretical models have been developed to characterize this type of crystallization.[86, 87]

Similar experimental studies of strain-induced crystallization have been carried out on stereoregular poly(methyl-3,3,3-trifluoropropylsiloxane).[47–50]

5.9 SOME ADDITIONAL UNUSUAL PROPERTIES OF PDMS

Table 5.4 summarizes some of the unusual physical properties exhibited by PDMS.[113, 240, 268, 271, 284] Atypically low values are exhibited for the characteristic pressure[206, 207] (a corrected internal pressure, which is used in the study of liquids), the bulk viscosity η, and the temperature coefficient of η.[252] Also, entropies of dilution and excess volumes on mixing PDMS with solvents are much lower than can be accounted for by theory.[268, 271] Finally, as has already been mentioned, PDMS has a surprisingly high permeability.

Another striking feature of siloxane polymers is their unusual surface properties.[54, 285–290] Fluorosiloxane polymers[291] have been studied most

Table 5.4. SOME UNUSUAL PROPERTIES OF PDMS[a]

Property	Experimental Result
Characteristic pressure	Unusually small
Bulk viscosity η	Unusually small
Temperature coefficient of η	Unusually small
Entropies of dilution	Significantly lower than theory
Excess volumes on mixing	Significantly lower than theory
Permeability	Unusually large

[a] Reproduced by permission of the American Chemical Society.

recently in this regard,[53, 285, 292] but the properties to be described are characteristic of a number of different members of the polysiloxane family. Surface properties can be exploited to serve in a variety of seemingly contradictory roles. For example, siloxanes can be both antifoaming agents and foam stabilizers, both paper-release coatings and pressure-sensitive adhesives, both water repellents and dewatering agents, and both emulsifiers and de-emulsifiers.[285, 293] This paradox is explained by the differing ways in which the siloxane chains interact with the other species present. For example, in foam technology it is critically important whether the siloxane dissolves in the liquid phase or stays at the liquid-gas interface. Similar questions arise in other applications, and specific properties are generated by an appropriate choice of side group; the addition of special polar groups, ionic groups, or reactive functional groups; or copolymerization with completely different classes of comonomer. Poly(methylphenylsiloxane) can undergo a reversible droplet-monolayer transition[294] that is relevant to the spreading of polymers on surfaces.

In the most general terms, the unusual surface properties of polysiloxanes are due to their low surface energies,[286, 293, 295] which are related to two important features of the chains themselves. The first is the very low intermolecular forces between the side chains, which are the methyl groups in PDMS, the commonest of the polymers having these unusual surface properties. The second is the remarkable flexibility of the siloxane backbone, which permits the chains to easily rearrange to place the methyl groups at the interface. A particularly interesting example of this is the ease with which a damaged polysiloxane surface quickly regenerates the surface characteristics of the original material. Some recent work has included the antifouling properties of polysiloxane hydrogels.[296, 297]

Although the molecular origin of the unusual properties of siloxane polymers remains controversial, a number of suggestions have been put forward. One involves low intermolecular interactions, as mentioned in the preceding paragraph. Others focus on differences between the nonpolar alkyl groups and the polar Si–O backbone,[293] or the very high rotational and oscillatory freedom of the methyl side groups in PDMS, the most important of the polysiloxanes.[298] Still another focuses on the chain's irregular cross section (very large at the substituted Si atom and very small at the unsubstituted O atoms.)[271] Another possible explanation is the fact that chains with unequal skeletal bond angles would not pack easily[299] since they would not be able to form the planar zig-zag conformations of polymers such as polyethylene.

Additional work in this area will almost certainly involve supramolecular structures.[300, 301]

REFERENCES

1. Brandrup, J.; Immergut, E. H.; Grulke, E. A., *Polymer Handbook*, 4th ed. Wiley: New York, 1999.
2. Mark, J. E., *Physical Properties of Polymers Handbook*, 2nd ed. Springer-Verlag: New York, 2007.
3. Mark, J. E., *Polymer Data Handbook, 2nd ed.* Oxford University Press: New York, 2009.
4. Smith, J. S.; Bedrov, D.; Smith, G. D.; Kober, E. M., Thermodynamics and Conformational Changes upon Stretching a Poly(dimethylsiloxane) Chain in the Melt. *Macromolecules* **2005**, *38*, 8101–8107.
5. Hanson, D. E.; Martin, R. L., How Far Can a Rubber Molecule Stretch Before Breaking? Ab Initio Study of Tensile Elasticity and Failure in Single-Molecule Polyisoprene and Polybutadiene. *J. Chem. Phys.* **2009**, *130*, 064903.
6. Durig, J. R.; Flanagan, M. J.; Kalasinsky, V. F., The Determination of the Potential Function Governing Low Frequency Bending Mode in Disiloxane. *J. Chem. Phys.* **1977**, *66*, 2775–2785.
7. Oberhammer, H.; Boggs, J. E., Importance of (p-d)π Bonding in the Siloxane Bond. *J. Am. Chem. Soc.* **1980**, *102*, 7241–7244.
8. Bock, H., Fundamentals of Silicon Chemistry: Molecular States of Silicon-Containing Compounds. *Angew. Chem. Int. Ed. Engl.* **1989**, *28*(12), 1627–1650.
9. Nicholas, J. B.; Winans, R. E.; Harrison, R. J.; Iton, L. E.; Curtiss, L. A.; Hopfinger, A. J., An ab Initio Investigation of Disiloxane Using Extended Basis Sets and Electron Correlation. *J. Phys. Chem.* **1992**, *96*, 7958–7965.
10. Sides, S. W.; Curro, J.; Grest, G. S.; Stevens, M. J.; Soddemann, T.; Habenschuss, A.; Londono, J. D., Structure of Poly(dimethylsiloxane) Melts: Theory, Simulation, and Experiment. *Macromolecules* **2002**, *35*, 6455–6465.
11. Frischknecht, A. L.; Curro, J., Improved United Atom Force Field for Poly(dimethylsiloxane). *Macromolecules* **2003**, *36*(6), 2122–2129.
12. Smith, J. S.; Borodin, O.; Smith, G. D., A Quantum Chemistry Based Force Field for Poly(dimethylsiloxane). *J. Phys. Chem. B* **2004**, *108*, 20340–20350.
13. Ismail, A. E.; Grest, G. S.; Heine, D. R.; Stevens, M. J.; Tsige, M., Interfacial Structure and Dynamics of Siloxane Systems: PDMS-Vapor and PDMS-Water. *Macromolecules* **2009**, *42*, 3186–3194.

14. Dubchak, I. L.; Pertsin, A. I.; Zhdanov, A. A., Calculation of Helical Chains of Polyorganosiloxanes. *Polym. Sci. U.S.S.R.* **1985**, *27*, 2340–2347.
15. Bercea, M.; Morariu, S.; Brunchi, C.-E.; Cazacu, M., Solution Behavior of Polysiloxanes. In *Recent Developments in Silicone-Based Materials*, Cazacu, M., Ed. Nova Science: New York, 2010; pp 35–76.
16. Flory, P. J., *Statistical Mechanics of Chain Molecules*. Wiley-Interscience: New York, 1969.
17. Mark, J. E., Conformations and Spatial Configurations of Inorganic Polymers. *Macromolecules* **1978**, *11*, 627–633.
18. Bahar, I.; Zuniga, I.; Dodge, R.; Mattice, W. L., PDMS Statistics 1. *Macromolecules* **1991**, *24*, 2986–2992.
19. Bahar, I.; Zuniga, I.; Dodge, R.; Mattice, W. L., PDMS Statistics 2. *Macromolecules* **1991**, *24*, 2993–2998.
20. Mattice, W. L.; Suter, U. W., *Conformational Theory of Large Molecules. The Rotational Isomeric State Model in Macromolecular Systems*. Wiley: New York, 1994.
21. Rehahn, M.; Mattice, W. L.; Suter, U. W., Rotational Isomeric State Models in Macromolecular Systems. *Adv. Polym. Sci.* **1997**, *131*, 1–18.
22. Polmanteer, K. E.; Hunter, M. J., Polymer Composition Versus Low-Temperature Characteristics of Polysiloxane Elastomers. *J. Appl. Polym. Sci.* **1959**, *1*, 3–10.
23. Albouy, P.-A., The Conformation of Poly(dimethylsiloxane) in the Crystalline State. *Polymer* **2000**, *41*, 3083–3086.
24. Mitchell, G. R.; Odajima, A., The Local Conformation of Poly(dimethylsiloxane). *Polym. J.* **1984**, *16*, 351–357.
25. Kim, C.; Gurau, M. C.; Cremer, P. S.; Yu, H., Chain Conformations of Poly(dimethyl siloxane) at the Air/Water Interface by Sum Frequency Generation. *Langmuir* **2008**, *24*, 10155–10160.
26. Evmenenko, G.; Mo, H.; Kewalramani, S.; Dutta, P., Conformational Rearrangements in Interfacial Region of Polydimethylsiloxane Melt Films. *Polymer* **2006**, *47*, 878–882.
27. Jiang, S.; Bae, S. C.; Granick, S., PDMS Melts on Mica Studied by Confocal Raman Scattering. *Langmuir* **2008**, *24*, 1489–1494.
28. Klingenfus, J.; Palmas, P., Weak Molecular Associations Investigated by ^{1}H-NMR Spectroscopy on a Neat Organosiloxane-Solvent Mixture. *Phys. Chem. Chem. Phys.* **2011**, *13*, 10661–10669.
29. Mark, J. E.; Erman, B., *Rubberlike Elasticity: A Molecular Primer.* Wiley-Interscience: New York, 1988.
30. Carmichael, J. B., Unperturbed Dimensions of Chains with Any Degree of Polymerization Application to Polydimethylsiloxane. *J. Polym. Sci., Part A: Polym. Chem.* **1965**, *3*, 4279–4283.
31. Kucukyavuz, Z.; Al-Ghezawi, N. N., An Experimental Study of the Dipole Moment of Poly(dimethylsiloxane). *Eur. Polym. J.* **1990**, *26*, 653–655.
32. Mark, J. E.; Ko, J. H., Random-Coil Dimensions of Poly(methylphenylsiloxane) and Their Dependence on Stereochemical Structure. *J. Polym. Sci., Part B: Polym. Phys.* **1975**, *13*, 2221–2235.
33. Wang, S.; Mark, J. E., An Experimental Study of the Unperturbed Chain Dimensions of Poly(Tetramethyl-*p*-silphenylene-siloxane). *Polym. Bull.* **1993**, *31*, 205–211.
34. Grigoras, S.; Lane, T. H., Conformational Analysis of Substituted Polysiloxane Polymers. In *Silicon-Based Polymer Science. A Comprehensive Resource*, Zeigler, J.

M.; Fearon, F. W. G., Eds. American Chemical Society: Washington, DC, 1990; Vol. 224.
35. Shemella, P. T.; Laino, T.; Fritz, O.; Curioni, A., Molecular Motion of Amorphous Silicone Polymers. *J. Phys. Chem. B* **2011**, *115*, 2831–2835.
36. Semlyen, J. A., Cyclic Polymers. *Pure Appl. Chem.* **1981**, *53* (9), 1797–1804.
37. Semlyen, J. A., *Cyclic Polymers.* Elsevier: London, 1986.
38. Semlyen, J. A., Cyclic Siloxane Polymers. In *Siloxane Polymers*, Clarson, S. J.; Semlyen, J. A., Eds. Prentice Hall: Englewood Cliffs, NJ, **1993**; pp 135–192.
39. Semlyen, J. A.; Wood, B. R.; Hodge, P., Cyclic Polymers: Past, Present and Future. *Polym. Adv. Technol.* **1994**, *5*, 473–478.
40. Dagger, A. C.; Semlyen, J. A., Studies of Cyclic and Linear Poly(dimethylsiloxanes): 34. Preparation, Fractionation and Characterization of the First Per-Deuterated Macrocyclic Poly(dimethylsiloxane). *Polym. Commun.* **1999**, *40*, 3243–3245.
41. Dagger, A. C.; Semlyen, J. A., Hydrogenated and Deuterated Cyclic Poly(dimethylsiloxanes). In *Silicones and Silicone-Modified Materials*, Clarson, S. J.; Fitzgerald, J. J.; Owen, M. J.; Smith, S. D., Eds. American Chemical Society: Washington, DC, 2000; Vol. 729, pp 38–80.
42. Bannister, D. J.; Semlyen, J. A., Studies of Cyclic and Linear Poly(dimethylsiloxanes): 6. Effect of Heat. *Polymer* **1981**, *22*, 377–381.
43. Lee, C.-L.; Emerson, F. A., Molecular Dimensions of Polydipropylsiloxamer. *J. Polym. Sci., Part A-2: Polym. Chem.* **1967**, *5*, 829–838.
44. Freire, J. J.; Pierola, I. F.; Horta, A., Conformational Analysis of Methyl-Phenyl-Siloxane Chains. *Macromolecules* **1996**, *29*, 5143–5148.
45. Clarson, S. J.; Dodgson, K.; Semlyen, J. A., Cyclic Polysiloxanes: 2. Neutron Scattering from Poly(phenylmethylsiloxane). *Polymer* **1987**, *28*, 189–192.
46. Odian, G., *Principles of Polymerization*, 4th ed.; Wiley-Interscience: New York, 2004.
47. Kuo, C.-M.; Saam, J. C.; Taylor, R. B., Stereoregularity in Poly[(methyl(3,3,3-trifluoropropyl)siloxane]. *Polym. Int.* **1994**, *33*, 187–195.
48. Battjes, K. P.; Kuo, C.-M.; Miller, R. L.; Saam, J. C., Strain-Induced Crystallization in Poly[methyl(3,3,3-trifluoropropyl)siloxane] Networks. *Macromolecules* **1995**, *28*, 790–792.
49. Kuo, C.-M.; Battjes, K. J.; Miller, R. L.; Saam, J. C., Strain-Induced Crystallization in Stereoregular Poly[(methyl(3,3,3-trifluoropropyl)siloxane] Networks. *Rubber Chem. Technol.* **1997**, *70*, 769–780.
50. Saam, J. C., Stereoregular Polysiloxanes via Ring-Opening Polymerization, a Review. *J. Inorg. Organomet. Polym.* **1999**, *9*, 1–16.
51. Kawakami, Y.; Imae, I., Functionality and Stereochemical Design of Silicon-Containing Polymers. *Macromol. Chem. Phys.* **1999**, *200*, 1245–1256.
52. Kawakami, Y., Control of the Stereochemical Structure of Silicon-Containing Polymers. *Macromol. Symp.* **2001**, *174*, 145–153.
53. Kobayshi, H.; Owen, M. J., Surface Tension of Liquid Polysiloxanes Having Fluorinated Alkyl Side-Chains. *Macromol. Chem.* **1993**, *194*, 1785–1792.
54. Kobayshi, H.; Owen, M. J., Surface Properties of Fluorosilicones. *Trends Polym. Sci.* **1995**, *3*, 330–335.
55. Patwardhan, D. V.; Zimmer, H.; Mark, J. E., Synthesis and Surface Properties of Polybutadiene Chains End-Capped with Two Types of Fluorosilicon Groups. *J. Macromol. Sci., Chem.* **1998**, *A35*, 1941–1955.

56. Zhao, Q.; Mark, J. E., Fluoro-Substituted Polysiloxanes. I. Synthesis of Poly(*p*-fluorophenylmethylsiloxane). *Macromol. Sci., Macromol. Rep.* **1992**, *A(29)*, 21–26.
57. Patwardhan, D. V.; Zimmer, H.; Mark, J. E., Synthesis of Some Fluorinated Phenylmethylsiloxane Polymers and Characterization of Their Surface Properties. *J. Inorg. Organomet. Polym.* **1997**, *7*, 93–109.
58. Falender, J. R.; Yeh, G. S. Y.; Chiu, D. S.; Mark, J. E., Thermodynamics of Fusion in Poly(diphenylsiloxane). *J. Polym. Sci., Part B: Polym. Phys.* **1980**, *18*, 389–391.
59. Babu, G. N.; Christopher, S. S.; Newmark, R. A., Poly(dimethylsiloxane-co-diphenylsiloxanes): Synthesis, Characterization, and Sequence Analysis. *Macromolecules* **1987**, *20*, 2654–2659.
60. Wang, S.; Mark, J. E., Reinforcement of Elastomeric Poly(dimethylsiloxane) by Glassy Poly(diphenylsiloxane). *J. Mater. Sci.* **1990**, *25*, 65–68.
61. Ibemesi, J.; Gvozdic, N.; Keumin, M.; Tarshiani, Y.; Meier, D. J., Synthesis and Properties of Copolymers of Diphenylsiloxane with Other Organosiloxanes. In *Polymer-Based Molecular Composites*, Schaefer, D. W.; Mark, J. E., Eds. Materials Research Society: Pittsburgh, PA, **1990**; Vol. 171, pp 105–115.
62. Papkov, V. S.; Gerasimov, M. V.; Buzin, M. I.; Il'ina, M. N.; Karzaryan, L. G., The Structure of Ordered Phases in Poly(diphenylsiloxane). *Polym. Sci. U.S.S.R.* **1996**, *38*, 1097–1102.
63. Buzin, M. I.; Gerasimov, M. V.; Obolonkova, E. S.; Papkov, V. S., Solid-State Polymerization of Hexaphenylcyclotrisiloxane. *J. Polym. Sci., Part A: Polym. Chem.* **1997**, *35*, 1973–1984.
64. Lebedev, B. V.; Smirnova, N. N.; Papkov, V. S.; Buzin, M. I.; Dubovik, I. I., Thermodynamics of Poly(diphenylsiloxane) in the Temperature Range 0–620 K. *Polym. Sci. U.S.S.R.* **2000**, *42*, 1111–1119.
65. Papkov, V. S.; Buzin, M. I.; Gerasimov, M. V.; Obolonkova, E. S., Thermoreversible Gelation in Poly(diphenylsiloxane)/Diphenyl Ether Systems Exhibiting Mesophase Behavior. *Macromolecules* **2002**, *35*, 1079–1090.
66. Lee, M. K.; Meier, D. J., Synthesis and Properties of Diarylsiloxane and (Aryl/Methyl)Siloxane Polymers: 1. Thermal Properties. *Polymer* **1994**, *34*, 4882–4892.
67. Lee, M. K.; Meier, D. J., Synthesis and Properties of Diarylsiloxane and (Aryl/Methyl)Siloxane Polymers: 2. Solution and Rheological Properties. *Polymer* **1994**, *35*, 3282–3284.
68. Kuo, P.-L.; Hou, S.-S.; Teng, C.-K.; Liang, W. J., Function and Performance of Silicone Copolymer. (VI). Synthesis and Novel Solution Behavior of Water-Soluble Polysiloxanes with Different Hydrophobicities. *Coll. Polym. Sci.* **2001**, *279*, 286–291.
69. Lin, G.; Zhang, X.; Kumar, S. R.; Mark, J. E., Improved Hydrophilicity from Poly(ethylene glycol) in Amphilic Conetworks with Poly(dimethylsiloxane). *Silicon* **2009**, *1*, 173–181.
70. Zhang, X.; Lin, G.; Kumar, S. R.; Mark, J. E., Hydrogels Prepared from Polysiloxane Chains by End Linking Them with Trifunctional Silanes Containing Hydrophilic Groups. *Polymer* **2009**, *50*, 5414–5421.
71. Mukbaniani, O.; Titvinidze, G.; Tatrishvili, T.; Mukbaniani, N.; Brostow, W.; Pietkiewicz, D., Formation of Polymethylsiloxanes with Alkyl Side Groups. *J. Appl. Polym. Sci.* **2007**, *104*, 1176–1183.

72. Villegas, J. A.; Cervantes, J., Unusual Behavior of Poly(methylhexylsiloxane) Short Chain Molecules in Solution. *J. Inorg. Organomet. Polym.* **2011**, *21*, 157–164.
73. Sacarescu, L., Non-Shrinking Polycyclocarbosiloxane Networks. *Polym. Bull.* **2008**, *59*, 873–882.
74. Hou, A.; Yu, J.; Shi, Y., Preparation and Surface Properties of the Polysiloxane Material Modified with Fluorocarbon Side Chains. *Eur. Polym. J.* **2008**, *44*, 1696–1700.
75. Ganjicz, T.; Kowalewska, A.; Stanczyk, W. A.; Butts, M.; Nye, S. A.; Rubinsztajn, S., A Novel Organometallic Route to Phenylethenyl-Modified Polysiloxanes. *J. Mater. Chem.* **2005**, *15*, 611–619.
76. Hui, C.; Xiaohui, T.; Yizhong, Y.; Jiaping, L., Preparation of Polysiloxane with Pendant Cr-Bound Carbazole Chromophore for Non-Linear Optics. *J. Appl. Polym. Sci.* **2008**, *107*, 4076–4081.
77. Mukbaniani, O.; Tatrishvili, T.; Titvinidze, G.; Patsatsia, S., Synthesis and Characterization of Polysiloxanes with Pendant Bicyclic Fragments. *J. Appl. Polym. Sci.* **2011**, *120*, 1572–1582.
78. Zhang, W.; Shiotsuki, M.; Masuda, T., Synthesis and Properties of Polymer Brush Consisting of Poly(phenylacetylene) Main Chain and Poly(dimethylsiloxane) Side Chains. *Polymer* **2007**, (48), 2548–2553.
79. Aranguren, M. I., Crystallization of Polydimethylsiloxane: Effect of Silica Filler and Curing. *Polymer* **1998**, *39*, 4897–4903.
80. Jones, B. A.; Torkelson, J. M., Crystallization and Enthalpy Relaxation of Physically Associating, End-Linked Polymer Networks: Telechelic Pyrene-Labeled Polydimethylsiloxane. *Polym. Bull.* **2004**, *51*, 411–418.
81. Chien, A.; Maxwell, R. S.; DeTeresa, S.; Thompson, L.; Cohenour, R.; Balazs, B., Effects of Filler-Polymer Interactions on Cold-Crystallization Kinetics in Crosslinked, Silica-Filled Polydimethylsiloxane/Polydiphenylsiloxane Copolymer Melts. *J. Polym. Sci., Part B: Polym. Phys.* **2006**, *44*, 1898–1908.
82. Lund, R.; Alegria, A.; Goitandia, L.; Colmenero, J.; Gonzales, M. A.; Lindner, P., Dynamical and Structural Aspects of the Cold Crystallization of Poly(dimethylsiloxane) (PDMS). *Macromolecules* **2008**, *41*, 1364–1376.
83. Wu, J.; McKenna, G. B., Anomalous Melting Behavior of Cyclohexane and Cyclooctane in Poly(dimethylsiloxane) Precursors and Model Networks. *J. Polym. Sci., Part B: Polym. Phys.* **2008**, *46*, 2779–2791.
84. Ohlberg, S. M.; Alexander, L. E.; Warrick, E. L., Crystallinity and Orientation in Silicone Rubber. I. X-Ray Studies. *J. Polym. Sci.* **1958**, *27*, 1–17.
85. Warrick, E. L., Crystallinity and Orientation in Silicone Rubber. I. Physical Measurements. *J. Polym. Sci.* **1958**, *27*, 19–38.
86. Sharaf, M. A.; Kloczkowski, A.; Mark, J. E., Networks Undergoing Strain-Induced Crystallization. Analysis in Terms of the Constrained-Junction Model. *Comput. Polym. Sci.* **1992**, *2*, 84–89.
87. Kloczkowski, A.; Mark, J. E.; Sharaf, M. A.; Erman, B., Theory of Strain-Induced Crystallization in Real Elastomeric Networks. In *Synthesis, Characterization, and Theory of Polymer Networks and Gels*, Aharoni, S. M., Ed. Plenum Press: New York, 1992; pp 227–241.
88. Mark, J. E.; Odian, G., *Polymer Chemistry Course Manual*. American Chemical Society: Washington, DC, 1984.
89. Brown, J. F., Double chain polymers and nonrandom crosslinking. *J. Polym. Sci., Part C: Polym. Symp.* **1963**, *1*(1), 83–97.

90. Helminiak, T. E.; Berry, G. C., Properties of the Ladder Polymer cis-syndiotactic poly(phenylsilsesquioxane) in Solution. *J. Polym. Sci., Part C: Polym. Symp.* **1978**, *65*, 107–123.
91. Billmeyer, F. W., *Textbook of Polymer Science*, 3rd ed. Wiley: New York, 1984.
92. Dunnavant, W. R., Silarylene and Silarylene-Siloxane Polymers. *Inorg. Macromol. Rev.* **1971**, *1*, 165–189.
93. Ballistreri, A.; Montaudo, G.; Lenz, R. W., Mass Spectral Characterization and Thermal Decomposition Mechanism of Alternating Silarylene-Siloxane Polymers. *Macromolecules* **1984**, *17*, 1848–1854.
94. Hani, R.; Lenz, R. W., Synthesis and Properties of Silphenylene-Siloxane Polymers. In *Silicon-Based Polymer Science. A Comprehensive Resource*, Zeigler, J. M.; Fearon, F. W. G., Eds. American Chemical Society: Washington, DC, 1990; Vol. 224. pp 741–752.
95. Ogawa, T.; Tachikawa, M.; Kushibiki, N.; Murakami, M., Poly(diarylsilmethylene)s. I. Synthesis and Characterization. *J. Polym. Sci., Part A: Polym. Chem.* **1995**, *33*, 2821–2831.
96. Ogawa, T.; Murakami, M., Poly(diarylsilmethylene)s. II. Thermal and Mechanical Properties. *J. Polym. Sci, Part B: Polym. Phys.* **1996**, *34*, 1317–1326.
97. Benouargha, A.; Boutevin, B.; Caporiccio, G.; Essassi, E.; Guida-Pietrasanta, F.; Ratsimihety, A., Hybrid Silalkylene Polysiloxanes: Synthesis and Thermal Properties. *Eur. Polym. J.* **1997**, *33*, 1117–1124.
98. Dorset, D. L.; McCourt, M. P., Direct Phase Determination for Polymer Fibre X-ray Data—The Structure of Poly(tetramethyl-*p*-silphenylene siloxane). *Polymer* **1997**, *38*, 1985–1988.
99. Mark, J. E., *Polymer Data Handbook*. Oxford University Press: New York, 1999.
100. Hanna, S.; Windle, A. H., Geometrical Limits to Order in Liquid Crystalline Random Copolymers. *Polymer* **1988**, *29*, 207–223.
101. Mark, J. E., Some Simulations on Elastomers and Rubberlike Elasticity. *Makromol. Symp.* **2001**, *171*, 1–9.
102. Madkour, T. M.; Mark, J. E., Sequence Distributions and Crystallinity in Random and Semi-Blocky Polysiloxane Copolymers. II. Searches in Multi-Chain Systems for Polymer Crystallinity. *Comput. Polym. Sci.* **1994**, *4*, 87–93.
103. Madkour, T. M.; Mark, J. E., Monte Carlo Simulations on the Stereoregularity, Crystallinity, and Related Physical Properties of Poly[methyl(3,3,3-trifluoropropyl)siloxane}. *Macromolecules* **1995**, *28*, 6865–6870.
104. Beatty, C. L.; Karasz, F. E., Transitions in Poly(diethylsiloxane). *J. Polym. Sci., Polym. Phys. Ed.* **1975**, *13*, 971–975.
105. Godovsky, Y. K.; Makarova, N. N.; Papkov, V. S.; Kuzmin, N. N., Crystalline Phases and a Mesophase in Poly(dipropylsiloxane). *Makromol. Chem.* **1985**, *6*, 443–450.
106. Godovsky, Y. K.; Papkov, V. S., Thermotropic Mesophases in Linear Polysiloxanes. *Makromol. Chem., Makromol. Symp.* **1986**, *4*, 71–87.
107. Friedrich, J.; Rabolt, J. F., Structural Studies of Phase Transitions in Poly(di-*n*-alkylsiloxanes). 1. Poly(dimethylsiloxane) and Poly(diethylsiloxane). *Macromolecules* **1987**, *20*, 1975–1979.
108. Godovsky, Y. K.; Volegova, I. A.; Valetskaya, L. A.; Rebrov, A. V.; Novitskaya, L. A.; Rotenburg, S. I., Mesophase Formation in Crosslinked Polydiethylsiloxanes Under Uniaxial Stretching. *Polym. Sci. U.S.S.R.* **1988**, *30*, 329–335.
109. Shibanov, Y. D., Optical Properties of Polydiethylsiloxane Mesophase. *Polym. Sci. U.S.S.R.* **1989**, *31*, 2653–2662.

110. Shul'gin, A. I.; Godovsky, Y. K., Effect of Mesophase on the Phase Transitions in Polydipropylsiloxane. *Polym. Sci. U.S.S.R.* **1989**, *29*, 2845–2847.
111. Papkov, V. S.; Godovsky, Y. K.; Svistunov, V. S.; Zhdanov, A. A., Uniaxial Stretching of Weakly Crosslinked Mesomorphous Polydiethylsiloxane Networks. *Polym. Sci. U.S.S.R.* **1989**, *31*, 1729–1737.
112. Moller, M.; Siffrin, S.; Kogler, G.; Oelfin, D., Mesomorphic Phases of Poly(di-*n*-alkylsiloxane)s. *Makromol. Chem., Makromol. Symp.* **1990**, *34*, 171–194.
113. Mark, J. E., Silicon-Containing Polymers. In *Silicon-Based Polymer Science. A Comprehensive Resource*, Zeigler, J. M.; Fearon, F. W. G., Eds. American Chemical Society: Washington, DC, 1990; Vol. 224, pp 47–68.
114. Miller, K. J.; Grebowicz, J.; Wesson, J. P.; Wunderlich, B., Conformations of Poly(diethylsiloxane) and Its Mesophase Transitions. *Macromolecules* **1990**, *23*, 849–856.
115. Antipov, E. M.; Kulichikhin, V. G.; Plate, N. A., Mesophase Structure of Some Flexible Chain Polymers. *Polym. Eng. Sci.* **1992**, *32*, 1188–1203.
116. Godovsky, Y. K., Rubber-Elastic Mesophase Siloxane Systems. *Angew. Makromol. Chemie* **1992**, *202*, 187–212.
117. Godovsky, Y. K.; Makarova, N. N., Self-Order in Flexible Non-Mesogenic Macromolecules. *Philos. Trans. R. Soc. London, Series A* **1994**, *348*, 45–57.
118. Out, G. J. J.; Turetskii, A. A.; Moller, M., Structural Correlation Between the Columnar Mesophase and the Melt in Poly(di-*n*-alkylsiloxane)s. *Makromol. Chem., Rapid Commun.* **1995**, *16*, 107–112.
119. Out, G. J. J.; Schlossels, F.; Turetskii, A. A.; Moller, M., Columnar Mesophase Behaviour of Poly(dibutylsiloxane) and Related Copolymers Containing Polar Substitutents. *Makromol. Chem. Phys.* **1995**, *196*, 2035–2046.
120. Turetskii, A. A.; Out, G. J. J.; Klok, H.-A.; Moller, M., Structural Transformations in Poly(di-*n*-alkylsiloxane)s with Alkyl Side Groups Containing 7 to 10 Carbon Atoms. *Polymer* **1995**, *36*, 1303–1308.
121. Out, G. J. J.; Turetskii, A. A.; Snijder, M.; Moller, M.; Papkov, V. S., Model Polydiethylsiloxane Networks: 1. Synthesis and Phase Behaviour. *Polymer* **1995**, *36*, 3213–3221.
122. Magonov, S. N.; Elings, V.; Papkov, V. S., AFM Study of Thermotropic Structural Transitions in Poly(diethylsiloxane). *Polymer* **1997**, *38*, 297–307.
123. Sautter, E.; Belousov, S. I.; Pechold, W.; Makarova, N. N.; Godovsky, Y. K., Langmuir-Blodgett Films of Novel Mesophase Polysiloxanes. *Polym. Sci. U.S.S.R.* **1996**, *38*, 49–55.
124. Molenberg, A.; Siffrin, S.; Moller, M., Well Defined Columnar Liquid Crystalline Polydiethylsiloxane. *Makromol. Chem., Makromol. Symp.* **1996**, *102*, 199–207.
125. Molenberg, A.; Moller, M.; Sautter, E., Columnar Mesophase Behavior of Poly(di-*n*-alkylsiloxane)s. *Prog. Polym. Sci.* **1997**, *22*, 1133–1144.
126. Lacey, D.; Beattie, H. N.; Mitchell, G. R.; Pople, J. A., Orientation Effects in Monodomain Nematic Liquid Crystalline Polysiloxane Elastomers. *J. Mater. Chem.* **1998**, *8*, 53–60.
127. Godovsky, Y. K.; Papkov, V. S., Polysiloxane Mesophases. In *Polymer Data Handbook*, Mark, J. E., Ed. Oxford University Press: New York, 1999; pp 394–395, 396–399, 400–401, 448–449, 453–454.
128. Belousov, S. I.; Buzin, A. I.; Godovsky, Y. K., Self-Organization of Polysiloxanes into Mono- and Multilayer Structures at Interphase Boundaries. *Polym. Sci. U.S.S.R.* **1999**, *41*, 303–324.

129. Inomata, K.; Yamamoto, K.; Nose, T., An X-Ray Diffraction Study on Crystalline and Mesomorphic Structures of Poly(diethylsiloxane). *Polym. J.* **2000**, *32*, 1044–1048.
130. Godovsky, Y. K.; Papkov, V. S.; Magonov, S. N., Atomic Force Microscopy Visualization of Morphology Changes Resulting from the Phase Transitions in Poly(di-*n*-alkoxysiloxane)s: Poly(diethylsiloxane). *Macromolecules* **2001**, *34*, 976–990.
131. Papkov, V. S.; Turetskii, A. A.; Out, G. J. J.; Moller, M., Uniaxial Extension of Poly(diethylsiloxane) Networks: Mesophase Formation Under Stress. *Int. J. Polym. Mat.* **2002**, *51*, 369–391.
132. Ganicz, T.; Stanczyk, W. A., Organosilicon Mesomorphic Polymer Systems. *Prog. Polym. Sci.* **2003**, *28*, 303–329.
133. Schneider, N. S.; Desper, C. R.; Beres, J. J., Mesomorphic Structure in Polyphosphazenes. In *Liquid-Crystalline Order in Polymers*, Blumstein, A., Ed. Academic Press: New York, 1978; pp 299–325.
134. Hammerschmidt, K.; Finkelmann, H., Stress-Optical and Thermomechanical Measurements on Liquid Crystalline Elastomers. *Makromol. Chem.* **1989**, *190*, 1089–1101.
135. Zentel, R., Liquid Crystalline Elastomers. *Angew. Chem. Int. Ed. Engl., Adv. Mater.* **1989**, *28*, 1407–1415.
136. Finkelmann, H.; Brand, H. R., Liquid-Crystalline Elastomers—A Class of Materials with Novel Properties. *Trends Polym. Sci.* **1994**, *2*, 222–226.
137. Poths, H.; Zentel, R., Dimesogenic Ferroelectric Polysiloxanes. *Macromol. Rapid Commun.* **1994**, *15*, 433–440.
138. Brehmer, M.; Zentel, R., Ferroelectric Liquid-Crystalline Elastomers with Short Switching Times. *Macromol. Rapid Commun.* **1995**, *16*, 659–662.
139. Disch, S.; Finkelmann, H.; Ringsdorf, H.; Schuhmacher, P., Macroscopically Ordered Discotic Columnar Networks. *Macromolecules* **1995**, *28*, 2424–2428.
140. Benne, I.; Semmler, K.; Finkelmann, H., Mechanically Induced Second Harmonic Generation in SC Elastomers. *Macromolecules* **1995**, *28*, 1854–1858.
141. Shilov, S. V.; Okretic, S.; Siesler, H. W.; Zentel, R.; Oge, T., Fourier-Transform Infrared Study of the Switching Process in a Ferroelectric Liquid Crystalline Polymer. *Macromol. Rapid Commun.* **1995**, *16*, 125–130.
142. Damewood, J. R., Jr.; West, R., Structure Calculations for Silane Polymers: Polysilane and Poly(dimethylsilylene). *Macromolecules* **1985**, *18*, 159–164.
143. Welsh, W. J.; DeBolt, L.; Mark, J. E., Conformational Energies and Unperturbed Chain Dimensions of Polysilane and Poly(dimethylsilylene). *Macromolecules* **1986**, *19*, 2978–2983.
144. Ferry, J. D., *Viscoelastic Properties of Polymers*, 3rd ed. Wiley: New York, 1980.
145. Bernardini, C.; Stoyanov, S. D.; Stuart, M. A. C.; Arnaudov, L. N.; Leermakers, F. A. M., PMMA Highlights the Layering Transition of PDMS in Langmuir Films. *Langmuir* **2011**, *27*, 2501–2508.
146. Heimenz, P. C., *Polymer Chemistry: The Basic Concepts*. Marcel Dekker: New York, 1984.
147. Mark, J. E., *Physical Properties of Polymers Handbook*. Springer-Verlag: New York, 1996.
148. Dvornic, P. R.; Jovanovic, J. D.; Govedarica, M. N., On the Critical Molecular Chain Length of Polydimethylsiloxane. *J. Appl. Polym. Sci.* **1993**, *49*, 1497–1507.

149. Di Landro, L.; Levi, M.; Nichetti, D.; Consolo, A., Experimental Determination of Rheological Properties of Polydimethylsiloxane. *Eur. Polym. J.* **2003**, *39*(1831–1838), 2559.
150. Takahashi, H.; Ishimuro, Y.; Watanabe, H., Nonlinear Mechanical Behavior of Scarcely Crosslinked Poly(dimethylsiloxane) Gel: Effect of Strand Length Polydispersity. *Polym. J.* **2007**, *39*, 465–474.
151. Rahalkar, R. R.; Lamb, J.; Harrison, G.; Barlow, A. J.; Hawthorn, W.; Semlyen, J. A.; North, A. M.; Pethrick, R. A., Viscoelastic Studies of Polydimethylsiloxanes. *Proc. R. Soc. London, Series A* **1984**, *394*, 207–222.
152. Arrighi, V.; Gagliardi, S.; Zhang, C.; Ganazzoli, F.; Higgins, J. S.; Ocone, R.; Telling, M. T. F., A Unified Picture of the Local Dynamics of Poly(dimethylsiloxane) Across the Melting Point. *Macromolecules* **2003**, *36*, 8738–8748.
153. Hartmann, L.; Kremer, F.; Pouret, P.; Leger, L., Molecular Dynamics in Grafted Layers of Poly(dimethylsiloxane). *J. Chem. Phys.* **2003**, *118*(6052–6058).
154. Hadjistamov, D., Determination of the Onset of Shear Thinning of Polydimethysiloxane. *J. Appl. Polym. Sci.* **2008**, *108*, 2356–2364.
155. Forgan, R. S.; Sauvage, J.-P.; Stoddart, J. F., Chemical Topology: Complex Molecular Knots, Links, and Entanglements. *Chem. Rev.* **2011**, *111*, 5434–5464.
156. Acosta, R. H.; Monti, G. A.; Villar, M. A.; Valles, E. M.; Vega, D. A., Transiently Trapped Entanglements in Model Polymer Networks. *Macromolecules* **2009**, *42*, 4674–4680.
157. Niamlang, S.; Sirivat, A., Electromechanical Response of a Crosslinked Polydimethylsiloxane. *Macromol. Symp.* **2008**, *264*, 176–183.
158. Hiamtup, P.; Sirivat, A.; Jamieson, A. M., Electromechanical Response of a Soft and Flexible Actuator Based on Polyaniline Particles Embedded in a Cross-Linked Poly(dimethylsiloxane) Network. *Mater. Sci. Eng., Part C* **2008**, *28*, 1044–1051.
159. Bell, W.; Daly, J.; North, A. M.; Pethrick, R. A.; Teik, P. B., Chain Entanglement and Normal Mode Motions in Poly(dimethylsiloxane) + Toluene Mixtures. *J. Chem. Soc., Faraday Trans. II* **1979**, *75*, 1452–1464.
160. Eastwood, A. E.; North, A. M.; Pethrick, R. A., Ultrasonic Relaxation Studies of Poly(methylphenylsiloxane) and Its Solutions in Toluene. *J. Chem. Soc., Faraday Trans. II* **1982**, *78*, 95–102.
161. Kawaguchi, M.; Kubota, K., Rheo-Optical Properties of Silicone Oil Emulsions in the Presence of Polymer Emulsifiers. *Langmuir* **2004**, *20*, 1126–1129.
162. Lorthioir, C.; Alegria, A.; Colmenero, J.; Deloche, B., Heterogeneity of the Segmental Dynamics of Poly(dimethylsiloxane) in a Diblock Lamellar Mesophase: Dielectric Relaxation Investigations. *Macromolecules* **2004**, *37*, 7808–7817.
163. Smith, D. A.; Thomas, J. K., Dynamic and Equilibrium Studies of Silicone Rubber Vulcanizates. *Eur. Polym. J.* **1971**, *7*, 433–443.
164. Warley, R. L.; Feke, D. L.; Manas-Zloczower, I., Effect of Peroxide Crosslinking on the Dynamic Modulus of Silicone Rubber. *J. Appl. Polym. Sci.* **2005**, *97*, 1504–1512.
165. Sun, Y.; Walker, G. C., Viscoelastic Response of Poly(dimethylsiloxane) in the Adhesive Interaction with AFM Tips. *Langmuir* **2005**, *21*, 8694–8702.
166. Kakiuchida, H.; Takahashi, M.; Tokuda, Y.; Masai, H.; Kuniyoshi, M.; Yoko, T., Viscoelastic and Structural Properties of Phenyl-Modified Polysiloxane Systems with a Three-Dimensional Structure. *J. Phys. Chem. B* **2006**, *110*, 7321–7327.

167. Cosgrove, T.; Roberts, C.; Choi, Y.; Schmidt, R. G.; Gordon, G. V.; Goodwin, A. J.; Kretschmer, A., Relaxation Studies of High Molecular Weight Poly(dimethylsiloxane)s Blended with Polysilicate Nanoparticles. *Langmuir* **2002**, *18*, 10075–10079.
168. Fragiadakis, D.; Pissis, P.; Bokobza, L., Modified Chain Dynamics in Poly(dimethylsiloxane)/Silica Nanocomposites. *J. Non-Cryst. Solids* **2006**, *352*, 4969–4972.
169. Warley, R. L.; Feke, D. L.; Manas-Zloczower, I., Transient Effects in Dynamic Modulus Measurements of Silicone Elastomers. 1. Zero Mean Strain Measurements. *J. Appl. Polym. Sci.* **2005**, *98*, 1001–1009.
170. Park, T.; O'Brien, E.; Lizotte, J. R.; Glass, T. E.; Ward, T. C.; Long, T. E.; Leo, D. J., Dielectric Relaxation Behavior of Poly(acrylonitrile-co-methacrylonitrile) microcapsules Dispersed in a Silicone Matrix. *Polymer* **2006**, *47*, 4218–4229.
171. Chiu, H.-T.; Wu, J.-H.; Shong, Z.-J., Dynamic Properties of Rubber Vibration Isolators and Antivibration Performance of a Nanoclay-Modified Silicone/ Poly(propylene oxide)-Poly(ethylene oxide) Copolymer with 20 Wt.% LiClO4 Blend System. *J. Appl. Polym. Sci.* **2006**, *101*, 3275–2381.
172. Warley, R. L.; Feke, D. L.; Manas-Zloczower, I., Transient Effects in Dynamic Modulus Measurements of Silicone Elastomers. 2. Effect of Mean Strains and Strain History. *J. Appl. Polym. Sci.* **2007**, *98*, 2197–2204.
173. Zhang, X. K.; Poojari, Y.; Clarson, S. J., A Comparison of Poly(methylphenylsiloxane) and Poly(dimethylsiloxane) Membranes for the Removal of Low Molecular Weight Organic Compounds from Water by Pervaporation. In *Science and Technology of Silicones and Silicone-Modified Materials*, Clarson, S. J.; Fitzgerald, J. J.; Owen, M. J.; Smith, S. D.; Van Dyke, M. E., Eds. American Chemical Society: Washington, DC, 2007; Vol. 964, pp 165–175.
174. Ye, L. Y.; Liu, Q. L.; Zhang, Q. G.; Zhu, A. M.; Zhou, G. B., Pervaporation Characteristics and Structure of Poly(vinyl alcohol)/Pol(ethylene glycol)/Tetraethoxysilane Hybrid Membranes. *J. Appl. Polym. Sci.* **2007**, *105*, 3640–3648.
175. Zhang, X. K.; Poojari, Y.; Drechsler, L. E.; Kuo, C. M.; Fried, J. R.; Clarson, S. J., Pervaporation of Organic Liquids from Binary Aqueous Mixtures Using Poly(t rifluoropropylmethylsiloxane) and Poly(dimethylsiloxane) Dense Membranes. *J. Inorg. Organomet. Polym. Mater.* **2008**, *18*, 246–252.
176. Chen, J.; Li, J.; Lin, Y.; Chen, C., Pervaporation Performance of Polydimethylsiloxane Membranes for Separation of Benzene/Cyclohexane Mixtures. *J. Appl. Polym. Sci.* **2009**, *112*, 2425–2433.
177. Garg, P.; Singh, R. P.; Pandey, L. K.; Choudhary, V., Pervaporative Studies Using Polyimide-Filled PDMS Membrane. *J. Appl. Polym. Sci.* **2010**, *115*, 1967–1974.
178. Vopicka, O.; Hynek, V.; Friess, K.; Izak, P., Blended Silicone-Ionic Liquid Membranes: Transport Properties of Butan-1-ol Vapor. *Eur. Polym. J.* **2010**, *46*, 123–128.
179. Modesti, G.; Zimmermann, B.; Borsch, M.; Herrmann, A.; Saalwachter, K., Diffusion in Model Networks as Studied by NMR and Fluorescence Correlation Spectroscopy. *Macromolecules* **2009**, *42*, 4681–4689.
180. Zhao, C.; Li, J.; Jiang, Z.; Chen, C., Measurement of the Infinite Dilution Diffusion Coefficient of Small Molecule Solvents in Silicone Rubber by Inverse Gas Chromatography. *Eur. Polym. J.* **2006**, *42*, 615–624.
181. Stern, S. A.; Bhide, B. D., Permeability of Silicone Polymers to Ammonia and Hydrogen Sulfide. *J. Appl. Polym. Sci.* **1989**, *38*, 2131–2147.

182. Jordon, S. M.; Koros, W. J., Permeability of Pure and Mixed Gases in Silicone Rubber at Elevated Pressures. *J. Polym. Sci., Polym. Phys. Ed.* **1990**, *28*(6), 795–809.
183. Stern, S.A.; Fried, J.R., "Permeability of Polymers to Gases and Vapors," in "Physical Properties of Polymers Handbook," 2nd Ed., J.A. Mark, Ed., Springer, NY, 2007, pp. 1009–1023.
184. Kamiya, Y.; Naito, Y.; Hirose, T.; Mizoguchi, K., Sorption and Partial Molar Volume of Gases in Poly (dimethyl siloxane). *J. Polym. Sci., Polym. Phys. Ed.* **1990**, *28*(8), 1297–1308.
185. Stern, S. A.; Krishnakumar, B.; Nadakatti, S. M., Permeability. In *Physical Properties of Polymers Handbook*, Mark, J. E., Ed. Springer-Verlag: New York, 1996; pp 687–700.
186. Kawakami, Y.; Karasawa, H.; Aoki, T.; Yamamura, Y.; Hsada, H.; Yamashita, Y., Polymers with Oligoorganosiloxane Side Chains as Materials for Oxygen Permeable Membranes. *Polym. J.* **1985**, *17*, 1159–1172.
187. Karunakaran, R.; Kennedy, J. P., Synthesis, Characterization and Properties of Novel Highly Oxygen Permeable Amphiphilic Membranes. *J. Polym. Sci. A, Polym. Chem.* **2007**, *45*, 308–316.
188. Rusina, T. P.; Smedes, F.; Klanova, J., Diffusion Coefficients of Polychlorinated Biphenyls and Polycyclic Aromatic Hydrocarbons in Polydimethylsiloxane and Low-Density Polyethylene Polymers. *J. Appl. Polym. Sci.* **2010**, *116*, 1803–1810.
189. Brijmohan, S. M.; Shaw, M. T., Proton Exchange Membranes Based on Sulfonated Crosslinked Polystyrene Micro Particles Dispersed in Poly(dimethylsiloxane). *Polymer* **2006**, *47*, 2856–2864.
190. Shi, S.; Du, Z.; Ye, H.; Zhang, C.; Li, H., A Novel Carbon Black Poly(dimethylsiloxane) Composite Membrane with High Flux for the Separation of Ethanol from Water by Pervaporation. *Polym. J.* **2006**, *38*, 949–955.
191. Ma, Z.; Wang, X.; Zhang, L., The Effect of Weight Fraction of H-PDMS on the Latex Membrane and the Stability of Composite Emulsion of H-PDMS/Acrylate. *J. Macromol. Sci., Pure Appl. Chem.* **2008**, *45*, 1078–1083.
192. De Figueiredo Pinto, R. C.; De Araujo Paschoal, C. W.; Paraguassu, W.; Castro Jr.,M. C.; Tanaka, A. A.; Filho, J. S. B.; Jose, N. M., Ionic Properties of an Organic-Inorganic Sol-Gel Hybrid Based on Polydimethylsiloxane and Tetraethoxysilane Doped with Sodium Dodecyl Sulfate. *J. Appl. Polym. Sci.* **2010**, *115*, 851–854.
193. Khounlavong, L.; Ganesan, V., Influence of Interfacial Layers upon the Barrier Properties of Polymer Nanocomposites. *J. Chem. Phys.* **2009**, *130*, 104901.
194. Adhikari, R.; Dao, B.; Hodgkin, J.; Mardel, J., Synthesis, Structure and Membrane Properties of Siloxane-Imide Co-polymers Produced by Aqueous Polymerization. *Eur. Polym. J.* **2011**, *47*, 1328–1337.
195. Di, C.; Jiang, Y.; Wang, R.; Yin, J., Multi-Responsive Nanoparticles from the Amphiphilic Poly(dimethylsiloxane) (PDMS)-Containing Poly(ether amine) (PDMS-gPEA) and Its Potential Applications for Smart Separation. *J. Mater. Chem.* **2011**, *21*, 4416–4423.
196. Juliac, E.; Arman, J.; Harran, D., An Ultrasonic Method to Study the Interdiffusion Phenomenon Between Two Polydimethylsiloxanes. *Eur. Polym. J.* **2002**, *38*, 2161–2169.
197. Huang, J.; Li, J.; Jhan, X.; Chen, C., A Modified Solution Diffusion Model and Its Application in the Pervaporization Separation of Alkane/Thiophenes Mixtures with PDMS Membrane. *J. Appl. Polym. Sci.* **2008**, *110*, 3140–3148.

198. Anim-Mensah, A.; Mark, J. E.; Krantz, W. R., Use of Solubility Parameters for Predicting the Separation Characteristics of Poly(dimethysiloxane) and Siloxane-Containing Membranes. In *Science and Technology of Silicones and Silicone-Modified Materials*, Clarson, S. J.; Fitzgerald, J. J.; Owen, M. J.; Smith, S. D.; Van Dyke, M. E., Eds. American Chemical Society: Washington, DC, 2007; Vol. 964, pp 203–219.
199. Sadrzadeh, M.; Shahidi, K.; Mohammadi, T., Synthesis and Gas Permeation Properties of a Single Layer PDMS Membrane. *J. Appl. Polym. Sci.* **2010**, *117*, 33–48.
200. Trohalaki, S.; DeBolt, L. C.; Mark, J. E.; Frisch, H. L., Estimation of Free Volume for Gaseous Penetrants in Elastomeric Membranes by Monte Carlo Simulations. *Macromolecules* **1990**, *23*, 813.
201. Madkour, T.; Kloczkowski, A.; Mark, J. E.; Roe, R. J., Molecular Dynamics Investigations into the High Permeability of Siloxane Polymers. In *Computational Methods in Materials Science*, Mark, J. E.; Glicksman, M. E.; Marsh, S. P., Eds. Materials Research Society: Pittsburgh, PA, 1992; Vol. 278, pp 33–40.
202. Charati, S. G.; Stern, S. A., Diffusion of Gases in Silicone Polymers: Molecular Dynamics Simulations. *Macromolecules* **1998**, *31*, 5529–5535.
203. Fried, J. R., *Polymer Science and Technology*, 2nd ed. Prentice Hall: Englewood Cliffs, NJ, 2003.
204. Fried, J. R., *Polymer Science and Technology*. Prentice Hall: Englewood Cliffs, NJ, 1995.
205. Masuda, T.; Isobe, E.; Higashimura, T.; Takada, K., Poly[1-(Trimethylsilyl)-1-Propyne]: A New High Polymer Synthesized with Transition-Metal Catalysts and Characterized by Extremely High Gas Permeability. *J. Am. Chem. Soc.* **1983**, *105*, 7473–7474.
206. Ichiraku, Y.; Stern, S. A.; Nakagawa, T., An Investigation of the High Gas Permeability of Poly(1-trimethylsilyl-1-propyne). *J. Membr. Sci.* **1987**, *34*, 5–18.
207. Masuda, T.; Higashimura, T., Polyacetylenes with Substituents: Their Synthesis and Properties. *Adv. Polym. Sci.* **1986**, *81*, 121–165.
208. Kita, H.; Sakamoto, T.; Tanaka, K.; Okamoto, K.-I., Preparation of Plasma-Polymerized Membranes from I-(Trimethylsilyl)-I-Propyne and Gas Permeability Through the Membranes. *Polym. Bull.* **1988**, *20*, 349–354.
209. Masuda, T.; Higashimura, T., Synthesis and Properties of Silicon-Containing Polyacetylenes. In *Silicon-Based Polymer Science. A Comprehensive Resource*, Zeigler, J. M.; Fearon, F. W. G., Eds. American Chemical Society: Washington, DC, 1990; Vol. 224. pp 641–661.
210. Baker, G. L.; Klausner, C. F.; Gozdz, A. S.; Shelburne III, J. A.; Bowmer, T. N., Brominated Poly(1-trimethylsilyl-1-propyne). In *Silicon-Based Polymer Science. A Comprehensive Resource*, Zeigler, J. M.; Fearon, F. W. G., Eds. American Chemical Society: Washington, DC, 1990; Vol. 224. ch 36, pp 663–678.
211. Nakagawa, T., The Highest Gas Permeable Membranes of Poly[1-(trimethylsilyl)-1-propyne]—Their Gas Permeability and Modification. In *Science and Technology of Polymers and Advanced Materials*, Prasad, P. N.; Mark, J. E.; Kandil, S. H.; Kafafi, Z. H., Eds. Plenum Press: New York, 1998; pp 821–834.
212. Fried, J. R.; Goyal, D. K., Molecular Simulation of Gas Transport in Poly[1-(trimethylsilyl)-1-propyne]. *J. Polym. Sci., Part B: Polym. Phys.* **1998**, (36), 519–536.
213. Zheng, T.; Fried, J. R., Monte Carlo Simulation of Sorption of Pure and Mixed Alkanes in Poly[1-(trimethylsilyl)-1-propyne]. *Sep. Sci. Technol.* **2001**, *36*, 959–974.

214. Rao, H.; Zhang, Z., Preparation, Characterization, and Permeation Property of a Liquid Crystal/PDMS Membrane Material. *J. Appl. Polym. Sci.* **2012**, *123*, 191–199.
215. Arkles, B.; Rediger, P., *Silicones in Biomedical Applications*. Technomic Publishing: Lancaster, PA, 1983.
216. Yannas, I. V.; Burke, J. F., Design of an Artificial Skin. I. Basic Design Principles. *J. Biomed. Mater. Res.* **1980**, *14*, 65.
217. Sweeney, D. F., *Silicone Hydrogels. Continuous-Wear Contact Lenses*, 2nd ed. Butterworth Heinemann: Edinburgh, 2004.
218. Sweeney, D. F., The Max Schapero Memorial Award Lecture 2004: Contact Lenses on and in the Cornea, What the Eye Needs. *Optometry Vis. Sci.* **2006**, *83*(3), 133–142.
219. Rajan, G. S.; Sur, G. S.; Mark, J. E.; Schaefer, D. W.; Beaucage, G., Preparation and Characterization of Some Unusually Transparent Poly(dimethylsiloxane) Nanocomposites. *J. Polym. Sci. B, Polym. Phys.* **2003**, *41*, 1897–1901.
220. Zheng, P.; McCarthy, T. J., Rediscovering Silicones: Molecularly Smooth, Low Surface Energy, Unfilled, UV/Vis-Transparent, Extremely Cross-Linked, Thermally Stable, Hard, Elastic PDMS. *Langmuir* **2010**, *26*, 18585–18590.
221. Gu, Q. G.; Zhou, Q. L., Preparation of High Strength and Optically Transparent Silicone Rubber. *Eur. Polym. J.* **1998**, *34*, 1727–1733.
222. Chalykh, A. E.; Avgonov, A.; Chaika, E. M.; Gerasimov, V. K.; Papkov, V. S.; Dubovik, I. I.; Rabkina, A. Y.; Zavin, B. G., Refractometry of Poly(organosiloxanes). *Polym. Sci. U.S.S.R., Ser. A* **2002**, *44*, 1048–1053.
223. Dasgupta, S.; Garg, S. K.; Smyth, C. P., Microwave Absorption and Molecular Structures in Liquids. LXIX. Atomic Polarization and Relaxation in Several Siloxanes and Hexamethyldisilazane. *J. Am. Chem. Soc.* **1967**, *89*, 2243–2246.
224. Rahalkar, R. R.; Chiu, D. S.; Mark, J. E., Dipole Moments of Polydiethylsiloxane Chains $[Si(CH_2CH_3)_2\text{-}O\text{-}]_x$. *J. Chem Phys.* **1978**, *69*, 938–940.
225. Baker, E. B.; Hunter, M. J., Dielectric Constants of Dimethyl Siloxane Polymers. *Ind. Eng. Chem. Res.* **1946**, *38*, 1117–1120.
226. Sauer, R. O.; Mead, D. J., Dipole Moments of Linear and Cyclic Polymethylsiloxanes. *J. Am. Chem. Soc.* **1946**, *68*, 1794–1797.
227. Dasgupta, S.; Smyth, C. P., Microwave Absorption and Molecular Structure in Liquids. LXX. Dipole Moment, Atomic Polarization, and Dielectric Relaxation in Polymethylpolysiloxanes. *J. Chem. Phys.* **1967**, *47*, 2911–2916.
228. Mark, J. E., Dipole Moments of Dimethylsiloxane Chains. *J. Chem. Phys.* **1968**, *49*, 1398–1402.
229. Sutton, C.; Mark, J. E., Dipole Moments of Dimethylsiloxane Oligomers and Polymers. *J. Chem. Phys.* **1971**, *54*, 5011–5014.
230. Adamski, P., Study of the Dipole Moments of Polydimethylsiloxane Under Theta-Conditions. *Vysokomol. Soedin.* **1972**, *A14*, 518–520.
231. Liao, S. C.; Mark, J. E., Effect of Excluded Volume on the Dipole Moments of Chain Molecules. *J. Chem. Phys.* **1973**, *59*, 3825–3830.
232. Mark, J. E., The Use of Dipole Moments to Characterize Configurations of Chain Molecules. *Acc. Chem. Res.* **1974**, *7*, 218–225.
233. Beevers, M. S.; Mumby, S. J.; Clarson, S. J.; Semlyen, J. A., Studies of Cyclic and Linear Poly(dimethylsiloxanes): 13. Static Dielectric Measurements and Dipole Moments. *Polymer* **1983**, *24*, 1565–1570.
234. Goodwin, A. A.; Beevers, M. S.; Clarson, S. J.; Semlyen, J. A., Studies of Cyclic and Linear Poly(dimethylsiloxanes): 32.Dielectric Relaxation Investigations. *Polymer* **1994**, *37*, 2603–2607.

235. Riande, E.; Mark, J. E., Dipole Moments of Some Poly(dimethylsiloxane) Linear Chains and Cyclics. *Eur. Polym. J.* **1984**, *20*, 517–518.
236. Gallone, G.; Carpi, F.; Rossi, D. D.; Levita, G.; Marchettk, A., Dielectric Constant Enhancement in a Silicone Elastomer Filled with Lead Magnesium Niobate-Lead Titanate. *J. Mater. Sci., Part C* **2007**, *27*, 110–116.
237. Kussmaul, S.; Risse, S.; Kofud, G.; Wache, R.; Wegener, M.; McCarthy, D. N.; Kruger, H.; Gerhard, R., Enhancement of Dielectric Permittivity and Electromechanical Response in Silicone Elastomers: Molecular Grafting of Organic Dipoles to the Macromolecular Network. *Adv. Funct. Mater.* **2011**, *21*, 4589–4594.
238. Zhang, B.; Zhuo, D.; Gu, A.; Liang, G.; Hu, J.-T.; Yuan, L., Preparation and Properties of Addition Curable Silicone Resins with Excellent Dielectric Properties and Thermal Resistance. *Polym. Eng. Sci.* **2011**, *52*, 259–267.
239. Lentz, C. W., It's Safe to Use Silicone Products in the Environment. *Ind. Res. & Dev.* **1980**, 139–143.
240. Hardman, B.; Torkelson, A., Silicones. In *Encyclopedia of Polymer Science and Engineering*, 2nd ed. Wiley-Interscience: New York, 1989; pp 204–308.
241. Wu, C.; Jin, J.; Yang, X.; Dong, H.; Xu, X.; Jiang, J., Effect of Structure and Additives on Thermal Stability of Silicone Fluids. *J. Appl. Polym. Sci.* **2010**, *117*, 2530–2537.
242. Hu, Y.; Hadziomerspahic, A.; Wang, Y., Role of Surfactants in Suppressing Aging of Silica-Poly(dimethylsiloxane) Gels. *Macromolecules* **2010**, *43*, 8233–8238.
243. Vasilakos, S. P.; Tarantili, P. A., The Effect of Pigments on the Stability of Silicone/Montmorillonite Prosthetic Nanocomposites. *J. Appl. Polym. Sci.* **2010**, *118*, 2659–2667.
244. Chinn, S. C.; Alviso, C. T.; Berman, E. S. F.; Harvey, C. A.; Maxwell, R. S.; Wilson, T. S.; Cohenour, R.; Saalwachter, K.; Chasse, W., MQ NMR and SPME Analysis of Nonlinearity in the Degradation of a Filled Silicone Elastomer. *J. Phys. Chem. B* **2010**, *114*, 9729–9736.
245. Lewicki, J. P.; Mayer, B. P.; Alviso, C. T.; Maxwell, R. S., Thermal Degradation Behavior and Product Speciation in Model Poly(dimethylsiloxane) Networks. *J. Inorg. Organomet. Polym. Mater.* **2012**, *22*, 636–645.
246. Huang, W.; Gao, X.; Xiong, J.; Xiong, L.; Xu, Y.; Fu, Y., Electron-Beam-Induced Radiation Effects on Siloxane Foam. *Macromol. Symp.* **2010**, *297*, 225–230.
247. Mozumder, A., Radiation Chemistry: Background, Current Status and Outlook. *J. Phys. Chem. Lett.* **2011**, *2*, 2994–2995.
248. Satti, A. J.; Nador, F.; Vitale, C.; Radivoy, G.; Andreucetti, N. N.; Ciolino, A. E.; Valles, E. M., Synthesis, Characterization, and Gamma Radiation Effects over Well-Defined Poly(vinylsiloxanes) Copolymers. *J. Appl. Polym. Sci.* **2012**, *124*, 832–839.
249. Fujitsuka, M.; Majima, T., Recent Approach in Radiation Chemistry Toward Material and Biological Science. *J. Phys. Chem. Lett.* **2011**, *2*, 2965–2971.
250. Huszank, R.; Szikra, D.; Simon, A.; Szilai, S. Z.; Nagy, I. P., $^{4}He^{+}$ Beam Irradiation Induced Modification of Poly(dimethylsiloxane). Characterization by Infrared Spectroscopy and Ion Beam Analytical Techniques. *Langmuir* **2011**, *27*, 3842–3848.
251. Patel, M.; Skinner, A. R., The Effect of Thermal Aging on the Non-Network Species in Room Temperature Vulcanized Polysiloxane Rubbers. In *Synthesis and Properties of Silicones and Silicone-Modified Materials*, Clarson, S. J.; Fitzgerald, J. J.; Owen, M. J.; Smith, S. D.; Van Dyke, M. E., Eds. American Chemical Society: Washington, DC, 2003; Vol. 838, pp 138–150.

252. Noll, W., *Chemistry and Technology of Silicones*. Academic Press: New York, 1968.
253. Hill, D. T.; Preston, C. M. L.; Whittaker, A. K.; Hunt, S. M., The Radiation Chemistry of Poly(dimethylsiloxane). *Macromol. Symp.* **2000**, *156*, 95–102.
254. Erdmenger, T.; Guerrero-Sanchez, C.; Vitz, J.; Hoogenboom, R.; Schubert, U. S., Recent Developments in the Utilization of Green Solvents in Polymer Chemistry. *Chem. Soc. Rev.* **2010**, *39*, 3317–3333.
255. Kostko, A. F.; McHugh, M. A.; Zanaten, J. H. V., Coil-Coil Interactions for Poly(dimethylsiloxane) in Compressible Supercritical CO_2. *Macromolecules* **2006**, *39*, 1657–1659.
256. Liu, Z.; Wang, J.; Dai, X.; Han, B.; Dong, Z.; Yang, G.; Zhang, X.; Xu, J., Synthesis of Composites of Silicone Rubber and Polystyrene Using Supercritical CO_2 as a Swelling Agent. *J. Mater. Chem.* **2002**, *12* (1), 2688–2691.
257. Horsch, S.; Serhatkulu, G.; Gulari, E.; Kannan, R. M., Supercritical CO_2 Dispersion of Nano-Clays and Clay/Polymer Nanocomposites. *Polymer* **2006**, *47*, 7485–7496.
258. Belinsky, M. R., *Supercritical Fluids*. Nova Science: New York, 2010.
259. Zhu, R.; Hoshi, T.; Sasaki, D.; Usui, R.; Hagiwara, T.; Yano, S.; Sawaguchi, T., Surface Properties and Depth Analysis of Polyethylene/Polydimethylsiloxane Composite Prepared by Using Supercritical Carbon Dioxide. *Polym. J.* **2010**, *42*, 562–566.
260. Shim, S. E.; Isayev, A. I., Ultrasonic Devulcanization of Precipitated Silica-Filled Silicone Rubber. *Rubber Chem. Technol.* **2001**, *74*, 303–316.
261. Shim, S. E.; Parr, J. C.; Von Meerwall, E.; Isayev, A. I., NMR Relaxation and Pulsed Gradient NMR Diffusion Measurements of Ultrasonically Devulcanized Poly(dimethylsiloxane). *J. Phys. Chem. B* **2002**, *106*, 12072–12078.
262. Shim, S. E.; Isayev, A. I.; Von Meerwall, E., Molecular Mobility of Ultrasonically Devulcanized Silica-Filled Poly(dimethylsiloxane). *J. Polym. Sci., Part B: Polym. Phys.* **2003**, *41*, 454–465.
263. Shim, S. E.; Isayev, A. I., Effects of the Presence of Water on Ultrasonic Devulcanization of Polydimethylsiloxane. *J. Appl. Polym. Sci.* **2003**, *88*, 2630–2638.
264. Varma-Nair, M.; Wesson, J. P.; Wunderlich, B., The Thermal Properties of Polysiloxanes Poly(dimethylsiloxane) and Poly(diethylsiloxane). *J. Therm. Anal.* **1989**, *35*, 1913–1939.
265. Andre, P.; Folk, S. L.; Adam, M.; Rubinstein, M.; DeSimone, J. M., Light Scattering Study of Polydimethyl Siloxane in Liquid and Supercritical Carbon Dioxide. *J. Phys. Chem. A* **2004**, *108*, 9901–9907.
266. Vera-Graziano, R.; Hernandez-Sanchez, F.; Cauich-Rodriguez, J. V., Study of Crosslinking Density in Polydimethylsiloxane Networks by DSC. *J. Appl. Polym. Sci.* **1995**, *55*, 1317–1327.
267. Soutzidou, M.; Panas, A.; Viras, K., Differential Scanning Calorimetry (DSC) and Raman Spectroscopy Study of Poly(dimethylsiloxane). *J. Polym. Sci., Polym. Phys. Ed.* **1998**, *36*, 2805–2810.
268. Shih, H.; Flory, P. J., Equation-of-State Parameters for Poly(dimethylsiloxane). *Macromolecules* **1972**, *5*, 758–761.
269. Dattelbaum, D. M.; Jensen, J. D.; Schwendt, A. M.; Kober, E. M.; Lewis, M. W.; Menikoff, R., A Novel Method for Static Equation-of-State Development: Equation of State of a Cross-Linked Poly(dimethylsiloxane) (PDMS) Network to 10 GPa. *J. Chem. Phys.* **2005**, *122*, 144903.
270. Coskun, S.; Cankurtaran, O.; Eran, B. B.; Sarac, A., A Study of Some Equation-of-State Parameters of Poly(methylhydrosiloxane)-co-dimethylsiloxane) with

Some Solvents by Gas Chromatography. *J. Appl. Polym. Sci.* **2007**, *104*, 1627–1631.

271. Flory, P. J.; Shih, H., Thermodynamics of Solutions of Poly(dimethylsiloxane) in Benzene, Cyclohexane, and Chlorobenzene. *Macromolecules* **1972**, *5*, 761–766.
272. Assary, R. S.; Curtiss, L. A.; Redfern, P. C.; Zhang, Z.; Amine, K., Computational Studies of Polysiloxanes: Oxidation Potentials and Decomposition Reactions. *J. Phys. Chem. C* **2011**, *115*, 12216–12223.
273. Schwaderer, P.; Funk, E.; Achenbach, F.; Weis, J.; Brauchle, C.; Wichaelis, J., Single-Molecule Measurement of the Strength of a Siloxane Bond. *Langmuir* **2008**, *24*, 1343–1349.
274. Niedermeyer, H.; Ab Rani, M. A.; Lickiss, P. D.; Hallett, J. P.; Welton, T.; White, A. J. P.; Hunt, P. A., Understanding Siloxane Functionalized Ionic Liquids. *Phys. Chem. Chem. Phys.* **2010**, *12*, 2018–2029.
275. Kuo, C. M.; Clarson, S. J., Investigation of the Interactions and Phase Behavior in Poly(dimethylsiloxane) and Poly(methylphenylsiloxane) Blends. *Macromolecules* **1992**, *25*, 2192–2195.
276. Beaucage, G.; Sukumaran, S.; Clarson, S. J.; Kent, M. S.; Schaefer, D. W., Symmetric, Isotopic Blends of Poly(dimethylsiloxane). *Macromolecules* **1996**, *29*, 8349–8356.
277. Kuo, C. M.; Clarson, S. J., An Investigation of the Thermodynamic Interactions of Oligomeric Cyclic Methylphenylsiloxanes in Siloxane Melts and Blends. *Polymer* **2000**, *41*, 5993–6002.
278. Clarson, S. J.; Stuart, J. O.; Selby, C. E.; Sabata, A.; Smith, S. D.; Ashraf, A., Surface Composition of Poly(methylphenylsiloxane)–Polystyrene Block Copolymers. *Macromolecules* **1995**, *28*, 674–677.
279. Spanos, C. G.; Ebbens, S. J.; Badyal, J. P. S., Surface Segregation and Plasma Oxidation of Poly(dimethylsiloxane)-Doped Polyolefins. *Macromolecules* **2001**, *34*, 8149–8155.
280. Ponjee, M. W. G.; Reijme, M. A.; Gon, A. W. D. V. D.; Brongersma, H. H.; Langeveld-Voss, B. M. W., Intermolecular Segregation of Siloxanes in P3HT: Surface Quantification and Molecular Surface-Structure. *Polymer* **2002**, *43*, 77–85.
281. Levin, V. Y.; Slonimsky, G. L.; Andrianov, K. A.; Zhdanov, A. A.; Godovski, Y. K.; Papkov, V. S.; Lyubavskaya, A. Y., Study of the Crystallization of Elongated Polyorganosiloxane Elastomers. *Polym. Sci. U.S.S.R.* **1973**, *15*, 256–266.
282. Erman, B.; Mark, J. E., *Structures and Properties of Rubberlike Networks*. Oxford University Press: New York, 1997.
283. Mark, J. E.; Erman, B., *Rubberlike Elasticity: A Molecular Primer*, 2nd ed. Cambridge University Press: Cambridge, 2007.
284. McLure, I. A., Thermodynamics of linear dimethylsiloxane–perfluoroalkane mixtures Part 1.—Liquid–liquid coexistence curves for hexamethyldisiloxane–, octamethyltrisiloxane– or decamethyltetrasiloxane–tetradecafluorohexane near the upper critical endpoint and upper coexistence temperatures for 21 other dimethylsiloxane–perfluoroalkane mixtures. *J. Chem. Soc., Faraday Trans.*, 1997, *93*, 249–256.
285. Owen, M. J.; L, G. J., Fluorosilicone Antifoams. *J. Appl. Polym. Sci.* **1990**, *40* (5–6), 789–797.
286. Owen, M. J., Analysis of Surfaces. In *The Analytical Chemistry of Silicones*, 2nd ed.; Smith, A. L., Ed. Wiley: New York, 1991; pp 97–116.

287. Owen, M. J., Surface Chemistry and Applications. In *Siloxane Polymers*, Clarson, S. J.; Semlyen, J. A., Eds. Prentice Hall: Englewood Cliffs, NJ, 1993; pp 309–372.
288. Owen, M. J.; Fritz, J. L., Surface Modification of Silicones. In *Polymers and Other Advanced Materials. Emerging Technologies and Business Opportunities*, Prasad, P. N.; Mark, J. E.; Ting, F. J., Eds. Plenum Press: New York, 1995; pp 201–211.
289. Owen, M. J., Surface Properties of Silicone High Voltage Insulators. In *Science and Technology of Polymers and Advanced Materials*, Prasad, P. N.; Mark, J. E.; Kandil, S. H.; Kafafi, Z. H., Eds. Plenum Press: New York, 1998; pp 99–106.
290. Fearon, F. W. G.; Owen, M. J., Silicone Surface Science Opportunities. In *Science and Technology of Polymers and Advanced Materials*, Prasad, P. N.; Mark, J. E.; Kandil, S. H.; Kafafi, Z. H., Eds. Plenum Press: New York, 1998; pp 873–879.
291. Cornelius, D. J.; Monroe, C. M., The Unique Properties of Silicone and Fluorosilicone Elastomers. *Polym. Eng. Sci.* **1985**, *25*, 467–473.
292. Owen, M. J., Surface and Interfacial Properties. In *Physical Properties of Polymers Handbook*, 2nd ed.; Mark, J. E., Ed. Springer-Verlag: New York, 1996; pp 669–676.
293. Owen, M. J., Siloxane Surface Activity. In *Silicon-Based Polymer Science. A Comprehensive Resource*, Zeigler, J. M.; Fearon, F. W. G., Eds. American Chemical Society: Washington, DC, 1990; Vol. 224, pp 705–739.
294. Granick, S.; Kuzmenka, D. J.; Clarson, S. J.; Semlyen, J. A., Spreading Transition of a Liquid Film. *Langmuir* **1989**, *5*(1), 144–147.
295. Uilk, J. M.; Mera, A. E.; Fox, R. B.; Wynne, K. J., Hydrosilation-Cured Poly(dimethylsiloxane) Networks: Intrinsic Contact Angles via Dynamic Contact Angle Analysis. *Macromolecules* **2003**, *36*, 3689–3694.
296. Ye, S.; Majumdar, P.; Chisholm, B.; Stafslien, S.; Chen, Z., Antifouling and Antimicrobial Mechanism of Tethered Quaternary Ammonium Salts in a Cross-Linked Poly(dimethylsiloxane) Matrix Studied Using Sum Frequency Generation Vibrational Spectroscopy. *Langmuir* **2010**, *26*, 16455–16462.
297. Wang, J. J.; Liu, F., Imparting Antifouling Properties of Silicone Hydrogels by Grafting Poly(ethylene glycol) Methyl Ether Acrylate Initiated by UV Light. *J. Appl. Polym. Sci.* **2012**, *125*, 548–554.
298. Rochow, E. G., *Silicon and Silicones*. Springer-Verlag: Berlin, 1987.
299. Mark, J. E., Overview of Siloxane Polymers. In *Silicones and Silicone-Modified Materials*, Clarson, S. J.; Fitzgerald, J. J.; Owen, M. J.; Smith, S. D., Eds. American Chemical Society: Washington, DC, 2000; Vol. 729, pp 1–10.
300. Aida, T.; Meijer, E. W.; Stupp, S. I., Functional Supramolecular Polymers. *Science* **2012**, *335*, 813–817.
301. Seiffert, S.; Sprakel, J., Physical Chemistry of Supramolecular Polymer Networks. *Chem. Soc. Rev.* **2012**, *41*, 909–930.
302. Allcock, H. R.; Lampe, F. W.; Mark, J. E., *Contemporary Polymer Chemistry*, 3rd ed. Prentice Hall: Englewood Cliffs, NJ, 2003.
303. Zeigler, J. M.; Fearon, F. W. G., *Silicon-Based Polymer Science. A Comprehensive Resource*. American Chemical Society: Washington, DC, 1990; Vol. 224.

CHAPTER 6

Surfaces

6.1 INTRODUCTION

Because of the great importance of the surface properties of the polysiloxanes, this topic is treated separately in this chapter.

6.2 INTERACTIONS WITH WATER

6.2.1 Hydrophilicity and Hydrophobicity

Hydrophobic polysiloxanes having simple aliphatic or aromatic side groups have surfaces that show essentially no attraction to water. In fact, polysiloxanes can serve as water repellants. This property is very useful for applications such as protective coatings on historical monuments[1] and for controlling the surfaces of other polymers,[2] sensors,[3] and quantum dots.[4] Hydrophobic surfaces can be readily regenerated if the surface becomes damaged. Regeneration occurs by rearrangements of the polysiloxane chains so that the hydrophobic methyl groups are once again covering the surface. The flexibility of the siloxane chain backbone facilitates this process.[5] It is also possible to prepare hydrophobic films using methyl-modified siloxane melting gels.[6] Glass surfaces[7] or wool fibers[8] can be coated with polydimethylsiloxane (PDMS) to make them more hydrophobic.

In some cases, it is necessary to modify a polysiloxane surface to make it hydrophilic or hydrophobic. Hydrophobization is one aspect of the general topic of modifying and managing the properties of polymer surfaces.[9–12] An important example involves soft contact lenses that contain

PDMS, which is often used because of its very high permeability to oxygen, which is required for metabolic processes within the eye.[13–15] Such lenses do not feel comfortable however because they do not float properly on the aqueous tears that coat the eye. There are a number of ways to modify the surfaces.[16–18] There is even a way to make "unreactive" silicones react with inorganic surfaces.[19]

In some applications it is useful to have hydrophilicity in the bulk of the polymer instead of just at the surface. One way of doing this is by simultaneously end linking hydrophilic poly(ethylene glycol) (PEG) chains and hydrophobic PDMS chains.[20] Another way is to make a PDMS network with a trifunctional organosilane $R'Si(OR)_3$ end linker that contains a hydrophilic R' side chain, such as a polyoxide.[21] Treating only the surfaces is another possibility, for example, by adding hydrophilic brushes by vapor deposition/hydrolysis cycles.[22] Such hydrophilic polysiloxanes can also serve as surfactants.[23] It has also been possible to make radially layered copolymeric dendrimers with hydrophilic polyamidoamine interiors and hydrophobic organosilicon exteriors.[24]

These ideas are being extended to materials that have high repellencies to a variety of liquids, including oils, solvents, and other low surface-energy liquids.[25] Such "superomniphobic" surfaces have been generated on PDMS substrates.[26]

6.2.2 Superhydrophilicity and Superhydrophobicity

There are also techniques for preparing surfaces of greater hydrophilicity ("superhydrophilicity")[27, 28] or increased hydrophobicity ("superhydrophobicity").[29–43] More specific terminology calls low-contact-angle materials hygrophilic and high-contact-angle materials hygrophobic.[32] Some of these materials display multiresponsive surfaces that can change reversibly between hydrophilicity or superhydrophilicity and superhydrophobicity.[27, 44, 45] Polymers showing transitions of this type are examples of stimuli-responsive materials, which are of increasing interest in a number of emerging applications.[46]

6.3 CHARACTERIZATION

6.3.1 Contact Angles

Measurement of advancing and receding contact angles of a water droplet on a polymer surfaces the standard way to characterize interactions

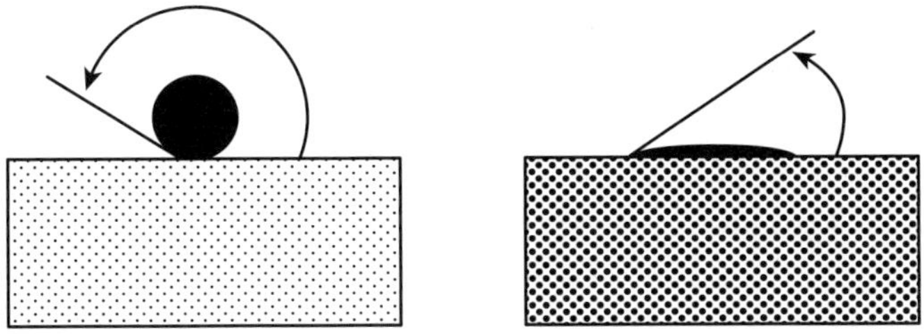

Figure 6.1:
Sketch showing the usual definition of the contact angle, for a very hydrophobic surface (high contact angle, left) and very hydrophilic (low contact angle right).

with water.[20, 21, 47, 48] Both static and dynamic techniques have been used.[47]

The left portion of figure 6.1 shows a hydrophobic surface (with a large contact angle between the water droplet and the surface), and the right portion shows a hydrophilic one with a small contact angle. A major problem is contamination of the water droplet from the polymer surface, making reproducibility very difficult.[47] Unusual increases in adhesion hysteresis and frictional forces have been observed when PDMS lenses were slid on smooth glassy surfaces after a period of aging.[49]

Applications in the biological area include stabilization of water in silicone oil emulsions by peptide-silicone hybrid polymers[50] and the mitigation of reactive human cell adhesion on PDMS by immobilized trypsin.[51]

6.3.2 Wettability

A polymer should be readily wet with water for applications such as contact lenses and for the preparation of hydrogels for a variety of biomedical applications. Increased wettability[52–55] is reflected by reduced values of the contact angle of the polymer with water. Another application is silicone surface modification of wool for self-cleaning.[56]

6.3.3 Spreading

In addition to the equilibrium properties there are important dynamic properties of surfaces. One example is the rate of spreading of PDMS on glass substrates.[57] A related study involved nanoviscosity of PDMS as measured by an anthracene probe.[58]

6.3.4 Surface Pressure

The relationships between surface pressure and surface or interfacial area can be used to gauge hydrophilicity. This method has been used, for example, to quantify the hydrophilicity of PDMS modified with ethylene oxide or propylene groups.[59]

6.3.5 Atomic Force Microscopy

Atomic force microscopy has been combined with nano-indentation measurements to map hardness variations on the surface of a $CaCO_3$-filled sample of PDMS.[60] In another application, PDMS-modified tips were used to obtain friction coefficients involving self-assembled monolayers.[61]

6.3.6 Nuclear Magnetic Resonance

A relevant example in this area involves self-diffusion measurements in a microemulsion consisting of water, oil, and ethoxylated polymethylsiloxane surfactant.[62]

6.3.7 Swelling

Surface modification of PDMS has been used to retard swelling in organic solvents, which is particularly important for fabricating microfluidic devices and for some types of lithography.[63]

6.3.8 Exposure to Seawater

PDMS has been widely used in antifouling coatings where it receives extensive exposures to seawater. The nature and extent to which the PDMS surface was modified can be determined in a variety of ways, including scanning electron microscopy and x-ray photoelectron spectroscopy.[64]

6.4 CHAINS BONDED TO OR EMBEDDED IN SURFACES

6.4.1 Tethering

Tethered chains can be placed at the surface of another polymer to change its surface characteristics. The placement can be through covalent bonding or by a physical process such as adsorption or embedding. It is also possible to tether quaternary ammonium salts to PDMS to enhance antifouling and antimicrobial characteristics.[65] The embedding approach is likely to be more robust.

One example is the silanization of the PDMS surface in the channels of a microfluidic devices such as biosensors.[66] The goal was to make the surface less hydrophobic and therefore less likely to absorb proteins. An organosilane was used to attach thiol groups on the surface, which were then reacted with modified DNA chains to give surface-immobilized chains of DNA oligomers. Oligomer reactions with target DNAs and an alkaline phosphatase conjugated with streptavidin then provides a platform for assaying some biologically important substances. DNA has also been bonded to maleimide-activated PDMS surfaces.[67]

PDMS coatings in the channels of microfluidic devices have also been modified with a perfluorinated alkoxysilane for selectivity toward fluorine-tagged peptides.[68] A more general approach employs a vinyl-terminated initiator with some H atoms on a dimethyl-methylhydrogen polysiloxane copolymer.[69] The polysiloxane surface activates to the extent that hydrophilic chains can be bonded onto it. Another biologically interesting material, heparin, has similarly been bonded to PDMS surfaces through a heterobifunctional poly(ethylene glycol) spacer to suppress the adsorption of molecules such as fibrinogen.[70]

In related work, acrylamide was grafted onto a PDMS surface using plasma polymerization. After this treatment, cell attachment to the PDMS surface became negligible.[71]

6.4.2 Grafting

One grafting approach involved peroxide-containing PDMS, using radical catalyzed thiol-ene chemistry to produce cross linkable polymers.[72] Another study focused on functionalization of surface-grafted poly(methylhydrosiloxane) thin films with alkyl side chains.[73] In some cases, the alkyl chains were long enough to undergo crystallization. Crystallization was also involved in the study of polypropylene-graft-polydiphenylsiloxane copolymers.[74]

An example of attachment by adsorption involves making PDMS surfaces more hydrophilic by addition of hydrophobins (a family of small, cysteine-rich and amphipathic fungal proteins).[75] A more complicated adsorption approach is illustrated by the adsorption of segments of a graft copolymer onto a PDMS surface that had been treated with an oxygen or ammonia plasma.[76, 77] The PDMS surface then contains hydroxyl-carboxyl groups that can attract the lysine segments of a poly(L-lysine-*graft*-poly(ethylene glycol)), leaving hydrophilic ethylene glycol segments covering the PDMS surface. This adsorption technique can be extended to an embedding approach, as illustrated in figure 6.2. The PDMS sequences of a PDMS-poly(ethylene oxide) block copolymer embed into a PDMS elastomer.[78] The ethylene oxide segments are then relegated to the surface, giving the PDMS a hydrophilic coating.

In a gas phase approach, undecenyl-trichlorosilane vapor was condensed onto a silicon wafer surface to form a self-assembled monolayer film whose surface was composed of olefin groups.[79] These groups react with SiH-functionalized PDMS in a hydrosilylation reaction to give an unusually thin coating of PDMS.

A hybrid approach involves adsorption followed by chemical bonding. Specifically, a vinyl-ended oligonucleotide was adsorbed onto a PDMS surface and then a hydrosilylation reaction used to bond it to the surface chemically.[80] This modification can greatly improve separation assays or the performance of biochips by enhancing target hybridization.

Whether bonded or physically adsorbed or embedded, surface polymers dangle from the polymer surface, as do the chains in any PDMS elastomer. Such chains are important with regard to mechanical properties such as adhesion, adhesion hysteresis,[81] and tack ("stickiness").[82] Their rheological and dielectric behavior have been modeled using molecular dynamics.[83]

Figure 6.2:
Modifying a surface using a diblock copolymer in which one block (thin lines) is the same as the substrate polymer, and the other block (rejected to the surface) has the desired amount of hydrophilicity or hydrophobicity.

6.5 RADIATION TREATMENTS

6.5.1 Plasmas and Photons

Modification of surfaces by radiation treatment can be difficult to control because of the high energies involved. The most extensively used treatment of this type involves modifications by chemical transformation of surface groups or adsorbed molecules, or induced polymerization of suitable monomers.

Transformation of PDMS can yield less hydrophobic surfaces, using radio-frequency plasmas,[84] argon or hydrogen plasmas,[85] oxygen plasmas,[86, 87] helium plasmas,[88] water plasmas,[89] or other environments.[90] The same treatments can be applied to molecules coated onto PDMS (e.g., imidazole and some of its derivatives) to facilitate attaching antimicrobial drugs onto the PDMS surface.[91] Other examples are layer-by-layer modifications using acrylamides, sulfonates, and trimethylammonium chloride salts.[92] A final example uses maleic anhydride to promote the formation of dicarboxylic acid groups on PDMS surfaces.[93] Plasma polymerization has also been used to modify fillers and curatives used in a variety of elastomers.[94]

An example of plasma polymerization is acrylic acid coating of PDMS. The polymerized coatings yielded hydrophilic or patterned hydrophilic/hydrophobic surfaces that were quite stable.[95] These materials were useful in PDMS microchannels in lab-on-a-chip devices or in soft microlithography. Maleic anhydride has also been used in these polymerizations.[89] Photons have also been used to modify surfaces.[96]

6.5.2 UV and UV/Ozone Treatments

UV (ultraviolet) irradiation of PDMS surfaces has been described and compared to the results from water-plasma treatments and plasma polymerization.[89] These treatments affect cracking of thin coatings of PDMS during stretching.

The use of UV in the presence of ozone is illustrated by the conversion of portions of PDMS surfaces to more hydrophilic SiO_x.[97, 98] UV/ozone treatments were also used to make poly(vinylmethylsiloxane) surfaces more hydrophilic, but only about one-twentieth the PDMS exposure time was required.[99] Such treatments have also improved the permeability of PDMS membranes.[100]

6.5.3 Ion Beams

Poly(hydroxymethylsiloxane) surfaces have been modified with 6 keV Ar^+ beams, and the results compared with those from oxygen plasma treatment.[101] The ion beam approach was much more successful in promoting proliferation of human dermal fibroblasts. This result was attributed to a peculiar electronic structure and related electrical properties in the ion-containing PDMS. Low-energy ion beams have also been used to form silica on PDMS surfaces.[102]

6.6 SOME ADDITIONAL CHEMICAL ASPECTS

6.6.1 Emulsions

PDMS oil/water emulsions have been stabilized by adsorbing hybrid siloxane polymers at the droplet surface.[103] Examples of such additives are PDMS backbones with side chains consisting of trimethylene spacers ending with amino groups, methylated amino groups, acid groups, or an ethylene oxide decamer. Microemulsions have also been prepared directly from silicone oil with an anionic/nonionic surfactant mixture.[104]

6.6.2 Radical Polymerization

Radical polymerization was used on acrylamide adsorbed onto a PDMS surface.[105] The hydrophilic surface thus generated had a twenty-fold improvement in resisting irreversible adsorption of lysozyme and was stable against reorganization to a hydrophobic surface for at least a month.

6.6.3 Copolymers with Polyurethanes

Segmented polyurethanes with PDMS blocks were found to have good environmental responsiveness.[106] The hydrophobic-to-hydrophilic transformation can be managed by varying the composition of the segments of the polymer. The associated switchable properties could be of interest in a number of applications including the release of soils, combating fouling, and selective adsorption–desorption of proteins. Surface microtopography studies of a thermoset siloxane-urethane system have been reported.[107, 108]

6.7 MIGRATION

6.7.1 Surface Segregation

Because of its low surface free energy, PDMS tends to migrate to surfaces and interfaces. For example, in blends of PDMS with low-density polyethylene[109] or linear low-density polyethylene,[109, 110] the surfaces are enriched with the hydrophobic methyl groups from the PDMS chains. The same is true for blends of PDMS with polypropylene[109] and poly(3-hexylthiophene).[111]

Similar enrichment of siloxane units occurs at the surfaces of block copolymers or even random copolymers containing PDMS sequences. Examples are poly(dimethylsiloxane-*b*-ethylene oxide)[112] and poly(dimethylsiloxane-*b*-imide) multi-block copolymers,[113] poly(dimethylsiloxane)/imine copolymers,[114] and poly(urethane-dimethylsiloxane) anionomers.[115] Other materials, such as long-chain alkenes[116] and highly branched polymer additives[117] exhibit surface segregation. Much of the information on these changes in surface composition has been obtained by measurements of contact angles.

6.7.2 Recovery and Restructuring

In some cases attempts to make PDMS surfaces more hydrophilic were only temporarily successful. Recovery or restructuring can occur, with methyl groups from the siloxane units migrating to the surface to lower the free energy of the system. Restructuring is facilitated by the high flexibility of the polysiloxane backbone.

Studies of this restructuring process have been carried out on PDMS with surfaces made hydrophilic by UV/ozone treatment,[118] oxygen plasma,[10] or corona discharge.[119] In these cases, changes in surface composition were determined using chemical force microscopy,[120] Fourier transform infrared spectroscopy and x-ray photoelectron spectroscopy,[10] or atomic force microscopy.[119]

Similar results have been obtained on PDMS elastomers. Examples are tetraethoxysilane-cured hydroxyl-terminated PDMS, methylhydrogensiloxane-cured vinyl-terminated PDMS, and methylhydrogensiloxane-cured vinyl-terminated poly(dimethylsiloxane-*co*-diphenylsiloxane) chemical copolymer.[121] Restructuring has also been observed in un-cross-linked copolymers, for example PDMS-*co*-polystyrene,[121] using sum frequency generation vibrational spectroscopy.

6.7.3 Self-Healing

For some applications, it is necessary to maintain a hydrophobic surface. If the surface is damaged then the recovery or restructuring is beneficial and is called self-healing. In this sense, PDMS is a "smart material," in that it responds constructively to a change in the environment.

Self-healing is often used in a broader sense, to mean reconstruction of the entire polymer instead of just its surface.[122] A relevant example here is a PDMS elastomer that contains microencapsulated PDMS resin and microencapsulated cross linker.[123] If this type of PDMS is damaged, both capsules rupture and the newly formed elastomer mends the damaged area.

6.8 INTERACTIONS WITH BIOMOLECULES

6.8.1 Trapped Biomolecules

Biologically active molecules are sometimes trapped in PDMS when end-functionalized PDMS chains are linked into a network structure. This method has been done, for example, with a lipase enzyme.[124] The PDMS plays a beneficial role as an activator or protective agent. Similar results were found for the enzyme α-chymotripsin, with some short-chain poly(ethylene oxide) used to enhance enzymatic activity.[125] It is also possible to generate microtopographic patterns that affect *Escherichia coli* biofilm formation on PDMS surfaces.

6.8.2 Controlled Release

Surface graft polymerization of poly(ethylene glycol) acrylate was used to modify the surface of PDMS.[126] Templates with channels were formed from this material and sol-gel chemistry was used to form amino-silane doped xerogel microarrays. These structures were then used to release nitric oxide at various rates, by control of micropattern dimensions, type and concentration of the amino-silane, and so on. This method parallels the use of polymers in controlled drug-delivery systems.[127–129]

6.8.3 Protein Adsorption

Proteins react with silicon-containing materials in a variety of ways.[130] For example, peptides in a phage library were exposed to the surface of several types of silsesquioxane cages.[131] The peptide units that bonded most readily were proline, histidine, and threonine. Another example of favorable interactions involves polysiloxanes in both the liquid and elastomeric forms in the presence of nonionic surfactants. These environments *increase* the activity of a number of enzymes, including α-chymotrypsin and a lipase. Studies have also been carried out on carbohydrate-modified siloxane surfactants.[132]

In many applications it is important to have surfaces that repel proteins. Hydrophilic poly(ethylene oxide) segments are effective in this regard. Such segments make polyoxide surfaces protein-repellant when either incorporated in the bulk material[133] or covalently bonded to its surface.[134, 135]

6.8.4 Cells and Antigen Molecules

The attachment and growth of several types of mammalian cells on fibronectin-coated PDMS have been reported.[136] All cells grew at the same rate on the PDMS irrespective of its degree of cross linking, but the compatibility of the cells on the surfaces did depend on the cell type. Similar results were obtained when micropatterns of hydrophilic regions were placed on PDMS by means of a gas plasma.[137] These modifications of the PDMS surfaces were found to increase the adhesion of fibroblast cells. On the other hand, grafting sulfobetaine onto PDMS surfaces decreases the adsorption of blood platelets.[138]

Geometric aspects can also be important, as illustrated by control of the spreading of mammalian cells on wavy PDMS surfaces.[139] Wavy surfaces were generated by stretching a sheet of the elastomer, exposing it to a oxygen plasma, and the allowing it to relax.

Another study focused on PDMS surfaces made hydrophilic by modification with hydrophobins (small, cysteine-rich and amphiphilic fungal proteins).[75] This approach was used to pattern antigen molecules, followed by immunoassays. For example, chicken immunoglobulin G was found to be compatible with the hydrophhobin-modified PDMS.

Some related work involved cellular interactions of collagen-immobilized PDMS surfaces,[140] and the immobilization of antibody fragments on polymer brushes supported by silicone nanofilaments.[141]

6.8.5 Biofouling

PDMS surfaces modified with poly(ethylene glycol) have a near-perfect resistance to nonspecific protein adsorption, making them effective for suppressing biofouling.[76, 142] The incorporation of diphenylsiloxane oils in PDMS is also effective. Methods have been developed to minimize depletion of these oils when the surfaces are exposed to freshwater and other marine environments.[143]

Coatings of the block copolymers, PDMS-polyurethane and PDMS-acrylic-polyurethane,[144] and PDMS-polyurea,[145] are also effective antibiofouling agents as are tethered quarternary ammonium salts in a cross-linked PDMS matrix.[65]

6.9 MECHANICAL ASPECTS

6.9.1 Friction and Lubricity

Frictional effects of PDMS sliding on gold surfaces,[146] glass,[147] silicon oxide surfaces,[148] or glassy polymers,[149, 150] have been investigated in detail. There has also been some interest in self-lubricating biomaterials.[151, 152]

6.9.2 Adhesion

Adhesion between polymers and various surfaces has long been of general interest.[153, 154] PDMS samples of varying degrees of cross linking were studied to determine the extent to which they adhered to one another (self-adhesion).[155] This type of adhesion was further explored by preparing PDMS model networks having known numbers of dangling chains.[81] The adhesion hysteresis increased with increase in the number of dangling chains because of the increase in the number of such chains bridging the interface.

Similar studies have been carried out on the adhesion of PDMS to polycarbonate using thermal gradients and x-ray photoelectron spectroscopy,[156] to mica using atomic force microscopy,[157] and to glass[158, 159] or to silicon[160] using measurements of force, contact angle, or other properties. There has also been considerable interest in the adhesion strength between coated particles and polymer matrixes.[161] Surface functionalization has been used to improve adhesion.[162, 163]

Modeling has been used to clarify the dependence of adhesive failure in PDMS on its viscosity, the adhesive thickness, the substrate properties, some rheological aspects, and peeling velocities.[164]

Interfacial tension between PDMS and another polymer such as liquid polybutadiene has also been studied.[165] The surface behavior of PDMS-filled nanoporous anodic alumina has been reported.[166]

6.9.3 Tribology

The low coefficients of friction shown by the polysiloxanes have been exploited in reducing abrasion or wear on a variety of materials. For example, PDMS thin coatings on mica were studied in this regard.[167] PDMS block copolymers have been blended with polypropylene and high-density polyethylene for this purpose.[168] Closely related is the scratch resistance of elastomeric PDMS coatings on materials such as stainless steel.[169]

6.10 SOME NOVEL MATERIALS

6.10.1 Dendrimers

Dendrimers,[170–173] including those based on organosilicon materials,[174] are of great relevance in siloxane surface science, particularly since their surfaces have functionalities that are numerous and controllable.

6.10.2 Ceramic Phases and Coatings

Work in this area involves applying polysiloxane coatings on inorganic substrates as has been done on silica,[175, 176] Al_2O_3/GaAs,[177] and the channels in microfluidic devices.[178] Variations include sol-gel blends of PDMS and titania[179] and titania coatings on PDMS.[180] PDMS cores have also been coated with poly(*N*-isopropylacrylamide) shells to give thermoresponsive microspheres.[181]

6.10.3 Micropatterning

Patterns with dimensions at the micron level have been placed on PDMS[182, 183, 184] and PDMS-acrylic interfaces.[185] Such materials have numerous applications, including soft lithography.

6.10.4 Nanofilaments and Molecular Wires

Polysiloxane nanofilaments can be superhydrophobic.[186] Such filaments can be subsequently patterned to present a "superfunctional" surface.[187] Activation is achieved by an oxygen plasma followed by reaction with a functionalized silane in the gas phase or in a solvent. One application is the selective enrichment of proteins.[188]

It is also possible to fabricate elastomeric wires by selective electroless metallization of PDMS.[189]

REFERENCES

1. Sadat-Shojai, M.; Ershad-Langroudi, A., Polymeric Coatings for Protection of Historic Monuments: Opportunities and Challenges. *J. Appl. Polym. Sci.* **2009**, *112*, 2535–2551.
2. Ward, L. J.; Schofield, W. C. E.; Badyal, J. P. S.; Goodwin, A. J.; Merlin, P. J., Atmospheric Pressure Glow Discharge Deposition of Polysiloxane and SiO_x Films. *Langmuir* **2003**, *19*, 2110–2114.
3. Shalal, T.; Melzak, K. A.; Lowe, C. R.; Gizeli, E., Poly(dimethylsiloxane)-Coated Sensor Devices for the Formation of Supported Lipid Bilayer and the Subsequent Study of Membrane Interactions. *Langmuir* **2008**, *24*, 11268–11275.
4. Moussodia, R.-O.; Balan, L.; Schneider, R., Synthesis and Characterization of Water-Soluble ZnO Quantum Dots Prepared Through PEG-Siloxane Coating. *New J. Chem.* **2008**, *32*, 1388–1393.
5. Mark, J. E., Some Interesting Things About Polysiloxanes. *Acc. Chem. Res.* **2004**, *37*, 946–953.
6. Jitianu, A.; Doyle, J.; Amatucci, G.; Klein, L. C., Methyl Modified Siloxane Melting Gels for Hydrophobic Films. *J. Sol-Gel Sci. Technol.* **2010**, *53*, 272–279.
7. Marinova, K. G.; Christova, D.; Tscholakova, S.; Efremov, E.; Denkov, N. D., Hydrophilization of Glass Surface by Adsorption of Poly(dimethylsiloxane). *Langmuir* **2005**, *21*, 11729–11737.
8. Canal, C.; Molina, R.; Bertran, E.; Erra, P., Polysiloxane Softener Coatings on Plasma-Treated Wool: Study of the Surface Interactions. *Macromol. Mater. Eng.* **2007**, *292*, 817–824.
9. Hutchings, L. R.; Narrianen, A. P.; Thompson, R. L.; Clarke, N.; Ansari, I., Modifying and Managing the Surface Properties of Polymers. *Polym. Int.* **2008**, *57*, 163–170.
10. Bodas, D.; Rauch, J.-Y.; Khan-Malek, C., Surface Modification and Aging Studies of Addition-Curing Silicone Rubbers by Oxygen Plasma. *Eur. Polym. J.* **2008**, *44*, 2130–2139.
11. Bodas, D.; Rauch, J.-Y.; Khan-Malek, C., Creation of a Stable Hydrophilic Poly(dimethylsiloxane) Surface by the Plasma-Induced Crosslinking of Monomers. *J. Appl. Polym. Sci.* **2011**, *120*, 1426–1430.
12. Lowe, R. D.; Pellow, M. A.; Stack, T. D. P.; Chidsey, C. E. D., Deposition of Dense Siloxane Monolayers from Water and Trimethoxyorganosilane Vapor. *Langmuir* **2011**, *27*, 9928–9935.
13. Lubick, N., Contact Lenses: The Hard and the Soft. *Sci. Am.* **2000**, *284*, 88–89.
14. Nicolson, P. C.; Vogt, J., Soft Contact Lens Polymers: An Evolution. *Biomaterials* **2001**, *22*, 3273–3283.
15. Tighe, B., Silicone Hydrogels: Structure, Properties and Behaviour. In *Silicone Hydrogels. Continuous-Wear Contact Lenses*, 2nd ed.; Sweeney, D., Ed. Butterworth Heinemann: Edinburgh, 2004; pp 1–27.

16. Benkova, Z.; Cordero, M. N. D. S., Molecular Dynamics Study of Water Interacting with Siloxane Surface Modified by Poly(ethylene oxide) Chains. *J. Phys. Chem. C* **2011**, *115*, 18740–18751.
17. Monticelli, O.; Fina, A.; Cozza, E. S.; Prato, M.; Bruzzo, V., POSS Vapor Phase Grafting: A Novel Method to Modify Polymer Films. *J. Mater. Chem.* **2011**, *21*, 18049–18054.
18. Benkova, Z.; Szefczyk, B.; Cordeiro, M. N. D. S., Molecular Dynamics Study of Hydrated Poly(ethylene oxide) Chains Grafted on Siloxane Surface. *Macromolecules* **2011**, *44*, 3639–3648.
19. Krumpfer, J. W.; McCarthy, T. J., Rediscovering Silicones; "Unreactive" Silicones React with Inorganic Surfaces. *Langmuir* **2011**, *27*, 11514–11519.
20. Lin, G.; Zhang, X.; Kumar, S. R.; Mark, J. E., Improved Hydrophilicity from Poly(ethylene glycol) in Amphilic Conetworks with Poly(dimethylsiloxane). *Silicon* **2009**, *1*, 173–181.
21. Zhang, X.; Lin, G.; Kumar, S. R.; Mark, J. E., Hydrogels Prepared from Polysiloxane Chains by End Linking Them with Trifunctional Silanes Containing Hydrophilic Groups. *Polymer* **2009**, *50*, 5414–5421.
22. Togulu, S.; Klok, H.-A., Surface Modification of Polydimethylsiloxane Substrates with Nonfouling Poly(ethylene glycol)methacrylate Brushes. *Macromol. Symp.* **2009**, *279*, 103–109.
23. Kuo, P.-L.; Hou, S.-S.; Teng, C.-K.; Liang, W. J., Function and Performance of Silicone Copolymer. (VI). Synthesis and Novel Solution Behavior of Water-Soluble Polysiloxanes with Different Hydrophobicities. *Coll. Polym. Sci.* **2001**, *279*, 286–291.
24. Dvornic, P. R.; Leuze-Jallouli, A. M. D.; Owen, M. J.; Perz, S. V., Radially Layered Poly(Amidoamine-organosilicon) Copolymeric Dedrimers and Their Networks Containing Controlled Hydrophilic and Hydrophobic Nanoscopic Domains. In *Silicones and Silicone-Modified Materials*, Clarson, S. J.; Fitzgerald, J. J.; Owen, M. J.; Smith, S. D., Eds. American Chemical Society: Washington, DC, 2000; Vol. 729, pp 241–269.
25. Wong, T.-S.; Kang, S. H.; Tang, S. K. Y.; Smythe, E. J.; Hatton, B. D.; Grinthal, A.; Alzenberg, J., Bioinspired Self-Repairing Slippery Surfaces with Pressure-Stable Omniphobicity. *Nature* **2011**, *477*, 443–447.
26. Dufour, R.; Harnois, M.; Coffinier, V.; Thomy, V.; Boukherroub, R.; Senez, V., Engineering Sticky Superomniphobic Surfaces on Transparent and Flexible PDMS Substrate. *Langmuir* **2010**, *26*, 17242–17247.
27. Xia, F.; Ge, H.; Hou, Y.; Sun, T.; Chen, L.; Zhang, G.; Jiang, L., Multiresponsive Surfaces Change Between Superhydrophilicity and Superhydrophobicity. *Adv. Mater.* **2007**, *19*, 2520–2524.
28. Drelich, J.; Chibowski, E., Superhydrophilic and Superwetting Surfaces: Definition and Mechanism of Control. *Langmuir* **2010**, *26*, 18621–18623.
29. Jin, M.; Feng, X.; Xi, J.; Zhai, J.; Cho, K.; Feng, L.; Jiang, L., Super-Hydrophobic PDMS Surface with Ultra-Low Adhesive Force. *Macromol. Rapid Commun.* **2005**, *26*, 1805–1809.
30. Crowe, J. A.; Efimenko, K.; Genzer, J., Manipulating Siloxane Surfaces: Obtaining the Desired Surface Function via Engineering Design. In *Science and Technology of Silicones and Silicone-Modified Materials*, Clarson, S. J.; Fitzgerald, J. J.; Owen, M. J.; Smith, S. D.; Van Dyke, M. E., Eds. American Chemical Society: Washington, DC, 2007; Vol. 964, pp 222–255.

31. Zhang, B. T.; Liu, B. L.; Deng, X. B.; Cao, S. S.; Hou, X. H.; Chen, H. L., Fabricating Superhydrophobic Surfaces by Molecular Accumulation of Polysiloxane on the Wool Textile Finishing. *Coll. Polym. Sci.* **2008**, *286*, 453–457.
32. Marmur, A., From Hygrophilic to Superhygrophobic: Theoretical Conditions for Making High-Contact-Angle Surfaces from Low-Contact-Angle Materials. *Langmuir* **2008**, *24*, 7173–7579.
33. Qian, T.; Li, Y.; Wu, Y.; Zheng, B.; Ma, H., Superhydrophobic Poly(dimethylsiloxane) via Surface-Initiated Polymerization with Ultralow Initiator Density. *Macromolecules* **2008**, *41*, 6641–6645.
34. Crick, C. R.; Parkin, I. P., A Single Step Route to Superhydrophobic Surfaces Through Aerosol Assisted Deposition of Rough Polymer Surfaces: Duplicating the Lotus Effect. *J. Mater. Chem.* **2009**, *19*, 1074–1076.
35. Chen, R.; Zhang, X.; Su, Z.; Gong, R.; Ge, X.; Zhang, H.; Wang, C., Perfectly Hydrophobic Silicone Nanofiber Coatings: Preparation from Methyltrialkoxysilanes and Use as Water-Collecting Substrate. *J. Phys. Chem. C* **2009**, *113*, 8350–8356.
36. De Givency, E. P. T.; Amigoni, S.; Martin, C.; Andrada, G.; Callier, L.; Geribaldi, S.; Guittard, F., Fabrication of Superhydrophobic PDMS Surfaces by Combining Acidic Treatment and Perfluorinated Monolayers. *Langmuir* **2009**, *25*, 6448–6453.
37. Hwang, J.; Hong, S.-H.; Lee, H., Mimicking the Nanostructure of Bamboo Leaves (Backside) for Hydrophobicity Using Polydimethylsiloxane Moulding and Nano-Imprint Lithography. *J. Nanosci. Nanotech.* **2009**, *9*, 3644–3647.
38. Lin, P.-C.; Yang, S., Mechanically Switchable Wetting on Wrinkled Elastomers with Dual-Scale Roughness. *Soft Matter* **2009**, *5*, 1011–1018.
39. Ke, Q.; Li, G.; Liu, Y.; He, T.; Li, X.-M., Formation of Superhydrophobic Polymerized *n*-Octadecylsiloxane Nanosheets. *Langmuir* **2010**, *26*, 3579–3584.
40. Peng, Y.-T.; Lo, K.-F.; Juang, Y.-J., Constructing a Superhydrophobic Surface on Polydimethylsiloxane via Spin Coating and Vapor-Liquid Sol-Gel Process. *Langmuir* **2010**, *26*, 5167–5171.
41. Lim, H. S.; Baek, J. H.; Park, K.; Shin, H. S.; Kim, J.; Cho, J. H., Multifunctional Hybrid Fabrics with Thermally Stable Superhydrophobicity. *Adv. Mater.* **2010**, *22*, 2138–2141.
42. Wohl, C. J.; Belcher, M. A.; Chen, L.; Connell, J. W., Laser Ablative Patterning of Copoly(imide siloxane)s Generating Superhydrophobic Surfaces. *Langmuir* **2010**, *26*, 11469–11478.
43. Nosonovsky, M., Slippery When Wetted. *Nature* **2011**, *477*, 412–413.
44. Crowe-Willoughby, J. A.; Genzer, J., Formation and Properties of Responsive Siloxane-Based Polymeric Surfaces with Tunable Surface Reconstruction Kinetics. *Adv. Mater.* **2009**, *21*, 460–469.
45. Blanco-Gomez, G.; Flendrig, L. M.; Cooper, J. M., Hysteresis and Reversibility of a Superhydrophobic Photopatternable Silicone Elastomer. *Langmuir* **2010**, *26*, 7248–7253.
46. Stuart, M. A. C.; Huck, W. T. S.; Genzer, J.; Muller, M.; Ober, C.; Stamm, M.; Sukhorukov, G. B.; Szleifer, I.; Tsukruk, V. T.; Urban, M.; Winnik, F.; Zauscher, S.; Luzinov, I.; Minko, S., Emerging Applications of Stimuli-Responsive Polymer Materials. *Nat. Mater.* **2010**, *9*, 101–113.
47. Uilk, J. M.; Mera, A. E.; Fox, R. B.; Wynne, K. J., Hydrosilation-Cured Poly(dimethylsiloxane) Networks: Intrinsic Contact Angles via Dynamic Contact Angle Analysis. *Macromolecules* **2003**, *36*, 3689–3694.

48. Bouteau, M.; Cantin, S.; Fichet, O.; Perrot, F.; Teyssie, D., Contribution Toward Comprehension of Contact Angle Values on Simple Polydimethylsiloxane and Poly(ethylene oxide) Polymer Networks. *Langmuir* **2010**, *26*, 17427–17434.
49. Kurian, A.; Prasad, S.; Dhinojwala, A., Unusual Surface Aging of Poly(dimethylsiloxane) Elastomers. *Macromolecules* **2010**, *43*, 2438–2443.
50. Sakai, K.; Ikeda, R.; Sharma, S. C.; Shrestha, R. G.; Ohtani, N.; Yoshioka, M.; Sakai, H.; Abe, M.; Sakamoto, K., Active Interfacial Modifier: Stabilization Mechanism of Water in Silicone Oil Emulsions by Peptide-Silicone Hybrid Polymers. *Langmuir* **2010**, *26*, 5349–5354.
51. Achyuta, A. K. H.; Stephens, K. D.; Lewis, H. G. P.; Murthy, S. K., Mitigation of Reactive Human Cell Adhesion on Poly(dimethylsiloxane) by Immobilized Trypsin. *Langmuir* **2010**, *26*, 4160–4167.
52. Svitova, T.; Theodoly, O.; Christiano, S.; Hill, R. M.; Radke, C. J., Wetting Behavior of Silicone Oils on Solid Substrates Immersed in Aqueous Electrolyte Solutions. *Langmuir* **2002**, *18*, 6821–6829.
53. Green, D. L.; Mewis, J., Connecting the Wetting and Rheological Behaviors of Poly(dimethylsiloxane)-Grafted Silica Spheres in Poly(dimethylsiloxane) Melts *Langmuir* **2006**, *22*, 9546–9553.
54. Kheirandish, S.; Jabbari, E., Effect and Surface Polarity on Wettability and Friction Coefficient of Silicone Rubber/Poly(acrylic acid)/Hydrogel Composite. *Coll. Polym. Sci.* **2006**, *284*, 1411–1417.
55. Schneemilch, M.; Quirke, N., Effect of Oxidation on the Wettability of Poly(dimethylsiloxane) Surfaces. *J. Chem. Phys.* **2007**, *127*, 114701.
56. Tung, W. S.; Daoud, W. A., Effect of Wettability and Silicone Surface Modification on the Self-Cleaning Functionalization of Wool. *J. Appl. Polym. Sci.* **2009**, *112*, 235–243.
57. Carre, A.; Woehl, P., Spreading of Silicone Oils on Glass in Two Geometries. *Langmuir* **2006**, *22*, 134–139.
58. Somoza, M. M.; Sluch, M. I.; Berg, M. A., Torsional Relaxation and Friction on the Nanometer Length Scale: Comparison of Small-Molecule Rotation in Poly(dimethylsiloxane) and Poly(isobutylene). *Macromolecules* **2003**, *36*, 2721–2732.
59. Mehta, S. C.; Somasundaran, P.; Maldaraelli, C.; Kulkarni, R., Effects of Functional Groups on Surface Pressure-Area Isotherms of Hydrophilic Silicone Polymers. *Langmuir* **2006**, *22*, 9566–9571.
60. Eaton, P.; Estarlich, F. F.; Ewen, R. J.; Nevell, T. G.; Smith, J. R.; Tsibouklis, J., Combined Nanoidentation and Adhesion Force Mapping Using the Atomic Force Microscope: Investigations of a Filled Polysiloxane Coating. *Langmuir* **2002**, *18*, 10011–10015.
61. Cho, J. H.; Lee, D. H.; Shin, H. S.; Pattanayek, S. K.; Ryu, C. Y.; Cho, K., Exploiting Poly(dimethylsiloxane)-Modified Tips to Evaluate Frictional Behavior by Friction Force Microscopy. *Langmuir* **2004**, *20*, 11499–11503.
62. Walderhaug, H., Structures in a Microemulsion System of an Ethoxylated Polydimethylsiloxane Surfactant, Water, and Oil Studied by NMR Self-Diffusion Measurements. *J. Phys. Chem. B* **2007**, *111*, 9821–9827.
63. Lee, J.; Kim, J.; Lee, H. H., Surface Modification of Poly(dimethylsiloxane) for Retarding Swelling in Organic Solvents. *Langmuir* **2006**, *22*, 2090–2095.
64. Arce, F. T.; Avci, R.; Beech, I. B.; Cooksey, K. E.; Cooksey, B. W., Modification of Surface Properties of Poly(dimethylsiloxane)-Based Elastomer, RTV11, upon Exposure to Seawater. *Langmuir* **2006**, *22*, 7217–7225.

65. Ye, S.; Majumdar, P.; Chisholm, B.; Stafslien, S.; Chen, Z., Antifouling and Antimicrobial Mechanism of Tethered Quaternary Ammonium Salts in a Cross-Linked Poly(dimethylsiloxane) Matrix Studied Using Sum Frequency Generation Vibrational Spectroscopy. *Langmuir* **2010**, *26*, 16455–16462.
66. Liu, D.; Perdue, R. K.; Sun, L.; Crooks, R. M., Immobilization of DNA onto Poly(dimethylsiloxane) Surfaces and Application to a Microelectrochemical Enzyme-Amplified DNA Hybridization Assay. *Langmuir* **2004**, *20*, 5905–5910.
67. Vaidya, A. A.; Norton, M. L., DNA Attachment Chemistry at the Flexible Silicone Elastomer Surface: Toward Disposable Microarrays. *Langmuir* **2004**, *20*, 11100–11107.
68. Wang, D.; Oleschuk, R. D.; Horton, J. H., Surface Modification of Poly(dimethylsiloxane) with a Perfluorinated Alkoxysilane for Selectivity Toward Fluorous Tagged Peptides. *Langmuir* **2008**, *24*, 1080–1086.
69. Wu, Y.; Huang, Y.; Ma, H., A Facile Method for Permanent and Functional Surface Modification of Poly(dimethylsiloxane). *J. Am. Chem. Soc.* **2007**, *129*, 7226–7227.
70. Chen, H.; Chen, Y.; Sheardown, H.; Brook, M. A., Immobilization of Heparin on a Silicone Surface Through a Heterobifunctional PEG Spacer. *Biomaterials* **2005**, *26*, 7418–7424.
71. Parvin, A.; Mirzadeh, H.; Khorasani, M. T., Physicochemical and Biological Evaluation of Plasma-Induced Graft Polymerization of Acrylamide onto Polydimethylsiloxane. *J. Appl. Polym. Sci.* **2008**, *107*, 2343–2349.
72. Lucas, P.; Fleury, E.; Estur, J.-F.; Lapinte, V.; Robin, J.-J., Peroxide-Grafted PDMS: Hydrosilylation Reaction and Thiol-Ene Chemistry as an Alternative Pathway. *Macromol. Chem. Phys.* **2009**, *210*, 1933–1941.
73. Thami, T.; Nasr, G.; Bestal, H.; Van Der Lee, A.; Bresson, B., Functionalization of Surface-Grafted Polymethylhydrosiloxane Thin Films with Alkyl Side Chains. *J. Polym. Sci., Part A: Polym. Chem.* **2008**, *46*, 3546–3562.
74. Wu, N., Study on the Crystallization Behavior of Polypropylene-Graft-Polydiphenylsiloxane Copolymer. *J. Macromol. Sci., Part B: Phys.* **2009**, *48*, 500–513.
75. Wang, R.; Yang, Y.-L.; Qin, M.; Wang, L.-K.; Yu, L.; Shao, B.; Qiao, M.-Q.; Wang, C.; Feng, X.-Z., Biocompatible Hydrophilic Modifications of Poly(dimethylsiloxane) Using Self-Assembled Hydrophobins. *Chem. Mater.* **2007**, *19*, 3227–3231.
76. Lee, S.; Voros, J., An Aqueous-Based Surface Modification of Poly(dimethylsiloxane) with Poly(ethylene glycol) to Prevent Biofouling. *Langmuir* **2005**, *21*, 11957–11962.
77. Cordeiro, A. L.; Zschoche, S.; Janke, A.; Nitschke, M.; Werner, C., Functionalization of Poly(dimethylsiloxane) Surfaces with Maleic Anhydride Copolymer Films. *Langmuir* **2009**, *25*, 1509–1517.
78. Yu, K.; Han, Y., A Stable PEO-Tethered PDMS Surface Having Controllable Wetting Property by a Swelling-Deswelling Process. *Soft Matter* **2006**, *2*, 705–709.
79. She, H.; Chaudhury, M. K.; Owen, M. J., Surface Properties of Thin Film Poly(dimethylsiloxane). In *Silicones and Silicone-Modified Materials*, Clarson, S. J.; Fitzgerald, J. J.; Owen, M. J.; Smith, S. D., Eds. American Chemical Society: Washington, DC, 2000; Vol. 729, pp 322–331.

80. Heyries, K. A.; Blum, L. J.; Marquette, C. A., Direct Poly(dimethylsiloxane) Functionalization with Vinyl Modified DNA. *Chem. Mater.* **2008**, *20*, 1251–1253.
81. Amouroux, N.; Leger, L., Effect of Dangling Chains on Adhesion Hysteresis of Silicone Elastomers, Probed by JKR Test. *Langmuir* **2005**, *21*, 1396–1401.
82. Mikrut, M.; Noordermeer, J. W. M.; Verbeek, G., The Influence of Loose and Semianchored Siloxane Polymer Chains on the Tack of Crosslinked Silicone Rubber. *J. Appl. Polym. Sci.* **2009**, *114*, 1357–1364.
83. Jancelewicz, J.; Nowaczyk, G.; Makrocka-Rydzyk, M.; Wypych, A.; Fojud, Z.; Jurga, S.; Maciejewski, H., Molecular Dynamics in Grafted Polydimethylsiloxanes. *J. Non-Cryst. Solids* **2010**, *356*, 669–675.
84. Chen, I.-J.; Lindner, E., The Stability of Radio-Frequency Plasma-Treated Polydimethylsiloxane Surfaces. *Langmuir* **2007**, *23*, 3118–3122.
85. Olander, B.; Wirsen, A.; Albertsson, A.-C., Silicone Elastomers with Controlled Surface Composition Using Argon or Hydrogen Plasma Treatment. *J. Appl. Polym. Sci.* **2003**, *90*, 1378–1383.
86. Olander, B.; Wirsen, A.; Albertsson, A.-C., Oxygen Microwave Plasma Treatment Silicone Elastomer: Kinetic Behavior and Surface Composition. *J. Appl. Polym. Sci.* **2003**, *91*, 4098–4104.
87. Befahy, S.; Lipnik, P.; Pardoen, T.; Nascimento, C.; Patris, B.; Bertrand, P.; Yunus, S., Thickness and Elastic Modulus of Plasma Treated PDMS Silica-Like Surface Layer. *Langmuir* **2010**, *26*, 3372–3375.
88. Rahman, M.; Amsarani, R.; Mooney, D. A.; MacElroy, J. M. D.; Dowling, D. P., Effect of Process Parameters on Chemistry, Growth Rate and Nano-Sized Particulate Formation of Atmospheric Plasma Deposited, nm Thick Siloxane Coatings. *J. Nanosci. Nanotech.* **2009**, *9*, 3506–3513.
89. Roucoules, V.; Ponche, A.; Geissler, A.; Siffer, F.; Vidal, L.; Ollivier, S.; Vallat, M. F.; Marie, P.; Voegel, J. C.; Schaaf, P.; Hemmerle, J., Changes in Silicone Elastomeric Surface Properties under Stretching Induced by Three Surface Treatments. *Langmuir* **2007**, *23*, 13136–13145.
90. Schneider, M. H.; Tran, Y.; Tabeling, P., Benzophenone Absorption and Diffusion in Poly(dimethylsiloxane) and Its Role in Graft Photo-polymerization for Surface Modification. *Langmuir* **2011**, *27*, 1232–1240.
91. Bae, W.-S.; Urban, M. W., Reactions of Antimicrobial Species to Imidazole-Microwave Plasma Reacted Poly(dimethylsiloxane) Surfaces. *Langmuir* **2004**, *20*, 8372–8378.
92. Bae, W.-S.; Convertine, A. J.; McCormick, C. L.; Urban, M. W., Effect of Sequential Layer-by-Layer Surface Modifications on the Surface Energy of Plasma-Modified Poly(dimethylsiloxane). *Langmuir* **2007**, *23*, 667–672.
93. Geissler, A.; Vallat, M.-F.; Vidal, L.; Voegel, J. C.; Hemmerle, J.; Schaaf, P.; Roucoules, V., Chemical Force Titration of Plasma Polymer-Modified PDMS Substrates by Using Plasma Polymer-Modified AFM Tips. *Langmuir* **2008**, *24*, 4874–4880.
94. Noordermeer, J. W. M.; Datta, R. N.; Dierkes, W. K.; Guo, R.; Mathew, T.; Talma, A. G.; Tiwari, M.; W., V. O., Surface Modification of Fillers and Curatives by Plasma Polymerization for Enhanced Performance for Single Rubbers and Dissimilar Rubber/Rubber Blends. *Adv. Polym. Sci.* **2011**, *239*, 167–218.
95. Barbier, V.; Tatoulian, M.; Li, H.; Arefi-Khonsari, F.; Adjari, A.; Tabeling, P., Stable Modification of PDMS Surface Properties by Plasma Polymerization:

Application to the Formation of Double Emulsions in Microfluidic Systems. *Langmuir* **2006**, *22*, 5230–5232.
96. Weibel, D. E., *Polymer Assisted Surface Modification by Photons*. Nova Science: New York, 2011.
97. Egitto, F. D.; Matienzo, L. J., Transformation of Poly(dimethylsiloxane) into Thin Surface Films of SiO_x by UV/Ozone Treatment. Part I: Factors Affecting Modification. *J. Mater. Sci.* **2006**, *41*, 6362–6373.
98. Matienzo, L. J.; Egitto, F. D., Transformation of Poly(dimethylsiloxane) into Thin Surface Films of SiO_x by UV/Ozone Treatment. Part II: Segregation and Modification of Doped Polymer Blends. *J. Mater. Sci.* **2006**, *41*, 6374–6384.
99. Efimenko, K.; Crowe, J. A.; Manias, E.; Schwark, D. W.; Fischer, D. A.; Genzer, J., Rapid Formation of Soft Hydrophilic Silicone Elastomer Surfaces. *Polymer* **2005**, *46*, 9329–9341.
100. Fu, Y.-J.; Qui, H.-Z.; Liao, K.-S.; Lue, S. J.; Hu, C.-C.; Lee, K.-R.; Lai, J.-Y., Effect of UV-Ozone Treatment on Poly(dimethylsiloxane) Membranes: Surface Characterization and Gas Separation Performance. *Langmuir* **2010**, *26*, 4392–4399.
101. Satriano, C.; Carnazza, S.; Guglielmino, S.; Marletta, G., Differential Cultured Fibroblast Behavior on Plasma and Ion-Beam-Modified Polysiloxane Surfaces. *Langmuir* **2002**, *18*, 9469–9475.
102. Hwang, B. H.; Choi, C. J.; Jo, M. K.; Kim, J. B.; Choe, H. M.; Chao, S. S.; Kim, Y. S.; Baik, H. K., Delicate Modification of Poly(dimethylsiloxane) Ultrathin Film by Low-Energy Ion Beam Treatment for Durable Intermediate Liquid Crystal Pretilt Angles. *Langmuir* **2010**, *26*, 5072–5076.
103. Mehta, S. C.; Somasundaran, P., Mechanism of Stabilization of Silicone Oil-Water Emulsions Using Hybrid Siloxane Polymers. *Chem. Mater.* **2008**, *20*, 4558–4563.
104. Wolf, L.; Hoffmann, H.; Watanabe, K.; Okamoto, T., Microemulsions from Silicone Oil with an Anionic/Nonionic Surfactant Mixture. *Phys. Chem. Chem. Phys.* **2011**, *13*, 3248–3256.
105. Xiao, D.; Zhang, H.; Wirth, M., Chemical Modification of the Surface of Poly(dimethylsiloxane) by Atom-Transfer Radical Polymerization of Acrylamide. *Langmuir* **2002**, *18*, 9971–9976.
106. Vaidya, A.; Chaudhury, M. K., Synthesis and Surface Properties of Environmentally Responsive Segmented Polyurethanes. *J. Coll. Interface Sci.* **2002**, *249*, 235–245.
107. Majumdar, P.; Webster, D. C., Influence of Solvent Composition and Degree of Reaction on the Formation of Surface Microtopography in a Thermoset Siloxane-Urethane System. *Polymer* **2006**, *46*, 4172–4181.
108. Majumdar, P.; Webster, D. C., Surface Microtopography in Siloxane-Polyurethane Thermosets: The Influence of Siloxane and Extent of Reaction. *Polymer* **2007**, *48*, 7499–7509.
109. Spanos, C. G.; Ebbens, S. J.; Badyal, J. P. S., Surface Segregation and Plasma Oxidation of Poly(dimethylsiloxane)-Doped Polyolefins. *Macromolecules* **2001**, *34*, 8149–8155.
110. Zhu, S.-H.; McManus, N. T.; Tzoganakis, C.; Penlidis, A., Effect of a Poly(dimethylsiloxane) Modified Polyolefin Additive on the Processing and Surface Properties of LLDPE. *Polym. Eng. Sci.* **2007**, *47*, 1309–1316.
111. Ponjee, M. W. G.; Reijme, M. A.; Gon, A. W. D. V. D.; Brongersma, H. H.; Langeveld-Voss, B. M. W., Intermolecular Segregation of Siloxanes in P3HT:

Surface Quantification and Molecular Surface-Structure. *Polymer* **2002**, *43*, 77–85.
112. Jukarainen, H.; Clarson, S. J.; Seppala, J. V., An Investigation of the Surface Properties and Phase Behavior of Poly(dimethylsiloxane-*b*-ethyleneoxide) Block Copolymers. In *Silicones and Silicone-Modified Materials*, Clarson, S. J.; Fitzgerald, J. J.; Owen, M. J.; Smith, S. D., Eds. American Chemical Society: Washington, DC, 2000; Vol. 729, pp 353–358.
113. Rimdusit, S.; Benjapan, W.; Assabumrungrat, S.; Takeichi, T.; Yokota, R., Surface Segregation of Siloxane Containing Component in Polysiloxane-Block-Polyimide and s-BPDA/ODA Polyimide Blends. *Polym. Eng. Sci.* **2007**, *47*, 489–498.
114. Cazacu, M.; Racles, C.; Alexandru, M.; Ignat, M.; Ioanid, A.; Vlad, A., Morphology and Surface Properties of Some Siloxane-Organic Copolymers. *Polym. Int.* **2008**, *58*, 697–702.
115. Krol, P.; Byczynski, L., Chemical Structure, Thermal Properties, and Free-Surface Energy Parameters of Coatings Synthesized from Poly(urethane-dimethylsiloxane) Anionomers. *J. Appl. Polym. Sci.* **2008**, *110*, 3488–3500.
116. Van Poll, M. L.; Khodabakhsh, S.; Brewer, P. J.; Shard, A. G.; Ramstedt, M.; Huck, W. T. S., Surface Modification of PDMS via Self-Organization of Vinyl-Terminated Small Molecules. *Soft Matter* **2009**, *5*, 2286–2293.
117. Qian, Z.; Minnikanti, V. S.; Archer, L. A., Surface Segregation of Highly Branched Polymer Additives in Linear Hosts. *J. Polym. Sci., Part B: Polym. Phys.* **2008**, *46*, 1788–1801.
118. Hillborg, H.; Tomczak, N.; Olah, A.; Schonherr, H.; Vancso, G. J., Nanoscale Hydrophobic Recovery: A Chemical Force Microscopy Study of UV/Ozone-Treated Cross-Linked Poly(dimethylsiloxane). *Langmuir* **2004**, *20*, 785–794.
119. Meincken, M.; Berhane, T. A.; Mallon, P. E., Tracking the Hydrophobic Recovery of PDMS Compounds Using the Adhesive Force Determined AFM Force Distance Measurements. *Polymer* **2005**, *46*, 203–208.
120. Hillborg, H.; Tomczak, N.; Olah, A.; Schonherr, H.; Vansco, C. J., Nanoscale Hydrophobic Recovery: A Chemical Force Microscopy Study of UV/Ozone-Treated Cross-Linked Poly(dimethylsiloxane). *Langmuir* **2004**, *20*, 785–794.
121. Chen, C.; Wang, J.; Chen, Z., Surface Restructuring Behavior of Various Types of Poly(dimethylsiloxane) in Water Detected by SFG. *Langmuir* **2004**, *20*, 10186–10193.
122. Yang, Y.; Urban, M. W., Self-Healing Polymeric Materials. *Chem. Soc. Rev.* **2013**, *42*(17), 7446–7467.
123. Keller, M. W.; White, S. R.; Sottos, N. R., A Self-Healing Poly(dimethylsiloxane) Elastomer. *Adv. Funct. Mater.* **2007**, *17*, 2399–2404.
124. Ragheb, A. M.; Brook, M. A.; Hrynyk, M., Highly Active, Lipase Silicone Elastomers. *Biomaterials* **2005**, *26*, 1653–1664.
125. Ragheb, A. M.; Hileman, O. E.; Brook, M. A., The Use of Poly(ethylene oxide) for the Efficient Stabilization of Entrapped α-Chymotripsin in Silicone Elastomers: A Chemometric Study. *Biomaterials* **2005**, *26*, 1653–1664.
126. Robbins, M. E.; Oh, B. K.; Hopper, E. D.; Schoenfisch, M. H., Nitric Oxide-Releasing Xerogel Microarrays Prepared with Surface-Tailored Poly(dimethylsiloxane) Templates. *Chem. Mater.* **2005**, *17*, 3288–3296.
127. Langer, R., Where a Pill Won't Reach. *Sci. Am.* **2003**, *288*(4), 50–57.

128. Gao, Z.; Nahrup, J. S.; Mark, J. E.; Sakr, A., Poly(dimethylsiloxane) Coatings for Controlled Drug Release. III. Drug Release Profiles and Swelling Properties of the Free-Standing Films. *J. Appl. Polym. Sci.* **2004**, *95*, 494–501.
129. Hoare, T. R.; Kohane, D. S., Hydrogels in Drug Delivery: Progress and Challenges. *Polymer* **2007**, *49*, 1993–2007.
130. Bartzoka, V.; McDermott, M. R.; Brook, M. A., Protein-Silicone Interactions. *Adv. Mater.* **1999**, *11*, 257–259.
131. Bassindale, A. R.; Codina-Barrios, A.; Frascione, N.; Taylor, P. G., The Use of Silsesquioxane Cages and Phage Display Technology to Probe Silicone-Protein Interactions. *New J. Chem.* **2008**, *32*, 240–246.
132. Guoyong, W.; Zhiping, D.; Qiuxiao, L.; Wei, Z., Carbohydrate-Modified Siloxane Surfactants and Their Adsorption and Aggregation Behavior in Aqueous Solution. *J. Phys. Chem. C* **2010**, *114*, 6872–6877.
133. Chen, H.; Brook, M. A.; Chen, Y.; Sheardown, H., Surface Properties of PEO-Silicone Composites: Reducing Protein Adsorption. *J. Biomater. Sci., Polymer Ed.* **2005**, *16*, 531–548.
134. Xia, N.; Hu, Y.; Grainger, D. W.; Castner, D. C., Functionalized Poly(ethylene glycol)-Grafted Polysiloxane Monolayers for Control of Protein Binding. *Langmuir* **2002**, *18*, 3255–3262.
135. Chen, H.; Zhang, Z.; Chen, Y.; Brook, M. A.; Sheardown, H., Protein Repellant Silicone Surfaces by Covalent Immobilization of Poly(ethylene oxide). *Biomaterials* **2005**, *26*, 2391–2399.
136. Lee, J. N.; Jianag, X.; Ryan, D.; Whitesides, G. M., Compatibility of Mammalian Cells on the Surfaces of Poly(dimethylsiloxane). *Langmuir* **2004**, *20*, 11684–11691.
137. Patrito, N.; McCague, C.; Norton, P. R.; Petersen, N. O., Spatially Controlled Cell Adhesion via Micropatterned Surface Modification of Poly(dimethylsiloxane). *Langmuir* **2007**, *23*, 715–719.
138. Yuan, Y.; Zang, X.; Ai, F.; Zhou, J.; Shen, J.; Lin, S., Grafting Sulfobetaine Monomer Onto Silicone Surface to Improve Haemocompatibility. *Polym. Int.* **2004**, *53*, 121–126.
139. Jiang, X.; Takayama, S.; Qian, X.; Ostuni, E.; Wu, H.; Bowden, N.; LeDuc, P.; Ingber, D. E.; Whitesides, G. M., Controlling Mammalian Cell Spreading and Cytoskeletal Arrangement with Conveniently Fabricated Continuous Wavy Features on Poly(dimethylsiloxane). *Langmuir* **2002**, *18*, 3273–3280.
140. Keranov, I.; Vladkova, T.; Minchev, M.; Kostadinova, A.; Altankov, G., Preparation, Characterization, and Cellular Interactions of Collagen-Immobilized PDMS Surfaces. *J. Appl. Polym. Sci.* **2008**, *110*, 321–330.
141. Iwasaki, Y.; Omichi, Y.; Iwata, R., Site-Specific Dense Immobilization of Antibody Fragments on Polymer Brushes Supported by Silicone Nanofilaments. *Langmuir* **2008**, *24*, 8427–8430.
142. Jiang, H.; Manolache, S.; Wong, A. C.; Denes, F. S., Synthesis of Dendrimer-Type Poly(ethylene glycol) Structures from Plasma-Functionalized Silicone Rubber Surfaces. *J. Appl. Polym. Sci.* **2006**, *102*, 2324–2337.
143. Burnell, T.; Carpenter, J.; Truby, K.; Serth-Guzzo, J.; Stein, J.; Weibe, D., Advances in Non-Toxic Silicone Biofouling Release Coatings. In *Silicones and Silicone-Modified Materials*, Clarson, S. J.; Fitzgerald, J. J.; Owen, M. J.; Smith, S. D., Eds. American Chemical Society: Washington, DC, 2000; Vol. 729, pp 180–193.

144. Krishnan, S.; Weinman, C. J.; Ober, C. K., Advances in Polymers for Anti-Biofouling Surfaces. *J. Mater. Chem.* **2008**, *18*, 3405–3413.
145. Fang, J.; Kelarakis, A.; Wang, D.; Giannelis, E. P.; Finlay, J. A.; Callow, M. E.; Callow, J. A., Fouling Release Nanostructured Coatings Based on PDMS-Polyurea Segmented Copolymers. *Polymer* **2010**, *51*, 2636–2642.
146. Elzein, T.; Galliano, A.; Bistac, S., Chain Anisotropy in PDMS Networks Due to Friction on Gold Surfaces. *J. Polym. Sci., Part B: Polym. Phys.* **2004**, *42*, 2348–2353.
147. Honig, C. D. F.; Ducker, W. A., Squeeze Film Lubrication in Silicone Oil: Experimental Test of the No-Slip Boundary Condition at Solid-Liquid Surfaces. *J. Phys. Chem. C* **2008**, *112*, 17324–17330.
148. Hsiao, E.; Kim, D.; Kim, S. H., Effects of Ionic Side Groups Attached to Polydimethylsiloxanes on Lubrication of Silicon Oxide Surfaces. *Langmuir* **2009**, *25*, 9814–9823.
149. Vorvolakes, K.; Chaudhury, M. K., The Effects of Molecular Weight and Temperature on the Kinetic Friction of Silicone Rubbers. *Langmuir* **2003**, *19*, 6778–6787.
150. Yurdumakan, B.; Nanjundiah, K.; Dhinojwala, A., Origin of Higher Friction for Elastomers Sliding on Glassy Polymers. *J. Phys. Chem. B* **2007**, *111*, 960–965.
151. Woolfson, A. D.; Malcolm, R. K.; Gorman, S. P.; Jones, D. S.; Brown, A. F.; McCullagh, S. D., Self-Lubricating Silicone Elastomer Biomaterials. *J. Mater. Chem.* **2003**, *13*, 2465–2470.
152. McCullagh, S. D.; Malcolm, R. K.; Woolfson, A. D.; Gorman, S. P.; Jones, D. S.; Cuddy, J., Release Kinetics of Olelyl Alcohol from a Self-Lubricating Silicone Biomaterial. *J. Mater. Chem.* **2004**, *14*, 1093–1098.
153. Awaja, F.; Gilbert, M.; Kelly, G.; Fox, B.; Pigram, P. J., Adhesion on Polymers. *Prog. Polym. Sci.* **2009**, *33*, 948–968.
154. Bowen, J.; Cheneler, D.; Andrews, J. W.; Avery, A. R.; Zhang, Z.; Ward, M. C. L.; Adams, M. J., Application of Colloidal Probe Atomic Force Spectroscopy to the Adhesion of Thin Films of Viscous and Viscoelastic Silicone Fluids. *Langmuir* **2011**, *27*, 11489–11500.
155. Vaenkatesan, V.; Li, Z.; Vellinga, W.-P.; De Jeu, W. H., Adhesion and Friction Behaviours of Polydimethylsiloxane—A Fresh Perspective on JKR Measurements. *Polymer* **2006**, *46*, 8317–8325.
156. Ahn, D.; Shephard, N. E.; Olney, P. A.; McMillan, C. S., Thermal Gradient Enabled XPS Analysis of PDMS Elastomer Adhesion to Polycarbonate. *Macromolecules* **2007**, *40*, 3904–3906.
157. Buzio, R.; Bosca, A.; Krol, S.; Marchetto, D.; Valeri, S.; Valbusa, U., Deformation and Adhesion of Elastomer Poly(dimethylsiloxane) Colloidal AFM Probes. *Langmuir* **2007**, *23*, 9293–9302.
158. Morashita, Y.; Morita, H.; Kaneko, D.; Doi, M., Contact Dynamics in the Adhesion Process Between Spherical Polydimethylsiloxane Rubber and Glass Substrate. *Langmuir* **2008**, *24*, 14059–14065.
159. Millare, B.; Thomas, M.; Ferreira, A.; Xu, H.; Holesinger, M.; Vullev, V. I., Dependence of the Quality of Adhesion Between Poly(dimethylsiloxane) and Glass Surfaces. *Langmuir* **2008**, *24*, 13218–13224.
160. Achyuta, A. K. H.; White, A. J.; Lewis, H. G. P.; Murthy, S. K., Incorporation of Linear Spacer Molecules in Vapor-Deposited Silicone Polymer Thin Films. *Macromolecules* **2009**, *42*, 1970–1978.

161. Lauke, B., Determination of Adhesion Strength Between a Coated Particle and Polymer Matrix. *Comput. Sci. Technol.* **2006**, *66*, 3153–3160.
162. Roth, J.; Albrecht, V.; Nitschke, M.; Bellmann, C.; Simon, F.; Zschoche, S.; Michel, S.; Luhmann, C.; Grundke, K.; Voit, B., Self Functionalization of Silicone Rubber for Permanent Adhesion Improvement. *Langmuir* **2008**, *24*, 12603–12611.
163. Lee, N. Y.; Chung, B. H., Novel Poly(dimethylsiloxane) Bonding Strategy via Room Temperature "Chemical Gluing." *Langmuir* **2009**, *25*, 3861–3866.
164. Verdier, C.; Ravilly, G., Peeling of Polydimethylsiloxane Adhesives: The Case of Adhesive Failure. *J. Polym. Sci., Polym. Phys. Ed.* **2007**, *45*, 2113–2122.
165. Juza, J., Thermodynamic Study of Surfaces of Liquid Polybutadienes and Their Interfaces with Poly(dimethylsiloxane). *J. Appl. Polym. Sci.* **2009**, *113*, 169–180.
166. Fang, T.-H.; Wong, T. H.; Kang, S. H., Nanomechanical and Surface Behavior of Polydimethylsiloxane Filled Nanoporous Anodic Alumina. *J. Mater. Sci.* **2009**, *44*, 1588–1593.
167. Yamada, S., Layering Transitions and Tribology of Molecularly Thin Films of Poly(dimethylsiloxane). *Langmuir* **2003**, *19*, 7399–7405.
168. Yilgor, E.; Sinmazcelik, T.; Yilgor, I., Modification of Polyolefins with Silicone Copolymers. II. Thermal, Mechanical and Tribological Behavior of PP and HDPE Blended with Silicone Copolymers. *J. Appl. Polym. Sci.* **2002**, *84*, 535–540.
169. Zhang, S. L.; Tsou, A. H.; Li, C. M., Scratching Behavior of Elastomeric Poly(dimethylsiloxane) Coatings. *J. Polym. Sci., Part B: Polym. Phys.* **2002**, *40*, 1530–1537.
170. Tomalia, D. A.; Frechet, J. M. J., Discovery of Dendrimers and Dendritic Polymers: A Brief Historical Perspective. *J. Polym. Sci., Part A: Polym. Chem.* **2002**, *40*, 2719–2728.
171. Frechet, J. M. J., Dendrimers and Other Dendritic Macromolecules: From Building Blocks to Functional Assemblies in Nanoscience and Nanotechnology. *J. Polym. Sci., Part A: Polym. Chem.* **2003**, *41*, 3713–3725.
172. Liang, C.; Frechet, J. M. J., Applying Key Concepts from Nature: Transition State Stabilization, Pre-concentration and Cooperativity Effects in Dendritic Biomimetics. *Prog. Polym. Sci.* **2005**, *30*, 385–402.
173. Tomalia, D. A., Birth of a New Macromolecular Architecture: Dendrimers as Quantized Building Blocks for Nanoscale Synthetic Polymer Chemistry. *Prog. Polym. Sci.* **2005**, *30*, 294–324.
174. Muzafarov, A. M.; Rebrov, E. A., From the Discovery of Sodiumoxyorganoalkoxysilanes to the Organosilicon Dendrimers. *J. Polym. Sci., Part A: Polym. Chem.* **2008**, *46*(2726–2733), 4935–4948.
175. Fadeev, A. Y.; Kazakevich, Y. V., Covalently Attached Monolayers of Oligo(dimethylsiloxane)s on Silica: A Siloxane Chemistry Approach for Surface Modification. *Langmuir* **2002**, *18*, 2665–2672.
176. Kazakevich, Y. V.; Fadeev, A. Y., Adsorption Characteristics of Oligo(dimethylsiloxane)-Modified Silicas: An Example of Highly Hydrophobic Surfaces with Non-Aliphatic Architecture. *Langmuir* **2002**, *18*, 3117–3122.
177. Song, M. M.; Ye, P. D.; Ivanisevic, A., Elastomeric Nanoparticle Composites Covalently Bound to Al_2O_3/GaAs Surfaces. *Langmuir* **2007**, *23*, 9472–9480.

178. Roman, G. T.; Culbertson, C. T., Surface Engineering of Poly(dimethylsiloxane) Microfluidic Devices Using Transition Metal Sol-Gel Chemistry. *Langmuir* **2006**, *22*, 4445–4451.
179. Shindou, T.; Katayama, S.; Yamada, N.; Kamiya, K., Effect of Composition on Surface Properties of Polydimethylsiloxane-Based Inorganic/Organic Hybrid Films. *J. Sol-Gel Sci. Technol.* **2004**, *30*, 229–237.
180. Girshevitz, O.; Nitzan, Y.; Sukenik, C. N., Solution-Deposited Amorphous Titanium Dioxide on Silicone Rubber: A Conformal, Crack-Free Antibacterial Coating. *Chem. Mater.* **2008**, *20*, 1390–1396.
181. Xiao, X.-C.; Zhuo, R.-X.; Xu, J.; Liu, B.; Liu, J.-Y., Preparation of Microspheres with Silicone Oil Cores and Poly(N-isopropylacrylamide) Shells by Physical Coating. *J. Appl. Polym. Sci.* **2006**, *102*, 5571–5576.
182. Lenz, P.; Ajo-Franklin, C. M.; Boxer, S. G., Patterned Supported Lipid Bilayers and Monolayers on Poly(dimethylsiloxane). *Langmuir* **2004**, *20*, 11092–11099.
183. Diaz-Quijada, G. A.; Wayner, D. D. M., A Simple Approach to Micropatterning and Surface Modification of Poly(dimethylsiloxane). *Langmuir* **2004**, *20*, 9607–9611.
184. Das, A. L.; Mukherjee, R.; Katiyer, V.; Kulkarni, M.; Ghatak, A.; Sharma, A., Generation of Sub-Micrometer-Scale Patterns by Successive Miniaturization Using Hydrogels. *Adv. Mater.* **2007**, *19*, 1943–1946.
185. Lamblet, M.; Verneuil, E.; Vilmin, T.; Boguin, A.; Silberzan, P.; Leger, L., Adhesion Enhancement Through Micropatterning at Polydimethylsiloxane-Acrylic Adhesive Interfaces. *Langmuir* **2007**, *23*, 6966–6974.
186. Arius, G. R. J.; Jung, S.; Zimmermann, J.; Gautschi, H.-P.; Marquardt, K.; Seeger, S., Silicone Nanofilaments and Their Application as Superhydrophobic Coatings. *Adv. Mater.* **2006**, *18*, 2758–2762.
187. Zimmermann, J.; Rabe, M.; Artus, G. R. J.; Seeger, S., Patterned Superfunctional Surfaces Based on a Silicone Nanofilament Coating. *Soft Matter* **2008**, *4*, 450–452.
188. Zimmermann, J.; Rabe, M.; Verdes, D.; Seeger, S., Functionalized Silicone Nanofilaments: A Novel Material for Selective Protein Enrichment. *Langmuir* **2008**, *24*, 1053–1057.
189. Miller, M. S.; Davidson, G. J. E.; Sahli, B. J.; Mailloux, C. M.; Carmichael, T. B., Fabrication of Elastomeric Wires by Selective Electroless Metallization of Poly(dimethylsiloxane). *Adv. Mater.* **2008**, *20*, 59–64.

CHAPTER 7

Elastomeric Networks

7.1 NETWORK FORMATION AND SOME ELASTOMERIC QUANTITIES

7.1.1 Gelation

Gelation is the cross-linking process that leads to the network structures required for rubberlike elasticity.[1–5] In some cases, gelation can be reversible.[6–8] There have been numerous studies involving theory and simulations exploring gelation and the mechanical properties of the resulting networks.[9–20]

Cross linking with free radicals is still quite common.[21–27] Radiation has often been used to carry out the cross linking,[28–36] as have new techniques known as "click" chemistry.[8, 37–40] Hydrosilylation is also popular.[41–51] Networks have even been designed with movable cross links.[52] Finally, reactive groups can be placed at the chain ends or within the chains themselves.[53–59]

Related studies have involved polydimethylsiloxane (PDMS)-based organogelators,[60] web-to-pillar transitions of gels,[61] and silica aerogels.[62] There has also been interest in polysiloxanes in interpenetrating hydrogels with high oxygen permeabilities[63] and viscoelastic magnetic gels.[64] Organic-inorganic hybrids with relatively low melting temperatures also exist,[65] some of which can be made to be self-healing.[66–68]

Gels are also formed in swelling experiments,[69, 70] which are useful for equilibrium experiments to characterize network structures.[71]

One of the recent topics in this area involves stimuli-responsive gels,[72–75] under the descriptive title of "self-walking gels,"[76] "wormlike motion of gels,"[77] and "peristaltic motion of gels."[78]

7.1.2 Cross Linking Under Unusual Conditions

The earliest studies of networks formed in solution were undertaken to investigate some subtle aspects of the elastic free energy expression—whether or not an additional term in the logarithm of the volume was required.[79–86]

Other studies focused on the properties of networks in general. As can be gathered from chapter 4, it is difficult to obtain information on the topology of a network. Some studies have therefore taken an indirect approach. Networks were prepared in a way as to simplify their topologies, and their properties were measured and interpreted in terms of reduced degrees of network-chain entanglement. [71, 82, 87–94]

The two techniques employed involved separating the chains prior to cross linking by either dissolution[84] or stretching.[95] After cross linking, the solvent is removed or the stretching force is relaxed, and the network is studied (unswollen) with regard to its stress-strain properties, typically in elongation.[85, 86]

Figure 7.1 schematically shows the preparation of networks by cross linking in solution followed by removal of the solvent.[96] Success in obtaining elastomers with fewer entanglements is supported by the observation that such networks come to elastic equilibrium much more rapidly than elastomers cross linked in the dry state. Table 7.1 shows results on PDMS networks cross linked in solution by means of γ radiation.[84, 97] Note the continual decrease in the time required to reach elastic equilibrium, t_{eq}, and in the extent of stress relaxation as measured by the ratio of equilibrium to initial values of the reduced stress, $[f^*]$, upon decrease in the volume

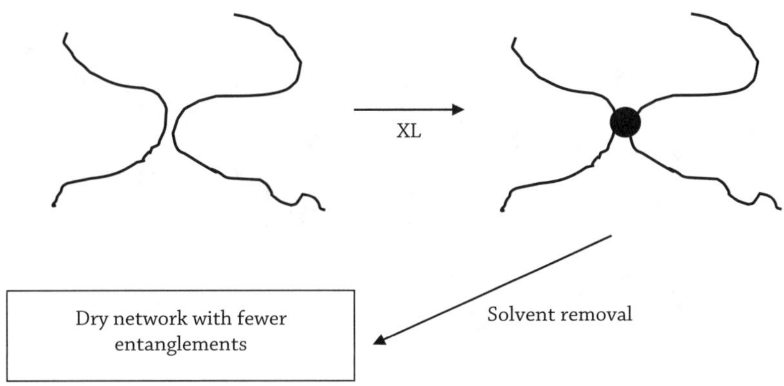

Figure 7.1:
Cross linking in solution to prepare networks of simpler topology. XL is the cross-linking step done in solution.

Table 7.1. PDMS NETWORKS COMPARED AT APPROXIMATELY CONSTANT MODULUS

v_{2S} [a]	t_{eq} (hr)	$[f^*]$(equil)/ $[f^*]$(init)	$2C_2$ (N mm^{-2})
1.00	0.70	0.95	0.062
0.75	0.48	0.98	0.057
0.62	0.10	0.99	0.059
0.55	0.30	0.99	0.062
0.48	0.02	1.00	0.067
0.40	0.03	1.00	0.039
0.30	0.00	1.00	0.031

[a]Volume fraction of polymer in the solution being irradiated during the cross-linking reaction, which is generally somewhat larger than the volume fraction, v_{2c}, actually incorporated in the gel.
Reprinted with permission from J. E. Mark et al., *Physical Properties of Polymers*. Copyright 1984 American Chemical Society.[84, 97]

fraction of polymer present during the cross linking. ($[f^*]$ is defined in section 7.1.3). Figure 7.2 qualitatively explains these observations. When a network is cross linked in solution and the solvent then removed, the chains collapse in such a way that there is reduced overlap in their configurational domains, which reduces chain-junction entangling embodied in the various constraint theories.

Solution cross-linked elastomers also exhibit stress-strain isotherms in elongation that are closer in form to those expected from the simplest molecular theories of rubberlike elasticity.[87] Specifically, there are large decreases in the Mooney-Rivlin $2C_2$ correction constant described in

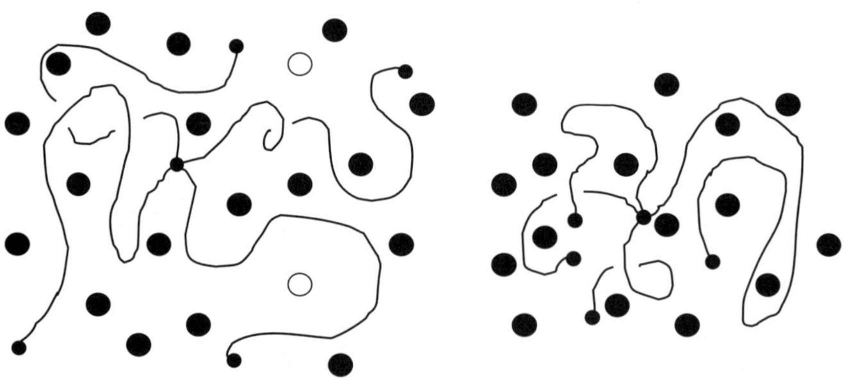

Figure 7.2:
Typical configurations of four chains emanating from a tetrafunctional cross link in a polymer network. In the left sketch, the network was prepared in the undiluted state, and in the right sketch, it was prepared in solution and then dried. When a network is cross linked in solution and the solvent then removed, the chains collapse in such a way that there is reduced overlap in their configurational domains.

figure 7.18. There have also been studies of the segmental orientation[90] and viscoelastic properties[88] of these networks.

In terms of the constrained-junction theory,[71, 96] the constraint parameter κ is reduced for solution cross-linked elastomers, indicating decreased chain-crosslink entangling. Specifically, the theory gives a good account of these results when κ was decreased though a reduction in the volume fraction v_{2c} of polymer present during the cross linking.[89, 91–93] The values of κ obtained are within the range obtained in other comparisons of theory and experiment, as are the values of an additional, relatively unimportant heterogeneity parameter ζ.[98] The values of κ generally decrease with decrease in the volume fraction v_{2c}, and with increase in degree of cross linking as represented by the Mooney-Rivlin constant $2C_1$. The dependence of κ on v_{2c} is significantly stronger than that suggested by theory however, indicating a particularly strong effect of dilution on the degree of network chain interpenetration. Simulations have been carried out to further clarify the structure-property relationships of these elastomers.[99]

Removal of the solvent has the additional effect of putting the chains into a "supercontracted" state (figure 7.3).[96] Experiments on strain-induced crystallization carried out on such solution cross-linked elastomers indicate that the decreased entangling is less important than the supercontraction of the chains. Crystallization required larger values of elongation compared to elastomers cross linked in the dry state.[100, 101] The most recent work in this area has focused on the unusually high extensibilities of such elastomers.[99, 102–107]

Unusual properties are also obtained for networks prepared and studied in the opposite way, specifically cross linking in the dry state and then swelling the network prior to the measurements of mechanical properties.[71] The approach to elastic equilibrium is more rapid and the stress-strain isotherms in elongation are closer to the form predicted by the

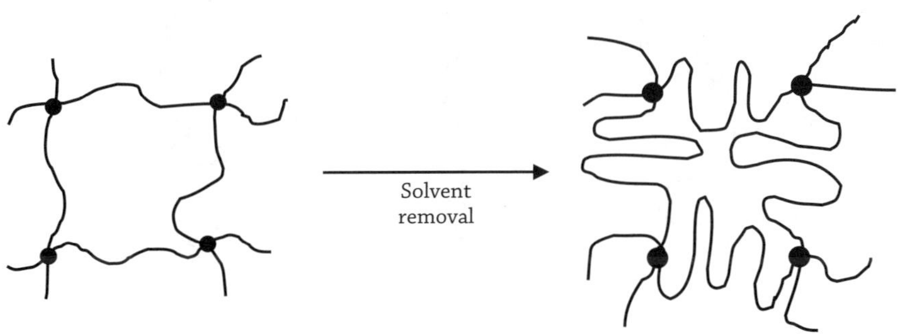

Figure 7.3:
"Supercompression" of the network chains when a network formed in solution is dried.

simplest molecular theories of rubberlike elasticity. Such networks do not have the unusually high extensibilities of networks formed in solution and studied in the dry state.

Networks can also be formed from chains in the deformed state.[71] In this approach a first network is generally introduced in the undeformed state, the resulting elastomer is elongated, and a second network is introduced in the stretched state. Release of the stress permits the network to retract, but the second network of this "double-network" structure prevents retraction down to the original dimensions (figure 7.4). The most interesting feature of the retracted network is the fact that it is anisotropic in structure and properties.

In some cases, double networks have shown increases in orientability and strain-induced crystallization,[108] as well as improved fatigue resistance.[109, 110] In fact, some results show that there may be less of a compromise between failure properties in general and the modulus,[111, 112] which may be due in part to the decreased hysteresis observed for some of these elastomers.[113] There have even been reports of improved thermal stability,[114] although it is hard to visualize how this would occur. Finally, electrical resistivity is more sensitive to strain in carbon-black reinforced double networks.[115] Better molecular understanding of these observations is being sought with, for example, extensive studies of residual strains and birefringence.[116]

Results on double networks in the unswollen state generally track the behavior of single networks. Recent results, however, now indicate that swelling double networks gives gels that have extremely high mechanical strength[117–119] and fracture energy.[120]

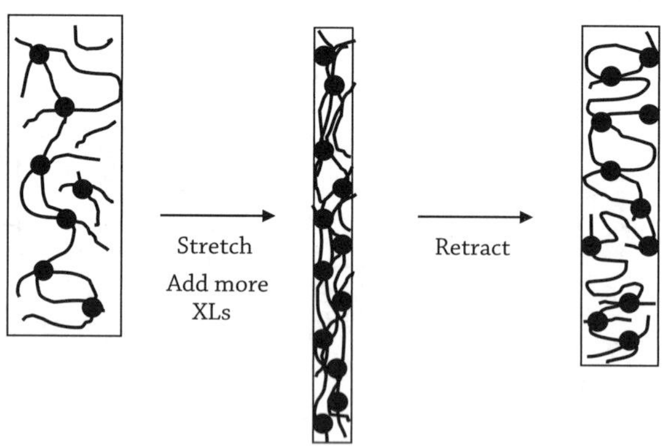

Figure 7.4:
Cross linking in the elongated state to prepare networks of simpler topology.

7.1.3 Some Elastomeric Quantities

The majority of studies of mechanical properties of elastomers have been carried out in elongation because of the simplicity of this type of deformation.[71, 121–123] Results are typically expressed in terms of the nominal or engineering stress $f^* = f/A^*$, which, in the simplest molecular theories, is given by

$$f^* = \nu kT(\alpha - \alpha^{-2}) \tag{1}$$

where ν is the number density of network chains, k is the Boltzmann constant, T is the absolute temperature, and α is the elongation or relative length of the stretched elastomer. The modulus or reduced stress is

$$[f^*] \equiv f^*/(\alpha - \alpha^{-2}) = \nu kT \tag{2}$$

$[f^*]$ is often fitted to the Mooney-Rivlin semi-empirical formula $[f^*] \equiv 2C_1 + 2C_2\alpha^{-1}$, where C_1 and C_2 are constants independent of deformation α.[123, 124] The simplest model assumes affine deformations, in which the cross links move in parallel with the macroscopic dimensions. This approach has been revisited recently.[125] There are a fewer studies using other types of deformation such as biaxial extension or compression, shear, and torsion.

Swelling is a nonmechanical property that is frequently used to characterize elastomeric materials.[71, 124, 126, 127] Swelling is an unusual deformation in that volume changes are of central importance, rather than being negligible. Swelling is a three-dimensional dilation in which the network absorbs solvent, reaching an equilibrium degree of swelling at which the free energy decrease due to the mixing of the solvent with the network chains is balanced by the free energy increase accompanying the stretching of the chains. The network is typically placed into an excess of solvent, which it imbibes until the dilational stretching of the chains prevents further absorption. The degree of cross linking can be determined from equilibrium extent of swelling, provided the polymer-solvent interaction parameter χ is known.[126] Conversely, if the degree of cross linking is known from an independent experiment, then the interaction parameter can be determined. The equilibrium degree of swelling and its dependence on various parameters are important tests of theory.[71, 96]

A number of specialized elastomeric quantities have also been investigated. PDMS networks have been particularly useful in investigating the "Mullins Effect," in which filler-reinforced elastomers exhibit a reduction in stress after the initial extension.[128–130] The same may be said for the "Payne Effect," in which increase in the shear strain of a filled elastomer causes a drop in the elastic modulus.[113, 131, 132] The tensile properties of PDMS[133] and the origins of the failure of silica-filled PDMS elastomers have also been elucidated from first principles, specifically using Car-Parrinello molecular dynamics.[134] In an unusual example of biomimicry, PDMS was molded into arrays of microlens shells that snapped from one curvature (concave) to another curvature (convex) in a way that is reminiscent of the trapping mechanism of the Venus flytrap plant.[135] In some cases, PDMS thin films showed extreme hardening due to high compressive strain and confined thickness.[136]

An example of a property that is less understood in molecular terms is the scratch resistance exhibited by polysiloxane elastomers.[137]

7.2 UNIMODAL MODEL ELASTOMERS

7.2.1 General Approach

Until recently, there was little reliable, quantitative information on the relationship of stress to structure, primarily because of the uncontrolled manner in which elastomeric networks were generally prepared.[3, 96, 124, 126, 138] Segments close together in space were linked irrespective of their locations along the chain trajectories, resulting in a random network structure in which the number and locations of the cross links were unknown. New synthetic techniques are now available, however, for the preparation of "model" polymer networks of known structure. If networks are formed by end linking functionally terminated chains then the required structural information is determined by the cross-linking reaction.[96, 139, 140] The functionality of the cross links is the same as that of the end-linking agent, and the molecular weight distribution between cross links matches that of the chains prior to end linking.

An example is the reaction shown in figure 7.5, in which hydroxyl-terminated chains of a polymer such as poly(dimethylsiloxane) (PDMS) are end-linked using tetraethyl orthosilicate (alternatively called tetraethoxysilane, TEOS). Networks have also been formed from PDMS chains that have either amide or urea groups at both ends.[141, 142] In these studies, hydrogen-bonded polymerizable end groups have been of particular interest. It is also possible to make networks from PDMS chains having a known number of potential cross-linking sites placed as side chains along the polymer backbone, as long as their distribution is known as well.[143]

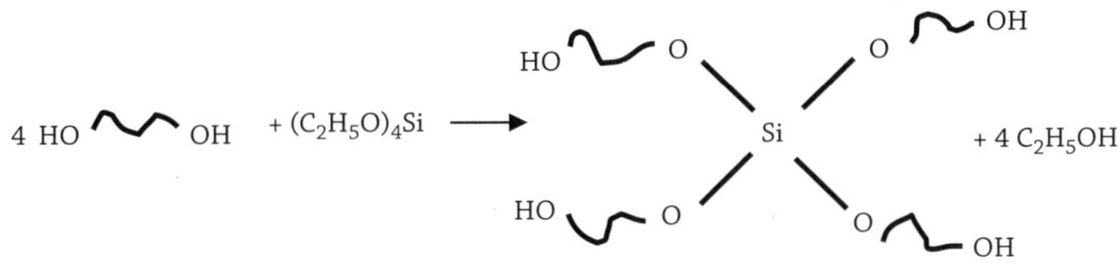

Figure 7.5:
End linking by a condensation reaction between hydroxyl groups at the ends of a polymer chain and the alkoxy groups on a tetrafunctional end-linking agent. The number-average molecular weight M_n of the precursor chains becomes the critically important molecular weight M_c between cross links. The distribution of M_n also characterizes the distribution of M_c (i.e., network chain lengths).

Because of their known structures, such model elastomers are now the preferred materials for the quantitative characterization of rubberlike elasticity. The properties of PDMS networks have been of interest to a variety of groups.[144–147] Such specific cross-linking reactions are also useful in the preparation of some of the liquid-crystalline elastomers,[148, 149] discussed in chapter 3.

7.2.2 Effects of Junction Functionality

Trifunctional and tetrafunctional PDMS networks can be prepared using the end-linking approach. Such networks have been used to test the molecular theories of rubber elasticity with regard to the increase in nonaffineness of the network deformation with increasing elongation. Figure 7.6 shows typical results.[150] The ratio $2C_2/2C_1$ decreases with increased cross-link functionality because cross links connecting four chains are more constrained than those connecting three.[151] Junction fluctuations, which lead to nonaffine behavior at large strain, are reduced for four-functional networks. The decrease in $2C_2/2C_1$ with decreasing network chain molecular weight is due to the fact that there is less configurational interpenetration for short network chains, which means that cross links are more easily displaced under deformation. As a result the deformation is nonaffine even at relatively small deformations.

A more thorough investigation of the effects of cross-link functionality requires more versatile chemical reactions, as illustrated in figure 7.7. Specifically, vinyl-terminated PDMS chains are end linked using a multifunctional silane. In the study summarized in figure 7.8,[152] this reaction was used to prepare PDMS model networks having functionalities ranging from three to eleven, with an unsuccessful attempt to achieve a functionality of thirty-seven. As shown in the figure, values of $2C_2$ relative to $2C_1$ both decrease, for the reasons described in the discussion of figure 7.6.

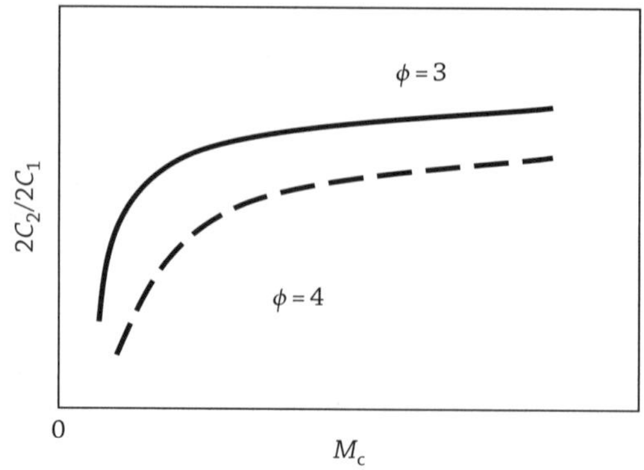

Figure 7.6:
Dependence of Mooney-Rivlin ratio, $2C2/2C1$, on the molecular weight between cross links. The factor $2C_2$ measures the departure from affineness as the elongation increases, and $2C_1$ approximates the high-deformation modulus. The ratio decreases with decrease in network chain molecular weight, and with increase in junction functionality, as predicted by theory.[98]

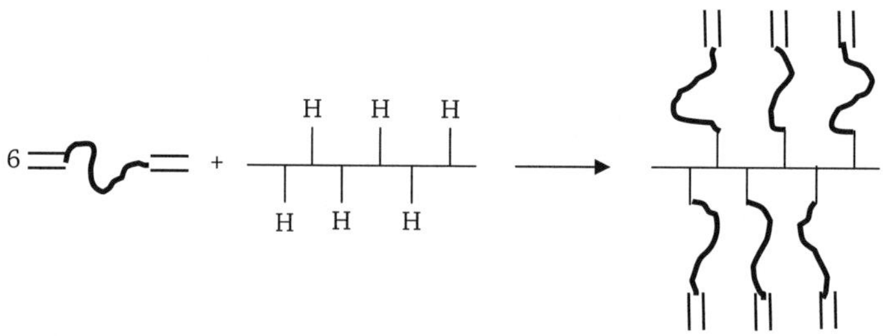

Figure 7.7:
End linking by an addition reaction between vinyl groups at the ends of a polymer chain and the active hydrogen atoms on silicon atoms in an oligomeric poly(methyl hydrogen siloxane). The case shown gives a junction functionality, ϕ, of six.

7.2.3 Effects of Entanglements

Model networks may also be used to test of molecular predictions for the modulus of a network of known degree of cross linking. Some experiments on model networks [10, 96, 153, 154] have given values of the elastic modulus in good agreement with theory. Others [155–160] have given values significantly larger than predicted.[96] The increase in modulus has been attributed to contributions from "permanent" chain entanglements of the type shown in figure 7.9. The issue, however, has not been completely resolved.

Aspects of greatest importance appear to be (i) investigating the effects of cross linking in solution, (ii) studying the effects of swelling on networks cross linked in the bulk (dry) state, (iii) building on the demonstration by

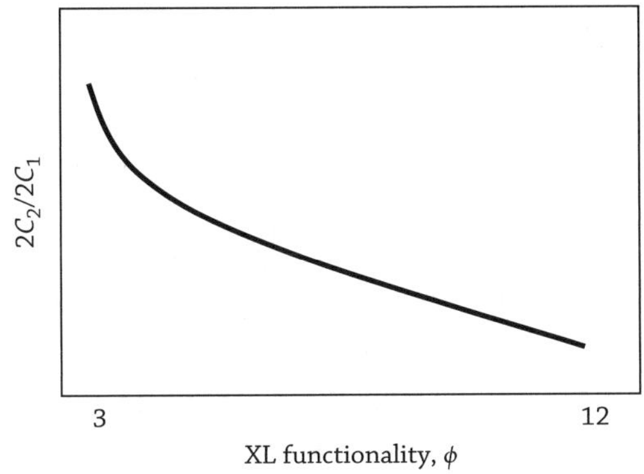

Figure 7.8:
The effect of cross-link functionality on $2C_2/2C_1$.

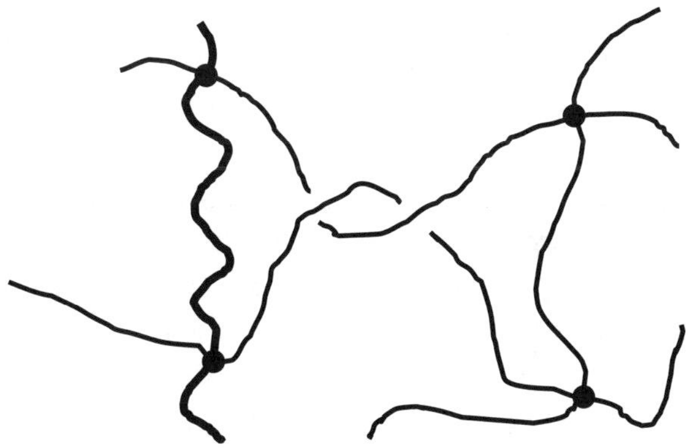

Figure 7.9:
Sketch of an interchain entanglement.

Vilgis and Erman[161] that the constraint models and slip-link models have much in common, (iv) elucidating the effects of cross-link functionality and degree of cross linking, (v) exploring a variety of elastomeric polymers, particularly those having very different values of the plateau modulus,[162] and (vi) generalizing rubber-elasticity models to include viscoelastic effects as well.

For purposes of illustration, it is useful to consider some studies of Oppermann and coworkers[163, 164] on end-linked PDMS networks of different functionalities, since some of the aforementioned issues arose in the analysis of their experimental data.[96] Here, studies of the small-strain shear modulus of end-linked PDMS networks with pentafunctional junctions were carried out as a function of network chain density. The data agree with predictions of the molecular theories at larger values of the chain or junction densities. At lower junction densities however the measured moduli are significantly larger than the predicted ones. Part of the difference, but probably not all, is due to problems in bringing such lightly cross-linked networks to elastic equilibrium in the unswollen state. Finally, the moduli tended to zero as the junction density went to zero as it should since PDMS in the liquid state above T_g.

The departure of the experimental results from the simple molecular theory when the network chains are long indicates the presence of additional contributions to the modulus. The question is whether the departures result from trapped entanglements or simply from the presence of other chains sharing the volume of a given network chain. One important difference between the two types of constraints involves their dependence on network swelling. The localized, permanent entanglements should be independent of swelling, while the more diffuse interchain interactions should decrease with increased swelling. The contributions from trapped entanglements should therefore persist even in highly swollen networks, and therefore contribute to the phantom modulus. For these reasons, it is quite important to carry out measurements of the modulus versus the degree of swelling, with the results at the highest degrees of swelling presumably being least complicated by nonequilibrium effects.

If one makes a network in the extremely dilute state so that there is no chain interpenetration during its formation, the result will be a "phantom-structure state" in which the chains can seemingly pass through one another. Experimentally, network formation in the dilute state decreases both the modulus and its dependence on elongation (the "C_2 effect").[96] It was observed that the number of entanglements resulting from the disperse interpenetration of chains in the cross-linked state were far more numerous than the specific localized points along the chain. Definitive experiments of this type would do much to resolve the nature and importance of chain entanglements in network structures at elastic equilibrium.

7.2.4 Interpretation of Ultimate Properties

Ultimate properties of unfilled elastomers at high elongations reveal how model networks can clarify elastomer properties. An upturn in modulus (as in figure 7.18-A) is frequently exhibited by elastomers at very high elongation.[121, 124, 126, 165] The upturn is very important since it corresponds to a significant toughening of the elastomer; its molecular origin, however, has been the source of some controversy.[166] The upturn has been widely attributed to the "limited extensibility" of the network chains—that is, to an inadequacy in the Gaussian distribution function. Specifically the Gaussian approximation does not assign a zero probability to a configuration unless the end-to-end separation is infinite. Since the increase in modulus had generally been observed only in networks that can undergo strain-induced crystallization, however, the increase could be due to the crystallites acting as additional cross links.

Reinforcement resulting from strain-induced crystallization is identified by the fact that the higher the temperature, the lower the extent of crystallization and the worse the ultimate properties. The effects of an increase in swelling were found to parallel those for an increase in temperature, as expected, since diluent also suppresses network crystallization.[167, 168] On the other hand, in those cases where the upturns are due to limited chain extensibility, increased temperature has little effect on the upturns.[96] Also, in these cases swelling can even make the upturns more pronounced because of the already-imposed stretching of the chains from the dilational effects of the swelling. Thus, the effects of temperature or swelling represent a way to determine whether the upturns are due to strain-induced crystallization or to a non-Gaussian contribution arising from limited-chain extensibility.

Attempts were made to observe upturns from non-Gaussian effects in noncrystallizable networks using end-linked, noncrystallizable model PDMS networks described section 5.3. These networks have high extensibilities, presumably because of the very low incidence of dangling-chain network irregularities. Extensibility is particularly high when networks are prepared from a mixture of very short chains (around a few hundred g mol^{-1}) with long chains (around 18,000 g mol^{-1}), giving a bimodal distribution of network chain lengths.[169] Apparently the short chains are important because of their limited extensibilities, and the long chains because of their ability to retard the rupture process. Such “bimodal” model networks are discussed further in section 7.3.2

Stress-strain measurements on bimodal PDMS networks exhibited upturns in modulus that were much less pronounced than those in crystallizable polymer networks. The upturns are independent of temperature and are not diminished by incorporation of solvent. These characteristics are expected for the case of limited chain extensibility.[166, 170]

7.2.5 Dangling-Chain Networks

Since dangling chains are imperfections in a network structure, their presence should have a detrimental effect on ultimate properties such as the tensile strength, as gauged by the nominal stress, f^*, at rupture, f^*_r.

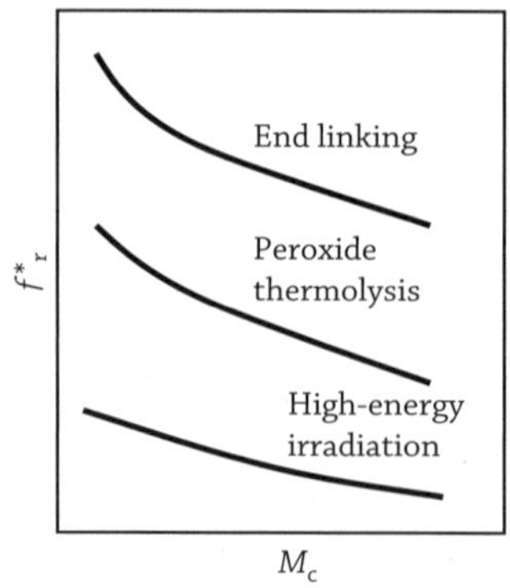

Figure 7.10:
Values of the ultimate strength shown as a function of the molecular weight between cross links for three cross-linking methods.

This expectation is confirmed by an extensive work on tetrafunctionally cross-linked PDMS networks. Some pertinent results are shown schematically as a function of the molecular weight between cross-links in figure 7.10.[171] The largest values f^*_r were observed for the networks prepared by selectively joining functional groups occurring as chain ends (or as side groups) on the chains, which leads to low incidence of dangling ends. As expected, the lowest values of the ultimate properties generally occur for the networks cured by radiation (ultraviolet light, high-energy electrons, and γ radiation). The peroxide-cured networks are intermediate to these two extremes, with the ultimate properties presumably depending on whether or not the free radicals generated by the peroxide are sufficiently reactive to cause some chain scission. Similar results were obtained for the maximum extensibility. These observations are certainly interesting, but they are somewhat deficient in that information on the number of dangling ends is generally not available.

Quantitative information has been obtained using the specific chemical reactions used to form ideal elastomers but modified to prepare intentionally nonideal networks containing known numbers and lengths of dangling-chain irregularities (figure 7.11).[172] If more chain ends are present than reactive groups on the end-linking molecules, then dangling ends result and their number is directly determined by the stoichiometric imbalance. The length of the dangling ends is of necessity the same as those of the elastically effective chains, as shown in part *a* of the figure. This constraint can be removed by separately preparing monofunctionally-terminated chains of any desired lengths and attaching them as shown in part *b*.

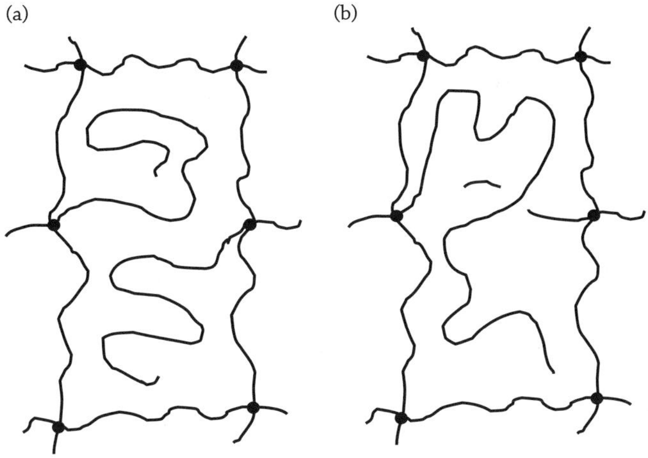

Figure 7.11:
Two end-linking techniques for preparing networks with known numbers and lengths of dangling chains. In part *a*, the dangling chains are produced by having more chain ends than complementary groups on the end-linking agent, and in part *b* by preparing some chains to have reactive groups on only one of the two chain ends.

More definitive come from a series of model networks prepared by end linking vinyl-terminated PDMS chains.[171, 173] The tetrafunctional end-linking agent was varied at levels below stoichiometric balance between its active hydrogen atoms and the chains' terminal vinyl groups. The ultimate properties of these networks, with known numbers of dangling ends, were then compared with those obtained on networks with negligible numbers of these irregularities.[171] Values of f^*_r of the networks containing the dangling ends was lower than that of the more perfect networks, with the largest differences occurring at high proportions of dangling ends (low $2C_1$), as expected.[171] These results, shown schematically in figure 7.12, thus confirm the results shown in figure 7.10. Maximum extensibility shows a similar dependence, as expected.

7.2.6 Interpenetrating Networks

If two types of chains have different end groups, then it is possible to end link them simultaneously into two networks that *interpenetrate* one another.[174] Such a network (figure 7.13) could, for example, be made by reacting hydroxyl-terminated PDMS chains with tetraethoxysilane (in a condensation reaction), while reacting vinyl-terminated PDMS chains

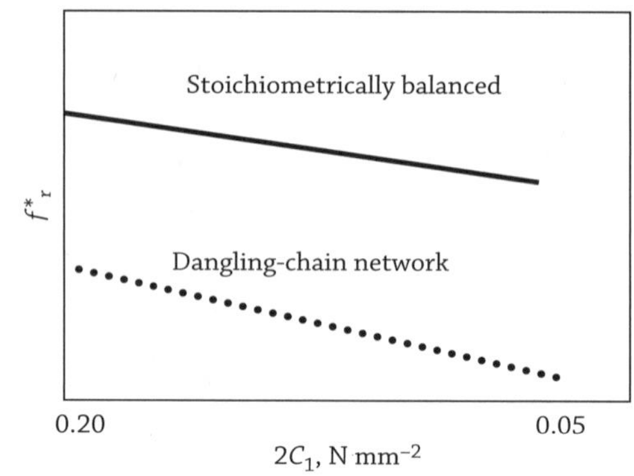

Figure 7.12:
The ultimate strength as a function of the high-deformation modulus for networks that either contain a negligible number of dangling ends (stoichiometrically balanced end linking) or contain dangling ends introduced by using less than the stoichiometrically required amount of end-linking agent. In the latter case, decrease in $2C_1$ corresponds to increase in the number of dangling ends.

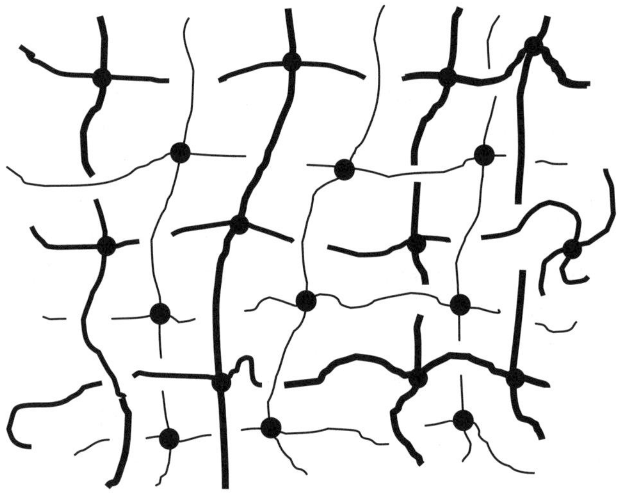

Figure 7.13:
An interpenetrating network structure in which one network is generated by a condensation end-linking reaction and the other by an additional end-linking reaction (heavier lines).

mixed into them with a multifunctional silane (in an addition reaction).[175] Interpenetrating networks (IPNs) can be unusual with regard to both equilibrium and dynamic mechanical properties. For example, such materials can have considerable toughness and unusual damping characteristics.

PDMS has been combined with a number of other polymers to form IPNs. Examples involve cellulose acetate butyrate,[176, 177] fluorinated acrylates,[178] polycarbonates,[179] epoxies,[180, 181] poly(vinyl alcohol),[182] and poly(*N*-vinyl-2-pyrrolidone).[183]

7.2.7 Sorption and Extraction of Diluents

7.2.7.1 General Approach

Swelling is an elastomeric deformation that specifically addresses the rate at which diluent absorbs into a network of known "pore size," and how rapidly it can subsequently be extracted.[184–187] The rate of absorption or extraction can be used to estimate diffusion coefficients. Extraction efficiencies are related to the extent of reaction in the end-linking procedure used to form the network and the degree to which the extractable chains are entangled with the network chains. Of interest is the dependence of these quantities on the molecular weight, M_c, of the network chains (as a measure of pore size), the molecular weight, M_d, of the diluent, the structure of the diluent (linear, branched, or cyclic), and whether or not the diluent is present during the end-linking process.

7.2.7.2 Linear Diluents

One way of obtaining a network swollen with diluent is to form the network in a first step and then absorb an unreactive diluent into it. Alternatively, the same diluent can be mixed into the reactive chains prior to end linking. In either case, oligomeric and polymeric diluents are of greatest interest. The diluent must be functionally inactive for it to "reptate" through the network rather than being bonded to it. Both types of networks can then be extracted to determine the ease with which various diluents can be removed, as a function of M_d and M_c.

The ease with which a diluent could be removed from a network was found to decrease with increase in M_d and with decrease in M_c, as expected. High molecular weight diluents are extremely hard to remove at values of M_c of interest in the preparation of model networks, complicating the analysis of soluble polymer fractions in terms of degrees of perfection of the network structure. The diluents added after the end linking are more easily removed, possibly because they were less entangled with the network structure.

7.2.7.3 Branched Diluents

The extraction of branched diluents is quite important. For example, in the preparation of networks by radiation cross linking there are presumably large numbers of soluble molecules formed that are highly branched (because of the essentially uncontrolled formation and coupling of free-radical species).[188] Another example occurs in the area of controlled release of drugs from cross-linked reservoirs.[189] Some of the drugs diffusing out of such delivery systems are assuredly "nonlinear," so understanding such systems is likely to become more important as interest shifts to larger and larger molecules (such as polypeptides). Finally, molecules with long branches can greatly affect the flow characteristics of a polymer,[190] to the extent that they can be used as processing aids in polymeric systems that will subsequently be cross linked. The ultimate in branched systems, "dendrimers,"[191–194] should also find applications in network structures.

7.2.7.4 Polar Diluents

There is a complication that can occur in the case of networks of polar polymers at relatively high degrees of swelling.[170, 195] The observation is that different solvents, at the same degree of swelling, can have significantly different effects on the elastic force, apparently due to a "specific solvent effect" on the unperturbed dimensions that appears in the various molecular forms of the elastic equations of state. Although frequently observed in studies of the solution properties of un-cross-linked polymers,[196, 197] the effect is not well understood. The effect is apparently partly due to the influence of the solvent's dielectric constant on the coulombic interactions between parts of a chain, but probably also to solvent-polymer segment interactions, which change the conformational preferences of the chain backbone.[195]

7.3 MULTIMODAL NETWORKS

7.3.1 Introduction

Multimodal networks represent a method to determine the effect of network chain-length distribution[124, 126] on rubberlike elasticity. Chain-length distribution has not received much attention even though manipulation of the chain-length distribution can give large improvements in mechanical properties. There are two primary reasons for this neglect. On the experimental side, the cross-linking techniques traditionally used to prepare the multimodal network structures have been uncontrolled.[71, 124, 126] The resulting network chain-length distributions are unimodal and probably very broad. On the theoretical side it turned out to be convenient, if not necessary, to assume a distribution of chain lengths that is not only unimodal but also *monodisperse*.[71, 124]

There are a number of reasons for developing techniques for controlling network chain-length distributions; one is to check the "weakest-link" theory[198] for elastomer rupture, which states that the shortest chains are the "culprits" in causing rupture. Due to their limited extensibility, short chains supposedly break at relatively small deformations and then act as rupture nuclei.

The preparation of controlled-structure networks requires special cross-linking reactions, such as end linking. In the case of a *bimodal* distribution, the network consists of a combination of unusually short chains (molecular weights of a few hundred) and the much longer chains typically associated with elastomeric behavior (molecular weights of 10,000 or 20,000) as in figure 7.14. Bimodal distributions clarify the dependence of ultimate properties on non-Gaussian effects arising from limited chain extensibility and elucidate synergistic effects that maximize the ultimate properties.

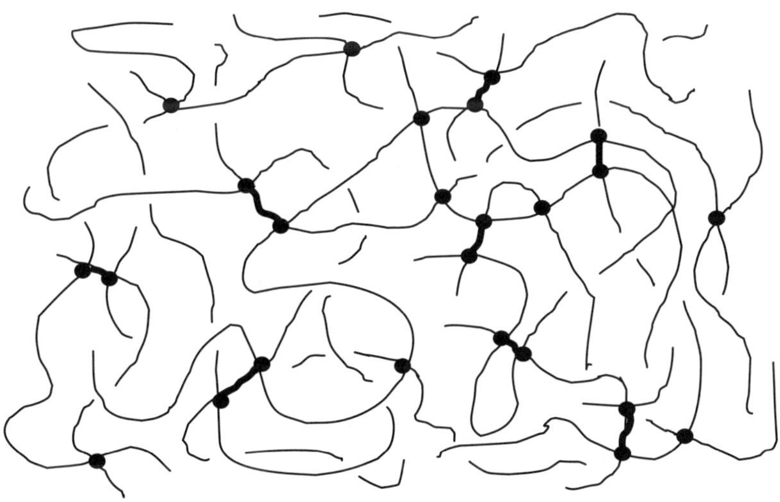

Figure 7.14:
A network having a bimodal distribution of network chain lengths. The short chains are shown by heavier lines, and the dots represent the cross links, typically resulting from the end linking of functionally terminated chains.

7.3.2 Bimodal Networks

7.3.2.1 Introduction

"Bimodal" elastomers prepared by end linking have very good ultimate properties, which underlies recent interest in such materials.[71, 96, 139, 199–204] There have been several patents in the area.[205–207] In additional to this experimental work, there are now theoretical studies addressing the novel properties of bimodal elastomers.[208–223]

7.3.2.2 Materials and Synthetic Techniques

Most bimodal networks have been prepared from PDMS, [–$Si(CH_3)_2O$–].[203, 204] PDMS is readily available with either hydroxyl or vinyl end groups and the reactions in these groups are relatively free of complicating side reactions. The end-linking reactions have generally involved hydroxyl-terminated chains, which are readily obtained from the usual ring-opening polymerization of the corresponding cyclic trimer or tetramer.[224] The ends of the chains react with the alkoxy groups in a multifunctional organosilicate, as described in chapter 3. In the application considered here, a mixture of short and long hydroxyl-terminated polymers is end linked. The end linking can take place in either the undiluted state or in solution.[225] Polyurethane elastomers have also been studied in this way.[226] In some cases, the end linking was carried out in two steps.[227] An alternative approach involves the addition reaction between vinyl groups at the ends of a polymer chain and the active hydrogen atoms on silicon atoms in the [$Si(CH_3)HO$-] repeat units in an oligomeric poly(methyl hydrogen siloxane).

One can also introduce short chains by using a trifunctional end linker with its fourth group able to associate with a similar group from another trifunctional end linker.[228] Yet another alternative is to have potential cross-linking sites that are closely spaced in one part of the chain backbone but widely spaced in another part.[143, 229] All these approaches can be extended to higher modalities (trimodal, etc.).

There is evidence of large-scale supramolecular structures in end-linked PDMS elastomers, particularly in the case of bimodal distributions.[230–233] Small-angle neutron scattering on bimodal networks of poly(tetrahydrofuran) suggests segregation of short and long chains.[234]

The distribution of network chain lengths in a bimodal elastomer can be much different from the usual unimodal distribution obtained in less-controlled methods of cross linking. Figure 7.15 shows a schematic distribution for the important example in which there is simultaneously a large *number* fraction of short chains and a large *weight* fraction of long chains. The major difference is the significant numbers of both very short chains and very long chains, which contrasts sharply with the *small* amounts of such chains in a typical unimodal distribution. The case shown, where the short chains predominate numerically, is of interest with regard to improvements in mechanical properties.[96] Studies have documented the effects of chain lengths, composition, and cross-link functionality.[235–239]

7.3.2.3 Testing of the Weakest Link

The weakest-link theory[198] was tested by preparing end-linked networks containing increasing amounts of short chains, on the order of 10–20 mol %.[71, 199] In striking disagreement with the suggested mode of elastomer failure, these elastomers showed no significant decrease in ultimate properties with increasing number of short chains. Networks are apparently much more resourceful than given credit for in the weakest-link theory. Apparently, the strain is continually being reapportioned during deformation, such that the more easily deformed long chains bear most of the burden of the deformation (figure 7.16). Thus, the short chains do not contribute significantly until just prior to rupture. The flaw in the weakest-link theory is the assumption that all parts of the network deform affinely, whereas the deformation is markedly nonaffine.[71, 96]

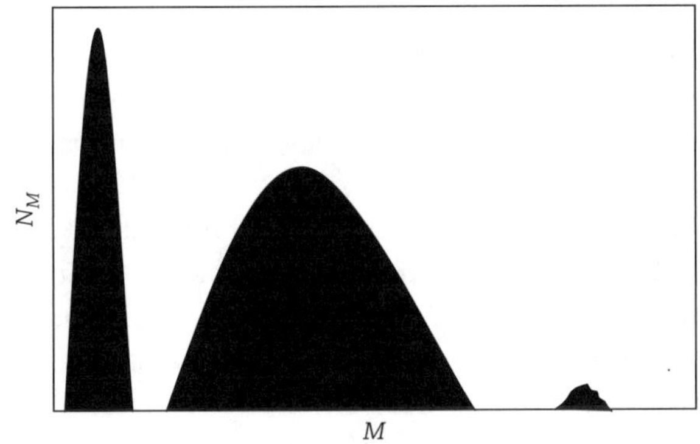

Figure 7.15:
Network chain-length distributions in which N_M is the number of chains in an infinitesimal interval around the specified value of the molecular weight M. For reference purposes, a unimodal distribution is shown between the two parts of the bimodal distribution.

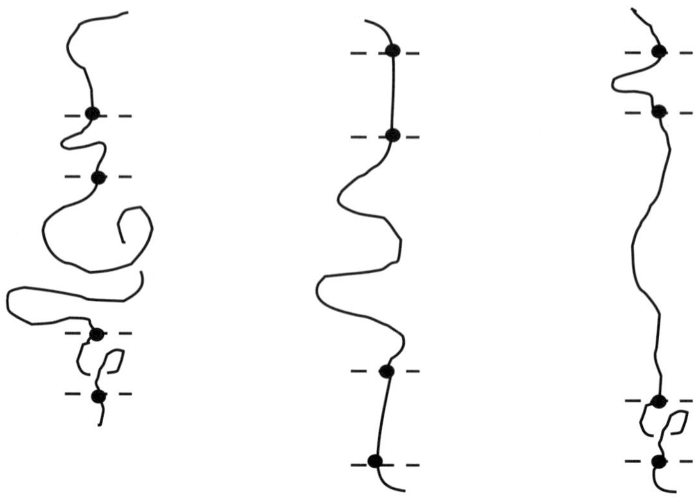

Figure 7.16:
The effect of deformation on an idealized network segment consisting of a long chain bracketed by two very short chains. The left sketch shows the undeformed segment, and the middle and right sketches show the segment deformed affinely and nonaffinely, respectively.

But what happens in the case of bimodal networks having such overwhelming numbers of short chains that they cannot be ignored? There is a synergistic effect leading to mechanical properties that are better than those obtainable from the usual unimodal distribution. The following sections describe these results in detail.

7.3.2.4 Elongation Results

Studies of bimodal networks have focused elongation because of the simplicity of this type of deformation. Examples of other deformations, equibiaxial extension (compression), shear, and torsion, are discussed later.

Many of the stress-strain isotherms of bimodal networks were obtained on PDMS elastomers in the vicinity of 25°C, a temperature sufficiently high to suppress strain-induced crystallization. These elongational isotherms show greatly *improved* the ultimate properties (figure 7.17).[203, 204] If the network consists entirely of short chains, then the material is brittle (which means that the maximum extensibility is very small). If the network consists of long chains, the ultimate strength is very low. As a result, neither unimodal material has a large area under its stress-strain curve and, thus, neither is a tough elastomer.

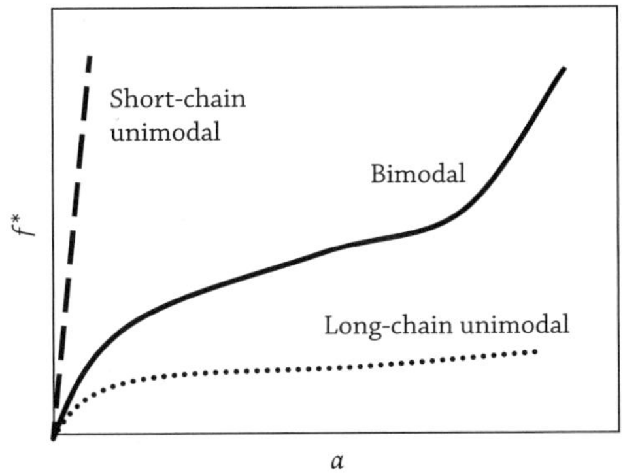

Figure 7.17:
Typical dependence of nominal stress against elongation for two unimodal networks having either all short chains or all long chains, and a bimodal network having some of both.

Bimodal networks can have high ultimate strengths without the usual diminished maximum extensibilities, corresponding to high values of the energy required for rupture, which makes them unusually tough elastomers. Apparently the short chains act primarily to increase the ultimate strength through their limited deformability, while the long chains somehow thwart the spread of rupture nuclei that would otherwise lead to catastrophic failure. This situation is analogous to what executives like to call a "delegation of responsibilities." Related improvements in mechanical properties have also been reported for other bimodal materials such as polyurethane elastomers.[240]

There are three requirements for bimodal enhancements. The ratio, M_S/M_L, of molecular weights of the short and long chains must be very small (i.e., that their molecular weights are very different). The second is that the short chains must be as short as possible (e.g., as low as 200 g/mol). Finally, there should be a large *number* concentration of the short chains, typically around 95 mol %.

Since this network approach to property improvement is of considerable practical and fundamental interest, a number of studies on PDMS elastomers were carried out to determine the impact of the molecular weights of the short chains, the proportions of short and long chains, and cross-link functionalities.[203, 204] The ultimate strength goes through a maximum with increasing amount of short chains, frequently in the vicinity of 95 mol % short chains. Too many short chains may make the elastomer brittle. These results are important in that they can be used to optimize improvements in mechanical properties.

The properties of several additional types of bimodal networks were also investigated. The first of these involves prereacting the short chains prior to incorporating the long ones.[207, 241, 242] The short-chain component can even be prepared separately, ground up, and then blended into another elastomeric network.[243] The resulting network can be spatially as well as compositionally heterogeneous, in that many of the short chains will be segregated into densely cross-linked domains that are only lightly cross linked to other such domains.[241] These networks could serve as models for inhomogeneously cross-linked elastomers, such as those cured by thermolysis of a partially miscible peroxide.

Another approach is based condensation and addition end-linking reactions carried out simultaneously and independently, which can give rise to the interpenetrating network structure.[96] The distribution of network chains can be made bimodal, even though the short chains and long chains communicate only through their entanglements. Finally, it is also possible to reinforce any type of bimodal network with filler particles, thereby further improving its mechanical properties.[96]

Measurements of the mechanical and optical properties of bimodal networks as a function of temperature and degree of swelling were used to test further the conclusion cited in section 7.3.2.4 that crystallization was not the origin of the improved properties.[96, 203, 204, 244,246] For example, stress-strain measurements on such bimodal PDMS networks exhibited upturns in modulus that were much less pronounced than those in crystallizable polymer networks such as natural rubber or *cis*-1,4-polybutadiene. The top portion of figure 7.18 illustrates upturns in the case of

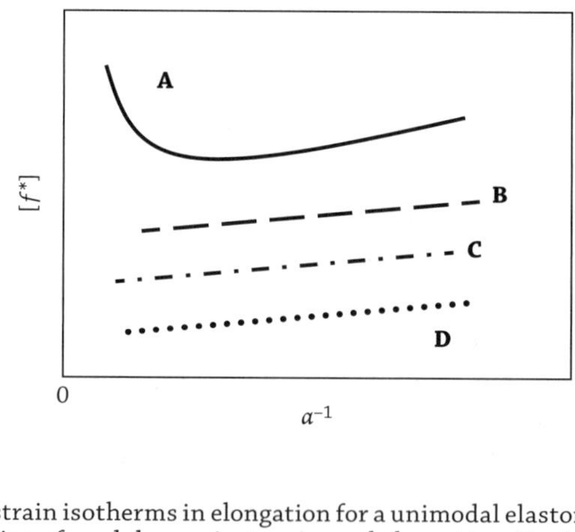

Figure 7.18:
Schematic stress-strain isotherms in elongation for a unimodal elastomer in the Mooney-Rivlin representation of modulus against reciprocal elongation. The isotherms are represented as the dependence of the reduced stress ($[f^*] = f^*/(\alpha - \alpha^{-2})$ on reciprocal elongation. ($f^* = f/A^*$, f = elastic force, A^* = undeformed area, α = elongation). The top three are for a crystallizable network: curve *A* for a relatively low temperature, *B* for an increased temperature, and *C* for the introduction of a swelling diluent. Isotherm *D* is for an unswollen unimodal network that is inherently noncrystallizable.

a crystallizable elastomer. Upturns decrease or disappear upon either an increase in temperature or addition of a diluent, as shown by two of the additional isotherms. The upturns due to crystallization are absent in the case of an elastomer that is inherently noncrystallizable, such as a stereoirregular polyacrylate.

In related experiments, temperature was found to have little effect on the Mooney-Rivlin isotherms for bimodal networks of (noncrystallizing) PDMS, as would be expected if limited chain extensibility causes the upturn (lower two isotherms in figure 7.19). Also, stress-temperature ("thermoelastic") and birefringence-temperature measurements showed no discontinuities or discernible changes of slope. Rather strikingly, swelling can even make the upturns in modulus *more pronounced* (upper isotherm in figure 7.19).[166, 170, 247] Apparently, the enhanced upturns are due to the chains being stretched in the solvent dilation process, prior to further stretching in the elongation experiments. In contrast, the upturns in crystallizable polymer networks *disappear* upon sufficient swelling.

A final experiment of relevance concerns the spatially heterogeneous PDMS networks in which the short chains are clustered. If the upturns in modulus were due to some type of intermolecular organization such as crystallization, then the behavior would presumably have been affected by this change in spatial heterogeneity. Instead, there was no discernible effect on the measured elastomeric properties. Also spectroscopy shows that bond-angle deformation was not significant in highly elongated PDMS elastomers.[248]

The foregoing findings argue against the presence of any crystallization or other type of intermolecular ordering. The upturns thus do seem to be

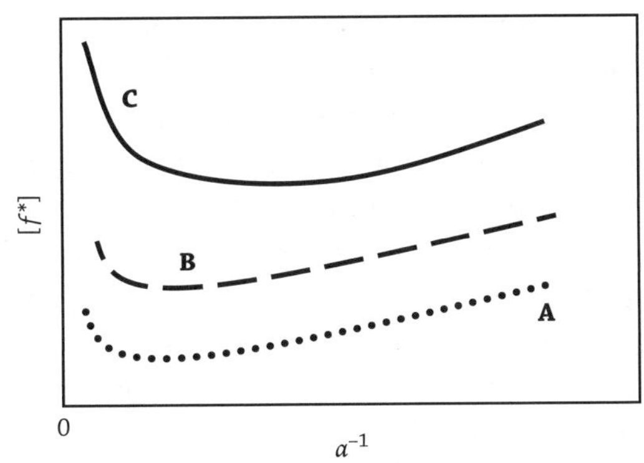

Figure 7.19:
Schematic Mooney-Rivlin isotherms for a noncrystallizable bimodal network: curve *A* for a relatively low temperature, *B* for an increased temperature, and *C* for the introduction of a swelling diluent.

intramolecular in origin. The observed increases in modulus and ultimate strength are due to the limited extensibility of the short network chains. In qualitative terms, the chains exhaust their spatial configurations, their entropies plummet, and the elastic force rises.

It is possible to characterize this non-Gaussian limited extensibility more quantitatively in a number of ways. The first involves the interpretation of limited chain extensibility in terms of the configurational characteristics of the PDMS chains making up the network structure.[170] The upturn in modulus generally begins at approximately 60–70% of maximum chain extensibility.[249] This value is approximately twice that estimated previously[124] from stress-strain isotherms of elastomers that may have been undergoing strain-induced crystallization.

More quantitative characterization of limited chain extensibility requires a non-Gaussian distribution function[124] for the end-to-end separation, *r*, of the short network chains. The Fixman-Alben distribution[250] was used[211] to calculate stress-strain isotherms in elongation for bimodal PDMS networks. Good agreement was found between theory and experiment. Other non-Gaussian distribution functions have also been successfully used.[96, 209, 210] The experimental isotherms can also be interpreted using the van der Waals theory of rubberlike elasticity.[96, 251]

Monte Carlo simulations utilize the wealth of information that rotational isomeric state theory provides on the spatial configurations of chain molecules. Monte Carlo calculations were used to simulate spatial configurations and thus distribution functions for the end-to-end separations of the chains.[71, 252] These distribution functions replaced the Gaussian function in the standard three-chain model[124] in the affine limit to give the desired non-Gaussian theory of rubberlike elasticity. Stress-strain isotherms calculated in this way were strikingly similar to the experimental isotherms obtained for the bimodal networks.[71, 96] The overall theoretical interpretation is thus quite satisfactory and would encourage other applications of these distributions, such as segmental orientation in networks containing very short chains. Such segmental orientation is of critical importance, for example, with regard to strain-induced crystallization.

A second important characteristic of an elastomeric network is the the elongation at rupture. Results on PDMS indicate that rupture occurs at approximately 80–90% of maximum chain extensibility.[96]

PDMS networks are unsuitable for characterizing the effects of bimodality on strain-induced crystallization, because of their very low crystallization temperatures. Poly(ethylene oxide), however, has a relatively high melting point (~ 65°C) [253, 255] and thus readily undergoes strain-induced crystallization. Decreasing temperature increases the extent to which the ultimate strength of the bimodal networks exceeds that of the unimodal ones (figure 7.20).[256] These results imply that bimodality facilitates strain-induced crystallization,[96] possibly through increased orientation of the longer, more easily crystallizable chains, into crystallization nuclei.[257] Similar conclusions have been reached in studies of bimodal networks of elongated poly(tetrahydrofuran),[258] but bimodality apparently had little effect in the undeformed state.[259]

In practical terms, the foregoing results demonstrate that short chains of limited extensibility may be bonded into a long-chain network to improve its toughness. It is also possible to achieve the converse effect. Thus, bonding a small number of relatively long elastomeric chains into a short-chain PDMS thermoset greatly improves both its energy of rupture and impact resistance (figure 7.21).[260] Approximately 95 mol % short chains gave the maximum effect for the molecular weights involved. Lower concentrations give smaller improvements and higher concentrations will gradually convert the composite to a more rubberlike material.

7.3.2.5 Results in Other Mechanical Deformations

There are numerous other deformations of interest, including compression, biaxial extension, shear and torsion.[124, 261] Equibiaxial extension was obtained by inflating sheets of unimodal and bimodal networks of PDMS.[225, 262] Upturns in the modulus occur at high biaxial extensions, as expected. Pronounced maxima precede the upturns (figure 7.22), which is yet to be explained by molecular theories.

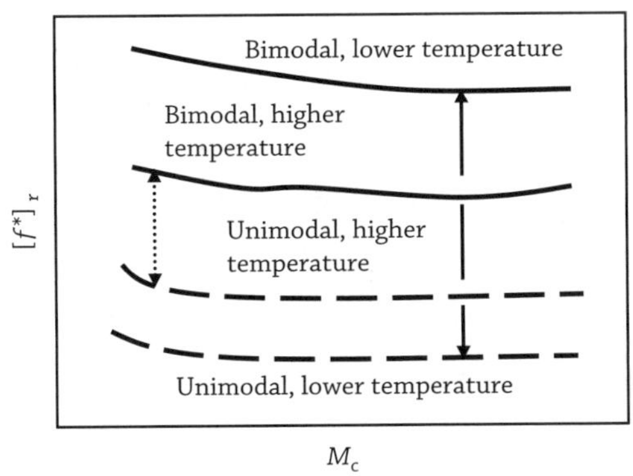

Figure 7.20:
Ultimate strength, as represented by the modulus at rupture, shown as a function of the molecular weight between cross links for a unimodal and bimodal elastomer compared at two temperatures. The improvement is larger at the lower temperature, presumably due to enhanced strain-induced crystallization.

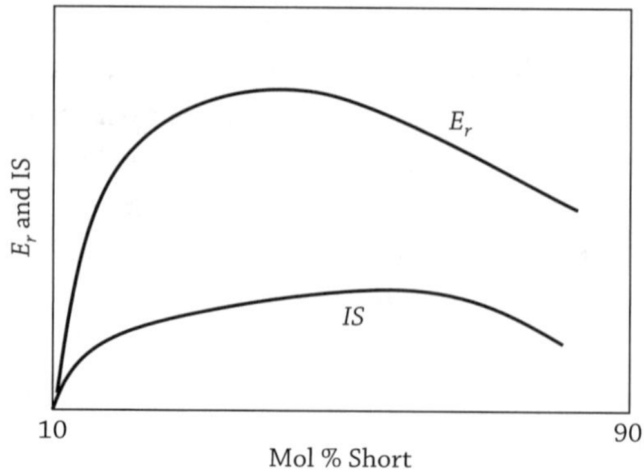

Figure 7.21:
The energy required for rupture E_r and the impact strength *IS* shown as a function of composition for typical bimodal networks that are sufficiently brittle for such testing.

Elastomeric networks can be studied in pure shear by stretching a sheet that has a large ratio of length to width, in the direction perpendicular to the length. Pure shear is compared to simple shear by Treloar (ref. 124, page 84). In shear measurements on some unimodal and bimodal networks of PDMS[263] the bimodal PDMS networks showed large upturns in the pure-shear modulus at high strains that were similar to those reported for elongation and biaxial extension.

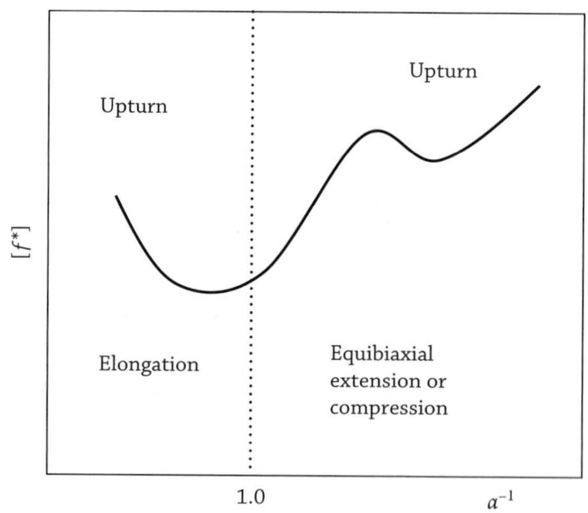

Figure 7.22:
Representative stress-strain isotherm for a bimodal elastomer in both uniaxial extension (left side) and biaxial extension (right side).

Very little work has been done on elastomers in torsion (twisting a cylindrical sample around its long axis). More results are presumably forthcoming, particularly on bimodal networks and on networks containing some of the unusual fillers introduced in situ by hydrolysis of organometallic compounds such as silicates or titanates.[96, 264] In any case, the same types of bimodal PDMS networks showed rather different behavior in torsion.[265] Specifically, ambiguous upturns in modulus were observed at large deformations. It has not yet been established whether the lack of upturn is due to the inability to reach sufficiently large torsions or whether there is some inherent difference in this type of deformation.

Tear tests have been carried out on bimodal PDMS elastomers[266–269] using the standard "trouser-leg" method. Tear energy is considerably increased for a bimodal distribution, with documentation of the effects of compositional and ratio of molecular weights of the short and long chains. The increase in tear energy did not seem to depend on tear rate,[266] an important observation that suggests that viscoelastic effects are not of paramount importance in explaining the observed improvements. Figure 7.23 shows some of these results schematically.

A subsequent series of shear tests[267] established the dependence of the tearing properties on the composition of the bimodal networks and the lengths of the chains used to prepare them. The maxima the tearing energy versus the amount of the short-chain component locates the composition giving the greatest increases in tear energy. Tensile strength depends on the ratio, M_S/M_L, of molecular weights of the two components (figure 7.24). The observed increase in strength with deceased molecular weight of the short chains must eventually decrease reverse when the chains become too short to have any elastic effectiveness at all. It is

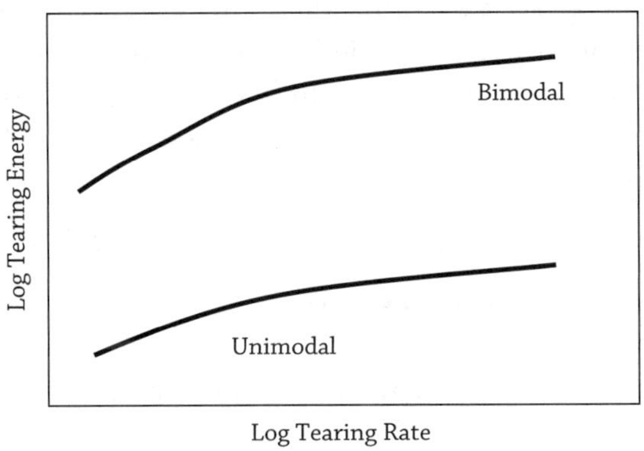

Figure 7.23:
Tearing energies for a unimodal and bimodal elastomer as a function of tearing rate

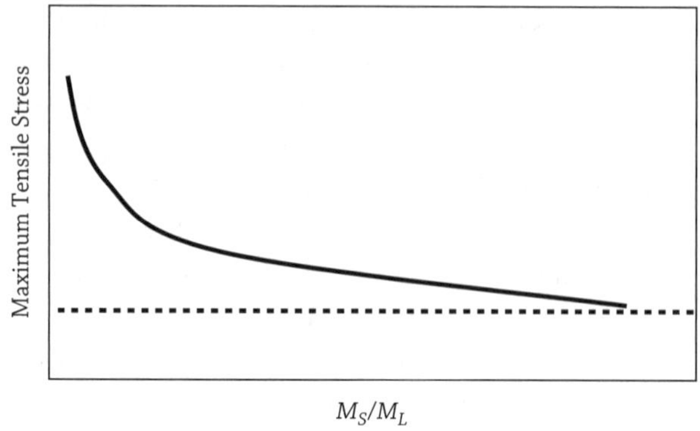

Figure 7.24:
Maximum tensile strengths in tearing for bimodal elastomers as a function of the ratio of the molecular weights of the short and long chains.

important to establish the molecular weight at which this decrease occurs for a variety of deformations and a number of different elastomers.

Some Rheovibron viscoelasticity results have been reported for bimodal PDMS networks.[270] Measurements are first carried out on unimodal networks consisting of the chains used in combination in the bimodal networks. One of the important observations was the dependence of crystallinity on the network chain-length distribution.

Some measurements have been made on the permanent set of PDMS networks in compressive cyclic deformation (figure 7.25).[271] There appeared to be less permanent set or creep in the case of the bimodal elastomers, consistent in a general way with some early results for polyurethane elastomers.[272] Specifically, cyclic elongation measurements on unimodal and bimodal networks indicated that the bimodal ones survived many more cycles before fatigue failure. The number of cycles to failure is approximately an order of magnitude higher for a bimodal network, at the same modulus (Mod10) at 10% deformation[96] Also viscoelastic effects in bimodal networks are not simple averages of contributions from the short and long chains.[270, 273]

7.3.2.6 Results on Nonmechanical Properties

Birefringence can be used to characterize non-Gaussian behavior in PDMS bimodal elastomers.[96, 274–278] A large decrease the stress-optical coefficient (ratio of birefringence to stress) was observed over a relatively small range in elongation,[78] presumably due to limited extensibility of the short chains.

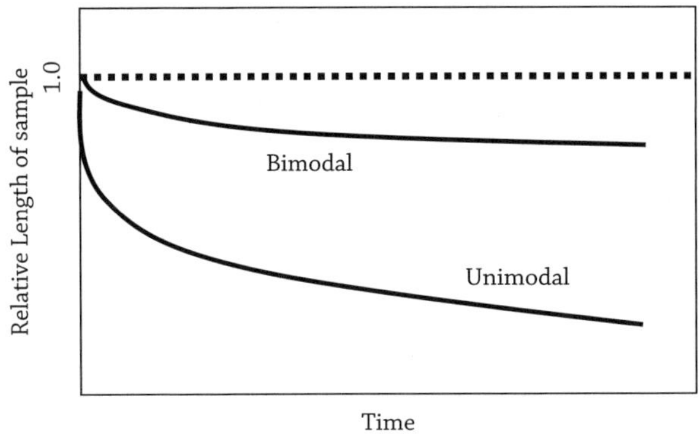

Figure 7.25:
Dependence of relative length of a sample on cyclic compressive stress as a function of time for a unimodal and bimodal elastomer.

Nuclear magnetic resonance (NMR) has also been used to characterize chain-segment orientation in PDMS bimodal materials.[279]

Another method, "thermoporometry," is based on the fact that solvent molecules constrained to small volumes form small crystallites and therefore exhibit lower crystallization temperatures.[280–286] Differential scanning calorimetry on solvent molecules constrained in the pores of PDMS elastomers shows several crystallization temperatures, which could be indicative of an unusual distribution of pore sizes.[287] The effects are most pronounced for trimodal networks, which are discussed later in this chapter.

Calorimetric measurements on bimodal poly(ethylene oxide) networks indicate that the short chains decrease crystallinity in the unstretched state.[257] This is an intriguing result since short chains *increase* the extent of crystallization in the stretched state.

When cyclic molecules are present during the end-linking reaction, the larger molecules tend to get trapped by being threaded with chains subsequently bonded into the network structure. Such experiments have also been carried out using a bimodal distribution of end-linkable PDMS chains.[288]

Additional insights into the dynamics and structure of bimodal elastomers come from experiments using dynamic light scattering,[289] NMR spectroscopy,[290, 292] neutron scattering,[213] and computer simulations on chain orientations and network mechanical properties.[214]

7.3.2.7 Inadvertent Bimodal Networks

Elastomers cured with sulfur frequently have improved mechanical properties when the curing conditions are chosen to give cross links that consist of *chains* of sulfur atoms.[293] If such polysulfidic cross links can themselves act as elastomeric network chains, then a bimodal network is produced, albeit inadvertently (figure 7.26). Calculations are in good agreement with experiment.[294]

A similar situation may occur in the case of networks end linked using the addition reaction involving vinyl chain ends and hydrogen atoms in an oligomeric poly(methyl hydrogen siloxane).[96, 154, 295–298] In the case of incomplete reactions, the segments between the reacted silicon atoms on the oligomer may be long enough to act as elastically effective chains in a bimodal structure. Finally, a bimodal chain-length distribution has also been proposed to explain some unusual properties of polysiloxane networks that were post cured,[299] elastomers prepared from two cross-linking systems,[300] and elastomers reclaimed by ultrasonic devulcanization.[301, 302]

7.3.2.8 Other Materials in Which Bimodality Might Be Advantageous

There are other cases where a bimodal distribution of chain length or some other physical property can be advantageous, possibly through the idea of a "delegation of responsibilities."[96]

In thermosets there seems to be an improvement in mechanical properties when the polymer being cured has a bimodal distribution of molecular weight.[303] In this case, the improvements may be due to different morphologies and degrees of inhomogeneity[304, 305] resulting from the fact that the long chains in a bimodal distribution could have considerably lower solubility than the short chains. Also, it is well known that the flow

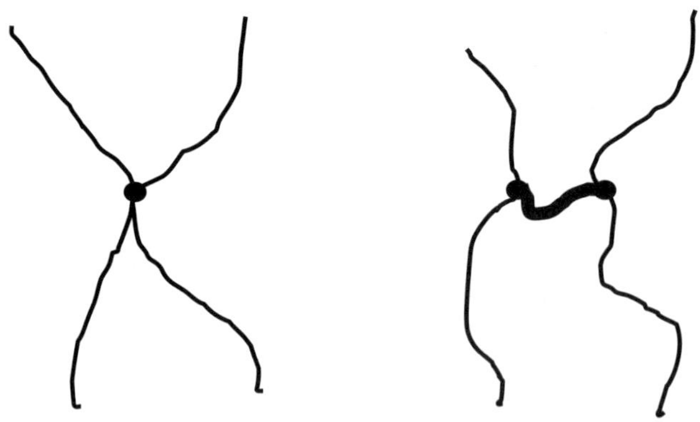

Figure 7.26:
Sketch showing the difference between a monosulfidic (left portion) and a polysulfidic cross link (right portion). In the latter case, the chains of sulfur atoms may act as additional, elastically effective chains in what is essentially a bimodal network.

characteristics of a polymer during processing[162] can be adjusted by the addition of a small amount of polymer of either very low or very high molecular weight. Bimodal polyethylene is an example.[306–308] In the case of branched polyethylenes, bimodal distributions of molecular weight affect thermal and crystallization behavior,[309] long-term creep failure,[310] and fracture.[311] In the case of brushlike structures, it has been of interest to produce bimodal layers of poly(ethylene glycol) on gaseous colloidal particles.[312]

Another example is rubber-toughened thermoplastics in which an elastomer is dispersed as domains within the thermoplastic matrix to improve its mechanical properties.[313, 314] A bimodal distribution of particle sizes gives the largest improvements.[315–322] Perhaps the small particles are most efficient at stopping one type of failure mechanism, and the large particles another type. There is the possibility that a mixture of two chemically different particles, such as silica (SiO_2) and titania (TiO_2),[323] could have significant advantages in elastomer reinforcement, with one functioning best at low and moderate temperatures and the other at elevated temperatures. In any case, the preparation of bimodal poly(organosiloxane) nanoparticles is of interest in its own right.[324]

These results can be brought into a broader context through applications not involving polymer networks. Example are the bimodal assemblies of diblock copolymer micelles and work on targeted bimodal imaging of pancreatic cancer.[325] There has also been considerable interest in preparing porous materials having bimodal distributions of pore sizes.[326–345] Frameworks having trimodal pore distributions have also been prepared.[346] Studies exist on bimodal magnetic-fluorescent nanostructures,[347–349] size distributions of nanoparticles,[350–352] bimodal arrays of nanoparticles on substrates,[353] nanocomposite scaffolds,[354] multimodal nanoparticles from miniemulsions,[355] proton transport in Nafion,[356] bimodal acid-base behavior on water-silica interfaces,[357] averaging effects in NMR attenuation in bimodal poly(methyl methacrylate) solutions,[358] and electrospinning of bimodal fiber meshes.[359] Other examples include mesoporous silica[360] and bulk ultrafine-grained nickel.[361] There have even been "ripening" studies of bimodally distributed AgCl nanoparticles.[362]

7.3.3 Trimodal Networks

Differential scanning calorimetry measurements exist on solvent molecules constrained in the pores of a variety of PDMS elastomers. Some results on trimodal networks have been reported.[287] The several crystallization temperatures were observed for benzene, which presumably reflect the pore sizes present.

Work on the mechanical properties of trimodal elastomers is limited. The problem is the large number of variables involved, three molecular weights and two independent composition variables (mol fractions). For this reason, early mechanical property experiments involved arbitrarily chosen molecular weights and compositions.[363] Not surprisingly, only modest improvements in mechanical properties were obtained over the bimodal materials. Recent results on both bimodal and trimodal PDMS elastomers showed significant improvement over unimodal PDMS elastomers with regard to some mechanical properties such as toughness, but less pronounced increases in fracture energy.[364] In summary, trimodal elastomers do show some improvements over the bimodal ones.

There has been some NMR work on trimodal PDMS elastomers, with regard to thermal degradation[365] and the presence of network heterogeneities.[366] The idea of employing trimodality has also been employed in some triblock copolymers.[367, 368]

Recent computational studies[214] indicate that it is possible to identify those molecular weights and compositions that should maximize improvement in mechanical properties. Such simulations are being extended to search for optimum properties of trimodal networks, specifically (i) the elastic modulus, (ii) maximum extensibility, (iii) tensile strength, and (iv) segmental orientability. Results[369] suggest that a trimodal network prepared by incorporating small numbers of very long chains into a bimodal network of long and short chains could significantly improve ultimate properties.

The interpretation of the attractive mechanical properties of bimodal networks has been in terms of the "delegation of responsibilities," with the short chains serving in one role and the long chains in another. If this is picture is true, then it would be interesting to study networks having extraordinarily broad molecular weight distributions, in that there would be network chains of all conceivable lengths, available for any possible mechanism that would improve properties.[96] Polymer prepared from a single polymerization would not have a broad enough distribution, but the combination of a series of samples of gradually differing average molecular weights could yield the desired distribution (figure 7.27). An elastomer of this type might have unusually attractive mechanical properties.

Trimodal distributions have also been studied in nonpolymeric systems (e.g., in aluminum metal-matrix composites).[370]

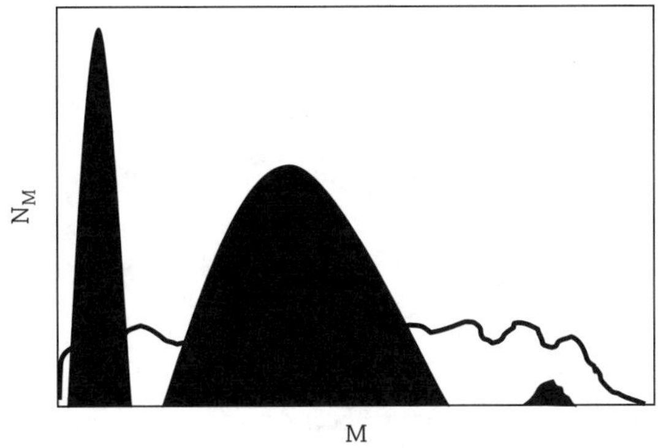

Figure 7.27:
The network chain-length distributions shown in figure 7.15, with the addition of the extremely broad "pseudo-unimodal" distribution obtainable by combining a number of samples of the same polymer made in different polymerizations.

7.4 TRAPPING OF CYCLIC OLIGOMERS WITHIN NETWORK STRUCTURES

7.4.1 Experimental Results

If cyclic molecules are present during the end linking of chains, some will be trapped because of threading by the linear chains prior to the latter being chemically bonded into the network structure (figure 7.28).[96, 288, 371, 373] The fraction trapped is estimated from solvent extraction studies. Figure 7.29 shows schematically the fraction trapped as a function of degree of polymerization of the cyclic.[374] The results were independent of intermingling time,[288] thus demonstrating the high mobility of the PDMS chains. As expected, very small cyclics don't get trapped at all, but the largest cyclics do. The following section describes the interpretation of these data in terms of the configurational characteristics of PDMS chains.

Cyclics can change the properties of the network. Since cyclics restrict motion of the network chains, they should increase the modulus of an elastomer. Small but significant increases in low-deformation modulus have in fact been observed.[372] Also, when PDMS cyclics are trapped in a thermoplastic, they can act as a plasticizer that is in a sense intermediate to the usual external (dissolved) and internal (copolymerized) varieties. Interesting changes in mechanical properties have been observed in materials of this type.[375, 376]

7.4.2 Theoretical Interpretation

The trapping process has been simulated using Monte Carlo methods based on a rotational isomeric state model [377–379] for the cyclic chains.[374]

Figure 7.28:
A tetrafunctional network containing cyclics (heavy lines). Cyclics *a* and *b* were trapped by linear chains that passed through them prior to end linking into a network structure.

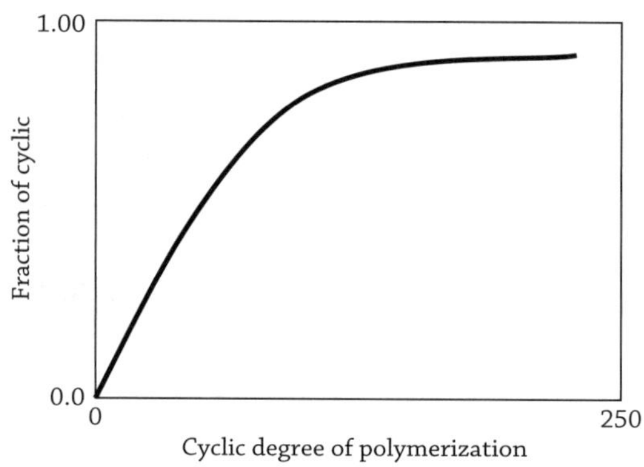

Figure 7.29:
Trapping efficiencies for cyclics as a function of cyclic size.

The first step is generation of a sufficient number of cyclic chains with the desired degree of polymerization having the known geometric features and conformational preferences. Matrix multiplication techniques common to rotational isomeric state theory are useful in the regard. Up to this point, the method is identical to that used to generate distribution functions for a non-Gaussian approach to rubberlike elasticity.[380] In the present application, however, a chain having an end-to-end distance less than a threshold value is considered to be a cyclic. The coordinates of each "cyclic" chain thus generated are stored for detailed examination of the chain's configurational characteristics such as the size of the "hole." Of particular interest is the size of this hole in comparison with the known diameter, 7–8 Å, of the PDMS chain.

The trapping process was simulated using a torus centered around each repeat unit in the cyclic.[374] Any empty torus was considered a pathway for a chain of specified diameter to thread and then incarcerate the cyclic once the end-linking process has been completed. Simulations were consistent with experimental trapping efficiencies. It is possible to interpret these experimental results in terms of a power law for the trapping probabilities and fractal cross sections for the PDMS chains.[381]

7.4.3 Olympic Networks

It may also be possible to use the trapping technique to prepare networks having no cross links whatsoever.[382] Mixed linear chains, with large amounts of cyclics, are *di*functionally end linked to yield an "Olympic" or "chain-mail" network (figure 7.30).[371, 383] Such materials are similar in some respects to the catenanes and rotaxanes that have long been of interest to a variety of scientists and mathematicians.[384–386] Computer simulations[387] could establish the conditions most likely to produce these novel structures.

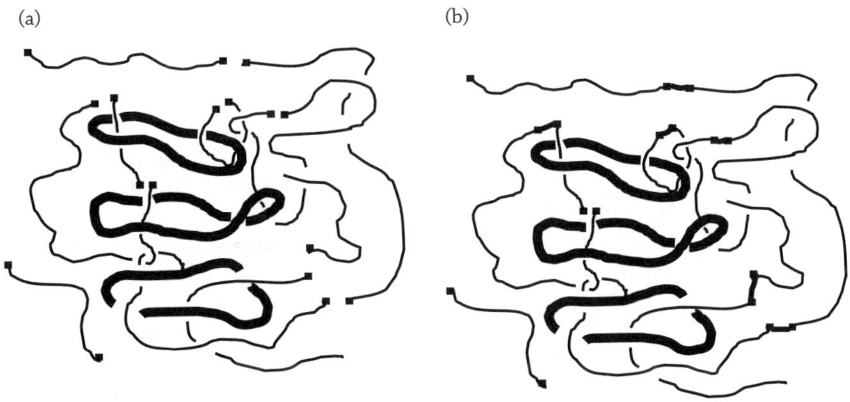

Figure 7.30:
Preparation of a chain-mail or Olympic network, which has no cross links at all. Linear chains (light lines) passing through the cyclics (heavy lines) in part *a* are di-functionally end linked to form a series of interpenetrating cyclics in part *b*.

7.5 ORIENTATION

When an elastomer is deformed, its network chains become oriented. Orientation is of both fundamental and practical importance.[71, 96, 388, 391] For example, since orientation decreases the entropy of the network chains, the melting points of the elastomer can increase significantly, which can induce crystallization. This strain-induced crystallization improves the mechanical properties of the elastomer.

In the case of polysiloxane elastomers, segmental orientation has been of interest for both elongation[392, 393] and compression.[394]

7.6 SOME VISCOELASTIC RESULTS

Examples of early studies in this area include studies of PDMS liquids,[395] PDMS solutions (e.g., in toluene),[396, 397] and PDMS elastomeric networks.[398]

REFERENCES

1. Yang, Y.; Ichise, N.; Li, Z.; Yuan, Q.; Mark, J. E.; Chan, E. K. M.; Alamo, R. G.; Mandelkern, L., The Use of Mechanical Property Measurements to Characterize Gels and Gelation Processes. In *Complex Fluids*, Sirota, E. B.; Weitz, D.; Witten, T.; Israelachvili, J., Eds. Materials Research Society: Pittsburgh, PA, 1992; Vol. 248, pp 325–331.
2. Ichise, N.; Yang, Y.; Li, Z.; Yuan, Q.; Mark, J. E.; Chan, E. K. M.; Alamo, R. G.; Mandelkern, L., Investigation of Gelation Processes and Gel Structures by Means of Mechanical Property Measurements. In *Synthesis, Characterization, and Theory of Polymeric Networks and Gels*, Aharoni, S. M., Ed. Plenum Press: New York, 1992; pp 217–225.
3. Alexandru, M., Siloxane-Organic Networks. In *Recent Developments in Silicone-Based Materials*, Cazacu, M., Ed. Nova Science: New York, 2010; pp 107–133.
4. Mrozek, R. A.; Cole, P. J.; Otim, K. J.; Shull, K. R.; Lenhart, J. L., Influence of Solvent Size on the Mechanical Properties and Rheology of Polydimethylsiloxane-Based Polymeric Gels. *Polymer* **2011**, *52*, 3422–3430.
5. Rambarran, T.; Gonzaga, F.; Brook, M. A., Generic Metal-Free Cross-Linking and Modification of Silicone Elastomers Using Click Ligation. *Macromolecules* **2012**, *45*, 2276–2285.
6. Klok, H.-A.; Rebrov, E. A.; Muzafarov, A. M.; Michelberger, W.; Moller, M., Reversible Gelation of Poly(dimethylsiloxane) with Ionic and Hydrogen-Bonding Substituents. *J. Polym. Sci., Part B: Polym. Phys.* **1999**, *37*, 485–495.
7. Yu, T.; Wakuda, K.; Blair, D. L.; Weiss, R. G., Reversible Cross-Linking Amino-Polysiloxanes by Simple Triatomic Molecules. Facile Methods for Tuning Thermal, Rheological, and Adhesive Properties. *J. Phys. Chem. C* **2009**, *113*, 11546–11553.

8. Lobez, J. M.; Swagger, T. M., Disassembly of Elastomers: Poly(olefin sulfone)—Silicones with Switchable Mechanical Properties. *Macromolecules* **2010**, *43*, 10422–10426.
9. Stepto, R. F. T., Intramolecular Interactions and Network Properties. In *Advances in Elastomers and Rubber Elasticity*, Lal, J.; Mark, J. E., Eds. Plenum Press: New York, 1986; pp 329–345.
10. Eichinger, B. E.; Akgiray, O., Computer Simulation of Polymer Network Formation. In *Computer Simulation of Polymers*, Colbourne, E. A., Ed. Longman: White Plains, NY, 1994; pp 263–302.
11. Stepto, R. F. T.; Taylor, D. J. R.; Partchuk, T.; Gottlieb, M., Poly(dimethylsiloxane) Gelation Studies. In *Silicones and Silicone-Modified Materials*, Clarson, S. J.; Fitzgerald, J. J.; Owen, M. J.; Smith, S. D., Eds. American Chemical Society: Washington, DC, 2000; Vol. 729, pp 194–203.
12. Stepto, R. F. T.; Cail, J. I.; Taylor, D. J. R.; Ward, I. M.; Jones, R. A., Understanding the Elastomeric Properties of Polymer Networks. *Macromol. Chem.* **2001**, *174*, 163–178.
13. Nosaka, M.; Takasu, M.; Katoh, K., Characterization of Gels by Monte Carlo Method Using a Model of Radical Polymerization with Cross Linkers. *J. Chem. Phys.* **2001**, *115*, 11333.
14. Stepto, R. F. T.; Cail, J.; Taylor, D. J. R., Formation, Structure and Elastomeric Properties of Polymer Networks: Theory and Modeling. *Macromol. Chem.* **2002**, *183*, 165–172.
15. Braun, J. L.; Mark, J. E.; Eichinger, B. E., Formation of Poly(dimethylsiloxane) Gels. *Macromolecules* **2002**, *35*, 5273–5282.
16. Stepto, R. F. T.; Cail, J. I.; Taylor, D. J. R., Prediction of the Elastomeric Modulus of Poly(dimethylsiloxane) Networks Formed by Endlinking. In *Synthesis and Properties of Silicones and Silicone-Modified Materials*, Clarson, S. J.; Fitzgerald, J. J.; Owen, M. J.; Smith, S. D.; Van Dyke, M. E., Eds. American Chemical Society: Washington, DC, 2003; Vol. 838, pp 151–162.
17. Stepto, R. F. T.; Cail, J. I.; Taylor, D. J. R.; Ward, I. M.; Jones, R. A., Developments in Understanding the Formation, Structure and Properties of Polymer Networks. *Macromol. Symp.* **2003**, *195*, 1–11.
18. Stepto, R. F. T.; Cail, J. I.; Taylor, D. J. R., Predicting the Modulus of End-Linked Networks from Formation Conditions. *Macromol. Symp.* **2003**, *200*, 255–264.
19. Cail, J. I.; Stepto, R. F. T., The Gel Point and Network Formation Including Poly(dimehylsiloxane) Polymerizations: Theory and Experiment. In *Science and Technology of Silicones and Silicone-Modified Materials*, Clarson, S. J.; Fitzgerald, J. J.; Owen, M. J.; Smith, S. D.; Van Dyke, M. E., Eds. American Chemical Society: Washington, DC, 2007; Vol. 964, pp 190–202.
20. Lopez, L. M.; Cosgrove, A. B.; Hernandez-Ortiz, J. P.; Osswald, T. A., Modeling the Vulcanization Reaction of Silicone Rubber. *Polym. Eng. Sci.* **2007**, *47*, 675–683.
21. Hartung, H. A.; Berger, S. E., Crosslinking of Vinylsilicone Resins by Dicumyl Peroxide. *J. Appl. Polym. Sci.* **1962**, *6*, 474–479.
22. De Buyl, F.; Comyn, J.; Shephard, N. E.; Subramaniam, N. P., Kinetics of Cure, Cross Link Density and Adhesion of Water-Reactive Alkoxysilicone Sealants. *Adhesion Sci. Technol.* **2002**, *16*, 1055–1071.
23. Baquey, G.; Moine, L.; Babot, O.; Degueil, M.; Maillard, B., Model Study of the Crosslinking of Polydimethylsiloxanes by Peroxides. *Polymer* **2005**, *38*, 6283–6292.

24. Baquey, G.; Moine, L.; Degueil-Castaing, M.; Lartigue, J.-C.; Maillard, B., Decomposition of Di-tert-butyl Peroxide in Siloxane: An Approach of the Free Radical Cross-Linking of Silicones. *Macromolecules* **2005**, *38*, 9571–9583.
25. Crowe-Willoughby, J. A.; Weiger, K. L.; Ozcam, A. E.; Genzer, J., Formation of Silicone Elastomer Network Films with Gradients in Modulus. *Polymer* **2010**, *51*, 763–773.
26. Mani, S.; Cassagnau, P.; Bousmina, M.; Chaumont, P., Rheological Modeling of the Free-Radical Crosslinking of PDMS Rubber in the Presence of TEMPO Nitroxide. *Polymer* **2010**, *51*, 3918–3925.
27. Chen, D.; Yi, S.; Wu, W.; Zhong, Y.; Liao, J.; Huang, C., Synthesis and Characterization of Novel Room Temperature Vulcanized (RTV) Silicone Rubbers Using Vinyl-POSS Derivatives as Cross Linking Agents. *Polymer* **2010**, *51*, 3867–3878.
28. Chu, H. K., UV Curable Silicones from Acryloxymethyldimethylacryloxysilane. In *Silicones and Silicone-Modified Materials*, Clarson, S. J.; Fitzgerald, J. J.; Owen, M. J.; Smith, S. D., Eds. American Chemical Society: Washington, DC, 2000; Vol. 729, pp 170–179.
29. Dworak, D. P.; Soucek, M. D., Synthesis of Cycloaliphatic Substituted Silane Monomers and Polysiloxanes for Photocuring. *Macromolecules* **2004**, *37*, 9402–9417.
30. Maxwell, R. S.; Chinn, S. C.; Solyom, D.; Cohenour, R., Radiation-Induced Cross-Linking in a Silica-Filled Silicone Elastomer as Investigated by Multiple Quantum ^{1}H NMR. *Macromolecules* **2005**, *38*, 7026–7032.
31. Kowalewska, A.; Kupcik, J.; Pola, J.; Stanczyk, W. A., Laser Irradiation of Oligosiloxane Copolymer Thin Films Functionalized with Side Chain Bulky Carbosilane Moities. *Polymer* **2008**, *49*, 857–866.
32. Ikeda, Y.; Higashitani, N.; Hijikata, K.; Kokubo, Y.; Morita, Y.; Shibayama, M.; Osaka, N.; Suzuki, T.; Endo, H.; Kohjiya, S., Vulcanization: New Focus on a Traditional Technology by Small-Angle Neutron Scattering. *Macromolecules* **2009**, *42*, 2741–2748.
33. Chakraborty, R.; Soucek, M. D., Mechanical and Film Properties of Thermally Curable Polysiloxane. *J. Appl. Polym. Sci.* **2010**, *115*, 358–369.
34. Satti, A. J.; Nador, F.; Vitale, C.; Radivoy, G.; Andreucetti, N. N.; Ciolino, A. E.; Valles, E. M., Synthesis, Characterization, and Gamma Radiation Effects over Well-Defined Poly(vinylsiloxanes) Copolymers. *J. Appl. Polym. Sci.* **2012**, *124*, 832–839.
35. Chen, J.; Vaino, A. R.; Smith, R. L.; Collins, S. C., Photomediated Crosslinking of Cinnamated PDMS for In-Situ Direct Photopatterning. *J. Polym. Sci., Part A: Polym. Chem.* **2008**, *46*, 3482–3487.
36. Wang, S.-J.; Fan, X.-D.; Kong, J.; Lu, J.-R., Synthesis, Characterization and UV Curing Kinetics of Hyperbranched Polysiloxanes from A2 and CB2 Type Monomers. *Polymer* **2009**, *50*, 3587–3594.
37. Gonzaga, F.; Yu, G.; Brook, M. A., Polysiloxane Elastomers via Room Temperature, Metal-Free Click Chemistry. *Macromolecules* **2009**, *42*, 9220–9224.
38. Mani, S.; Cassagnau, P.; Bousmina, M.; Chaumont, P., Cross-Linking Control of PDMS Rubber at High Temperatures Using TEMPO Nitroxide. *Macromolecules* **2009**, *42*, 8460–8467.
39. Hoyle, C. E.; Bowman, C. N., Thiol-Ene Click Chemistry. *Angew. Chem. Int. Ed. Engl.* **2010**, *49*, 1540–1573.
40. Kade, M. J.; Burke, D. J.; Hawker, C. J., The Power of Thiol-ene Chemistry. *J. Polym. Sci., Part A: Polym. Chem.* **2010**, *48*, 743–750.

41. Barrall II, E. M.; Flandera, M. A.; Logan, J. A., A Thermodynamic Study of the Crosslinking of Methyl Silicone Rubber. *Thermochim. Acta* **1973**, *5*, 415–432.
42. James, P. M.; Barrall, E. M.; Dawson, B.; Logan, J. A., Crosslinking of Methyl Silicone Rubbers. 2. Analysis of Extractables from Samples Crosslinked Under Various Conditions. *J. Macromol. Sci.: Part A—Chem.* **1974**, *8*(1), 135–155.
43. Steffanut, P.; Osborn, J. A.; DeCian, A.; Fisher, J., Efficient Homogeneous Hydrosilylation of Olefins by Use of Complexes of Pt^0 with Selected Electron-Deficient Olefins as Ligands. *Chem. Eur. J.* **1998**, *4*, 2008–2017.
44. Racles, C.; Cazacu, M.; Hitruc, G.; Hamaide, T., On the Feasibility of Chemical Reactions in the Presence of Siloxane-Based Surfactants. *Coll. Polym. Sci.* **2009**, *287*, 461–470.
45. Hotta, Y.; Komatsu, K.; Wang, F. W., Novel Fluorescence Method for Cure Monitoring of Hydrosilation-Curable Silicones. *J. Appl. Polym. Sci.* **2009**, *112*, 2441–2444.
46. Wu, Y.; McGarry, F. J.; Zhu, B.; Keryk, J. R.; Katsoulis, D. E., Temperature Effects on Mechanical Properties of Toughened Silicone Resins. *Polym. Eng. Sci.* **2005**, *45*, 1522–1531.
47. Esteves, A. C. C.; Brokken-Zijp, J.; Laven, J.; Huinink, H. P.; Reuvers, N. J. W.; Van, M. P.; With, G. D., Influence of Cross-Linker Structure Concentration on the Cross-Linking of PDMS and the Network Structures Formed. *Polymer* **2009**, *50*, 3955–3966.
48. Matloka, P. R.; Kean, Z.; Wagener, K. B., Chain Internal/Chain End Latent Crosslinking in Thermoset Polymer Systems. *J. Polym. Sci., Part A: Polym. Chem.* **2010**, *48*, 1866–1877.
49. White, C. C.; Hunston, D. L.; Williams, R. S., Studies on the Effect of Movement During the Cure on the Mechanical Properties of a Silicone Joint Sealant. *Polym. Eng. Sci.* **2010**, *50*, 113–119.
50. Stiubianu, G.; Racles, C.; Cazacu, M.; Simionescu, B. C., Silicone-Modified Cellulose. Crosslinking of Cellulose Acetate with Poly[dimethyl(methyl-H)siloxane] by Pt-Catalyzed Dehydrogenative Coupling. *J. Mater. Sci.* **2010**, *45*, 4141–4150.
51. McConville, C.; Andrews, G. P.; Laverty, T. P.; Woolfson, A. D.; Malcolm, R. K., Rheological Evaluation of the Isothermal Cure Characteristics of Medical Grade Silicone Elastomers. *J. Appl. Polym. Sci.* **2010**, *116*, 2320–2327.
52. Miki, K.; Inamoto, Y.; Inoue, S.; Uno, T.; Itoh, T.; Kubo, M., Preparation of Three-Dimensional Poly(dimethylsiloxane) (PDMS) with Movable Cross-Linking. *J. Polym. Sci. A, Polym. Chem.* **2009**, *47*, 5882–5890.
53. Boileau, S.; Bouteiller, L.; Kowalewska, A., Polysiloxanes Containing Cross-linkable Diacetylene Units in the Main Chain. *Polym. Int.* **2004**, *53*, 191–197.
54. Matloka, P. P.; Sworen, J. C.; Zuluaga, F.; Wagener, K. B., Chain-End and Chain-Internal Crosslinking in "Latent Reactive" Silicon Elastomers. *Macromol. Chem. Phys.* **2005**, *206*, 218–228.
55. Dai, L.; Zhang, Z.; Zhao, Y.; Xie, Z., Polymeric Curing Agent Reinforced Silicone Rubber Composites with Low Viscosity and Low Volume Shrinkage. *J. Appl. Polym. Sci.* **2008**, *110*, 1624–1631.
56. Han, Y.; Zhang, J.; Yang, Q.; Shi, L.; Qi, S.; Jin, R., Novel Polymethoxysiloxane-Based Crosslinking Reagent and Its In-Situ Improvement for Thermal and Mechanical Properties of Siloxane Elastomer. *J. Appl. Polym. Sci.* **2008**, *107*, 3788–3795.

57. Yin, Y.; Zhang, Y.; Chi, Z.; Xu, J., Self-Catalytic Cross-Linking Reaction and Reactive Mechanism Studies of α-Aminomethyl Triethoxysilanes for α,ω-Dihydroxy Poly(dimethylsiloxane) *Eur. Polym. J.* **2008**, *44*, 2284–2294.
58. Dai, L.; Zhang, Z.; Zhao, Y.; Liu, H.; Xie, Z., Effects of Polymeric Curing Agent Modified with Silazanes on the Mechanical Properties of Silicone Rubber. *J. Appl. Polym. Sci.* **2009**, *111*, 1057–1062.
59. Longuet, C.; Ratsimihety, A.; Andre, S.; Boutevin, G.; Guida-Pietrasanta, F.; Deschamps, B.; Ramonda, M.; Joly-Duhamel, C.; Boutevin, B.; Ganachaud, F., Physically Crosslinked Fluorosilicone Elastomers Obtained by Self-Assembly and Template Polycondensation of Tailored Building Blocks. *J. Mater. Chem.* **2010**, *20*, 10269–10276.
60. Susuki, M.; Owa, S.; Shirai, H.; Hanabusa, K., Poly(dimethylsiloxane)-Based Polymer Organogelators with L-Lysine Derivatives as an Organogelation-Causing Segment. *J. Polym. Sci., Part A: Polym. Chem.* **2006**, *44*, 3817–3824.
61. Kanamori, K.; Nakanishi, K.; Hirao, K.; Jinnai, H., Interface-Directed Web-to-Pillar Transition of Microphase-Separated Gels. *Langmuir* **2003**, *19*, 9101–9103.
62. Garrido, L.; Mark, J. E.; Wang, S.; Ackerman, J. L.; Vevea, J. M., Polymer Distribution in Silica Aerogels Impregnated with Siloxanes by ^{1}H NMR Imaging. *Polymer* **1992**, *33*, 1826–1830.
63. Wang, J.; Li, X., Preparation and Characterization of Interpenetrating Polymer Network Silicone Hydrogels with High Oxygen Permeability. *J. Appl. Polym. Sci.* **2010**, *116*, 2749–2757.
64. Matsumura, K.; Yamaguchi, H., Elastic Modulus of Viscoelastic Magnetic Silicone Gel Body. *Polym. Eng. Sci.* **2010**, *50*, 857–862.
65. Klein, L. C.; Jitianu, A., Organic-Inorganic Hybrid Melting Gels. *J. Sol-Gel Sci. Technol.* **2010**, *55*, 86–93.
66. Cho, S. H.; Andersson, H. M.; White, S. R.; Sottos, N. R.; Braun, P. V., Polydimethylsiloxane-Based Self-Healing Materials. *Adv. Mater.* **2006**, *18*, 997–1000.
67. Keller, M. W.; White, S. R.; Sottos, N. R., A Self-Healing Poly(dimethylsiloxane) Elastomer. *Adv. Funct. Mater.* **2007**, *17*, 2399–2404.
68. Mangun, C. L.; Mader, A. C.; Sottos, N. R.; White, S. R., Self-Healing of a High Temperature Cured Epoxy Using Poly(dimethylsiloxane) Chemistry. *Polymer* **2010**, *51*, 4063–4068.
69. Falcao, A. N.; Pedersen, J. S.; Mortensen, K.; Boue, F., Polydimethylsiloxane Networks at Equilibrium Swelling: Extracted and Nonextracted Networks. *Macromolecules* **1996**, *29*, 809–818.
70. Braden, M.; Latham, D.; Patel, M. P., Observations on the Swelling of Cross-Linked Poly(dimethylsiloxane) Networks by Solvents. *Eur. Polym. J.* **2005**, *41*, 3069–3071.
71. Mark, J. E.; Erman, B., *Rubberlike Elasticity: A Molecular Primer*, 2nd ed. Cambridge University Press: Cambridge, 2007.
72. Kaneko, Y.; Noguchi, K.; Kadokawa, J.-I., Synthesis of Temperature-Responsive Organic-Inorganic Hybrid Hydrogel by Free-Radical Polymerization of Methacrylamide Using Water-Soluble Rigid Polysiloxane Having Acrylamide Side-Chains as a Cross-linking Agent. *Polym. J.* **2007**, *39*, 1078–1081.
73. Lendlein, A., Progress in Actively Moving Polymers. *J. Mater. Chem.* **2010**, *20*, 3332–3334.

74. Crowe-Willoughby, J. A.; Stevens, D. R.; Genzer, J.; Clarke, L. I., Investigating the Molecular Origins of Responsiveness in Functional Silicone Elastomer Networks. *Macromolecules* **2010**, *43*, 5043–5051.
75. Kimura, T.; Umehara, Y.; Kimura, F., Magnetic Field Responsive Silicone Elastomer Loaded with Short Steel Wires having Orientation Distribution. *Soft Matter* **2012**, *8*(23), 6206–6209.
76. Maeda, S.; Hara, Y.; Sakai, T.; Yoshida, R.; Hashimoto, S., Self-Walking Gel. *Adv. Mater.* **2007**, *19*, 3480–3484.
77. Shinohara, S.-I.; Seki, T.; Sakai, T.; Yoshida, R.; Takeoka, Y., Photoregulated Wormlike Motion of a Gel. *Angew. Chem. Int. Ed. Engl.* **2008**, *47*, 9039–9043.
78. Maeda, S.; Hara, Y.; Yoshida, R.; Hashimoto, S., Peristaltic Motion of Polymer Gels. *Angew. Chem. Int. Ed. Engl.* **2008**, *47*, 6690–6693.
79. Rijke, A. M.; Prins, W., The Swelling of Cellulose Acetate Networks Obtained by Crosslinking in Solution. *J. Polym. Sci.* **1962**, *59*, 171–190.
80. Mukherji, R.; Prins, W., Applicability of Polymer Network Theories to Gels Obtained by Crosslinking a Polymer in Solution. *J. Polym. Sci., Part A: Polym. Chem.* **1964**, *2*, 4367–4390.
81. Dusek, K.; Prins, W., Structure and Elasticity of Non-Crystalline Polymer Networks. *Adv. Polym. Sci.* **1969**, *6*, 1–102.
82. Mark, J. E., An Experimental Comparison of the Theories of Elasticity of Polymer Networks. *J. Am. Chem. Soc.* **1970**, *92*, 7252–7257.
83. Froelich, D.; Crawford, D.; Rozek, T.; Prins, W., Ideal Network Behavior of Anionically Prepared Poystyrene Gels. *Macromolecules* **1972**, *5*, 100–102.
84. Johnson, R. M.; Mark, J. E., Properties of Poly(dimethylsiloxane) Networks Prepared in Solution, and Their Use in Evaluating the Theories of Rubberlike Elasticity. *Macromolecules* **1972**, *5*, 41–45.
85. Vasiliev, V. G.; Rogovina, L. Z.; Slonimsky, G. L., Dependence of Properties of Swollen and Dry Polymer Networks on the Conditions of Their Formation in Solution. *Polymer* **1985**, *26*, 1667–1676.
86. Rogovina, L. Z.; Vasil'ev, V. G.; Slonimskii, G. L., Influence of the Nature and Concentration of the Solvent on the Formation and Properties of Polydimethylsiloxane Networks. *Polym. Sci. U.S.S.R.* **1990**, *32*, 1997–2004.
87. Mark, J. E., The Volume Dependence of the C2 Correction in the Mooney-Rivlin Elastic Equation of State. *J. Polym. Sci., Part C: Polym. Symp.* **1970**, *31*, 97–106.
88. Cohen, R. E.; Severson, S. D.; Yu, C. U.; Mark, J. E., Viscoelastic Properties of Poly(dimethylsiloxane) Networks Prepared by Cross-Linking the Chains in Solution. *Macromolecules* **1977**, *10*, 663–667.
89. Erman, B.; Mark, J. E., Interpretation of Stress-Strain Isotherms for Elastomers Cross-Linked in Solution. *Macromolecules* **1987**, *20*, 2892–2897.
90. Erman, B.; Mark, J. E., Segmental Orientation in Networks Cross Linked in Solution. *Macromolecules* **1989**, *22*, 480–481.
91. Queslel, J. P.; Mark, J. E., Theoretical Equilibrium Moduli and Swelling Extents for Elastomers Cross-Linked in Solution. *Rubber Chem. Technol.* **1990**, *63*, 46–55.
92. Xu, P.; Mark, J. E., Biaxial and Uniaxial Extension Studies of Polydimethylsiloxane Networks Prepared by End Linking in Solution. *Makromol. Chem.* **1991**, *192*, 567–578.
93. Erman, B.; Mark, J. E., Stress-Strain Isotherms for Elastomers Cross-Linked in Solution. 2. Interpretation in Terms of the Constrained-Chain Model. *Macromolecules* **1992**, *25*, 1917–1921.

94. Mark, J. E., The Use of Polysiloxanes to Elucidate Molecular Aspects of Rubberlike Elasticity. *J. Inorg. Organomet. Polym. Mater.* **2012**, *22*, 560–563.
95. Greene, A.; Smith, K. J.; Ciferri, A., Elastic Properties of Networks Formed from Oriented Chain Molecules Part-2. Composite Networks. *Trans. Faraday Soc.* **1965**, *61*, 2772–2783.
96. Erman, B.; Mark, J. E., *Structures and Properties of Rubberlike Networks*. Oxford University Press: New York, 1997.
97. Mark, J. E., The Rubber Elastic State. In *Physical Properties of Polymers*, 1st ed.; Mark, J. E.; Eisenberg, A.; Graessley, W. W.; Mandelkern, L.; Koenig, J. L., Eds. American Chemical Society: Washington, DC, 1984; pp 1–54.
98. Flory, P. J.; Erman, B., Theory of Elasticity of Polymer Networks. 3. *Macromolecules* **1982**, *15*, 800–806.
99. Cifra, P.; Bleha, T., Simulations of Elongation of Supercoiled and Biopolymer Networks. *J. Polym. Sci., Part B: Polym. Phys.* **1999**, *37*, 2013–2023.
100. Premachandra, J.; Mark, J. E., Effects of Dilution During Cross Linking on Strain-Induced Crystallization in *cis*-1,4-Polyisoprene Networks. 1. Experimental Results. *J. Macromol. Sci., Pure Appl. Chem.* **2002**, *39*, 287–300.
101. Premachandra, J.; Kumudinie, C.; Mark, J. E., Effects of Dilution During Cross Linking on Strain-Induced Crystallization in *cis*-1,4-Polyisoprene Networks. 2. Comparison of Experimental Results with Theory. *J. Macromol. Sci., Pure Appl. Chem.* **2002**, *39*, 301–320.
102. Urayama, K.; Kohjiya, S., Uniaxial Elongation of Deswollen Polydimethylsiloxane Networks with Supercoiled Structure. *Polymer* **1997**, *38*, 955–962.
103. Kohjiya, S.; Urayama, K.; Ikeda, Y., Poly(siloxane) Network of Ultra-High Elongation. *Kautschuk Gummi Kunststoffe* **1997**, *50*, 868–872.
104. Urayama, K.; Kohjiya, S., Extensive Stretch of Polysiloxane Network Chains with Random and Super-Coiled Conformations. *Eur. Phys. J. B* **1998**, *2*, 75–78.
105. Urayama, K.; Yokoyama, K.; Kohjiya, S., Low-Temperature Behavior of Deswollen Poly(dimethylsiloxane) Networks. *Polymer* **2000**, *41*, 3273–3278.
106. Urayama, K., Network Topology-Mechanical Properties Relationships of Model Elastomers. *Polym. J.* **2008**, *40*, 669–678.
107. Urayama, K.; Kawamura, T.; Kohjiya, S., Structure-Mechanical Property Correlations of Model Siloxane Elastomers with Controlled Network Topology. *Polymer* **2008**, *50*, 347–356.
108. Roland, C. M.; Warzel, M. L., Orientation Effects in Rubber Double Networks. *Rubber Chem. Technol.* **1990**, *63*, 285–297.
109. Santangelo, P. G.; Roland, C. M., Role of Strain Crystallization in the Fatigue Resistance of Double Network Elastomers. *Rubber Chem. Technol.* **2003**, *76*, 892–898.
110. Yang, Z. H.; Detwiler, A. T.; Lesser, A. J., Prestressed Double Network Themoset: Preparation and Characterization. *J. Mater. Sci.* **2012**, *47*, 4251–4261.
111. Santangelo, P. G.; Roland, C. M., The Mechanical Behavior of Double Network Elastomers. *Rubber Chem. Technol.* **1994**, *67*, 359–365.
112. Santangelo, P. G.; Roland, C. M., Failure Properties of Natural Rubber Double Networks. *Rubber Chem. Technol.* **1995**, *68*, 124–131.
113. Wang, J.; Hamed, G. R.; Umetsu, K.; Roland, C. M., The Payne Effect in Double Network Elastomers. *Rubber Chem. Technol.* **2005**, *78*, 76–83.
114. Aprem, A. S.; Joseph, K.; Thomas, S., Studies of Double Networks in Natural Rubber Vulcanizates. *J. Appl. Polym. Sci.* **2004**, *91*, 1068–1076.
115. Roland, C. M.; Peng, K. L., Electrical Conductivity in Rubber Double Networks. *Rubber Chem. Technol.* **1991**, *64*, 790–800.

116. Mott, P. H.; Roland, C. M., Mechanical and Optical Properties of Double Network Rubbers. *Macromolecules* **2000**, *33*, 4132–4137.
117. Gong, J. P.; Katsuyama, Y.; Kurokawa, T.; Osada, Y., Double Network Hydrogels with Extremely High Mechanical Strength. *Adv. Mater.* **2003**, *15*, 1155–1158.
118. Na, Y.-H.; Kurokawa, T.; Katsuyama, Y.; Tsukeshiba, H.; Gong, J. P.; Osada, Y.; Okabe, S.; Karino, T.; Shibayama, M., Structural Characteristics of Double Network Gels with Extremely High Mechanical Strength. *Macromolecules* **2004**, *37*, 5370–5374.
119. Naficy, S.; Razal, J. M.; Whitten, P. G.; Wallace, G. G.; Spinks, G. M., A pH-Sensitive, Strong Double-Network Hydrogel: Poly(ethylene glycol) Methyl Ether Methacrylates-Poly(acrylic acid). *J. Polym. Sci., Part B: Polym. Phys.* **2011**, *50*, 423–430.
120. Tanaka, Y.; Kuwabara, R.; Na, Y.-H.; Kurokawa, T.; Gong, J. P.; Osada, Y., Determination of Fracture Energy of High Strength Double Network Hydrogels. *J. Phys. Chem. B.* **2005**, *109*, 11559–11562.
121. Mark, J. E., Rubber Elasticity. *J. Chem. Educ.* **1981**, *58*, 898–903.
122. Mark, J. E., New Developments and Directions in the Area of Elastomers and Rubberlike Elasticity. *Macromol. Symp., Kyoto issue* **2003**, *201*, 77–83.
123. Mark, J. E., Some Recent Theory, Experiments, and Simulations on Rubberlike Elasticity. *J. Phys. Chem. B* **2003**, *107*, 903–913.
124. Treloar, L. R. G., *The Physics of Rubber Elasticity*, 3rd ed. Clarendon Press: Oxford, 1975.
125. Stepto, R. F. T.; Taylor, D. J. R., Molecular Modeling of the Elastic Behavior of Polymer Chains in Networks. *J. Chem. Soc., Faraday Trans.* **1995**, *91*, 2639–2647.
126. Flory, P. J., *Principles of Polymer Chemistry.* Cornell University Press: Ithaca NY, 1953.
127. Erman, B.; Mark, J. E., Rubber-like elasticity. *Ann. Rev. Phys. Chem.* **1989**, *40*, 351–374.
128. Clement, F.; Bokobza, L.; Monnerie, L., On the Mullins Effect in Silica-Filled Polydimethylsiloxane Networks. *Rubber Chem. Technol.* **2001**, *74*, 847–870.
129. Hanson, D. E.; Hawley, M.; Houlton, R.; Chitanvis, K.; Rae, P.; Orler, E. B.; Wrobleski, D. A., Stress Softening Experiments in Silica-Filled Polydimethylsiloxane Provide Insight Into a Mechanism for the Mullins Effect. *Polymer* **2005**, *46*, 10989–10995.
130. Diani, J.; Fayolle, B.; Gilormini, P., A Review of the Mullins Effect. *Eur. Polym. J.* **2009**, *45*, 601–612.
131. Clement, F.; Bokobza, L.; Monnerie, L., Investigation of the Payne Effect and Its Temperature Dependence on Silica-Filled Polydimethylsiloxane Networks. Part I. Experimental Results. *Rubber Chem. Technol.* **2005**, *78*, 211–231.
132. Clement, F.; Bokobza, L.; Monnerie, L., Investigation of the Payne Effect and Its Temperature Dependence on Silica-Filled Polydimethylsiloxane Networks. Part II. Test *Rubber Chem. Technol.* **2005**, *78*, 232–244.
133. Lupton, E. M.; Achenbach, F.; Weis, J.; Brauchle, C.; Frank, I., Modified Chemistry of Siloxanes Under Tensile Stress: Interaction with Environment. *J. Phys. Chem. B* **2006**, *110*, 14557–14563.
134. Lupton, E. M.; Achenbach, F.; Weis, J.; Brauchle, C.; Frank, I., Origins of Material Failure in Siloxane Elastomers from First Principles. *ChemPhysChem* **2009**, *10*, 119–123.

135. Holmes, D. P.; Crosby, A. J., Snapping Surfaces. *Adv. Mater.* **2007**, *19*, 3589–3593.
136. Xu, W.; Chahine, N.; Sulchek, T., Extreme Hardening of PDMS Thin Film Due to High Compressive Strain and Confined Thickness. *Langmuir* **2011**, *27*, 8475–8477.
137. Kohl, J. G.; Randall, N. X.; Schwarzer, N.; Ngo, T. T.; Schockley, J. M., An Investigation of Scratch Testing of Silicone Elastomer Coatings with a Thickness Gradient. *J. Appl. Polym. Sci.* **2012**, *124*, 2978–2986.
138. Hernandez-Ortiz, J. P.; Osswald, T. A., Modeling Processing of Silicone Rubber: Liquid Versus Hard Silicone Rubbers. *J. Appl. Polym. Sci.* **2011**, *119*, 1864–1871.
139. Mark, J. E., Some Interesting Things About Polysiloxanes. *Acc. Chem. Res.* **2004**, *37*, 946–953.
140. Mark, J. E.; Ngai, K. L.; Graessley, W. W.; Mandelkern, L.; Samulski, E. T.; Koenig, J. L.; Wignall, G. D., *Physical Properties of Polymers*, 3rd ed. Cambridge University Press: Cambridge, 2004.
141. Leir, C. M.; Galkiewicz, R. K.; Kantner, S. S.; Mazurek, M., Telechelic Siloxanes with Hydrogen-Bonded Polymerizable End Groups. I. Liquid Rubbers and Elastomers. *J. Appl. Polym. Sci.* **2010**, *117*, 756–766.
142. Mazurek, M.; Leir, C. M.; Galkiewicz, R. K., Telechelic Siloxanes with Hydrogen-Bonded Polymerizable End Groups. II. IR Studies of End-Groups Interactions—Model Amides and Ureas. *J. Appl. Polym. Sci.* **2010**, *117*, 982–995.
143. Falender, J. R.; Yeh, G. S. Y.; Mark, J. E., The Effect of Chain Length Distribution on Elastomeric Properties. 2. Comparisons Among Networks of Varying Degrees of Randomness. *Macromolecules* **1979**, *12*, 1207–1209.
144. Heine, D. R.; Grest, G. S.; Lorenz, C. D.; Tsige, M.; Stevens, M. J., Atomistic Simulations of End-Linked Poly(dimethylsiloxane) Networks: Structure and Relaxations. *Macromolecules* **2004**, *33*, 3857–3864.
145. Bienzobaz, P. F.; Malacarne, L. C.; Mendes, R. S.; Lenzi, E. K., An Expansion Approach in Rubber Elasticity. *Polymer* **2008**, *49*, 1968–1971.
146. Liu, Q.; Wang, L.; Yu, W.; Xiao, A.; Yu, H.; Huo, J., Hypercrosslinked Polystyrene Microspheres with Bimodal Pore Size Distribution and Controllable Macroporosity. *J. Appl. Polym. Sci.* **2010**, *116*, 84–92.
147. Wang, J. J.; Feng, L.-J.; Lei, A.-L.; Yan, A.-J.; Wang, X.-J., Thermal Stability and Mechanical Properties of Room Temperature Vulcanized Silicone Rubbers. *J. Appl. Polym. Sci.* **2012**, *125*, 505–511.
148. Disch, S.; Finkelmann, H.; Ringsdorf, H.; Schuhmacher, P., Macroscopically Ordered Discotic Columnar Networks. *Macromolecules* **1995**, *28*, 2424–2428.
149. Brehmer, M.; Zentel, R., Ferroelectric Liquid-Crystalline Elastomers with Short Switching Times. *Macromol. Rapid Commun.* **1995**, *16*, 659–662.
150. Mark, J. E.; Rahalkar, R. R.; Sullivan, J. L., Model Networks of End-Linked Polydimethylsiloxane Chains. III. Effect of the Functionality of the Cross-Links. *J. Chem. Phys.* **1979**, *70*, 1794–1797.
151. Mark, J. E., Some Topological Issues in the Area of Elastomeric Polymeric Networks. *New J. Chem.* **1993**, *17*, 703–709.
152. Llorente, M. A.; Mark, J. E., Model Networks of End-Linked Polydimethylsiloxane Chains. VIII. Networks Having Cross-Links of Very High Functionality. *Macromolecules* **1980**, *13*, 681–685.
153. Sharaf, M. A., The Effects of Network Imperfections on the Small-Strain Moduli of Polydimethylsiloxane Elastomers Having High Functionality Cross-Links. *Int. J. Polym. Mater.* **1992**, *18*(3–4), 237–252.

154. Sharaf, M. A.; Mark, J. E., Elastomers Having High Functionality Cross Links Viewed as Bimodal Networks. An Alternative Interpretation in Terms of Bimodal Chain-Length Distributions. *J. Polym. Sci., Polym. Phys. Ed.* **1995**, *33*, 1151–1165.
155. Gottlieb, M.; Macosko, C. W.; Benjamin, G. S.; Meyers, K. O.; Merrill, E. W., Equilibrium Modulus of Model Poly(dimethylsiloxane) Networks. *Macromolecules* **1981**, *14*, 1039–1046.
156. Miller, D. R.; Macosko, C. W., Molecular Weight Relations for Crosslinking of Chains with Length and Site Distribution. *J. Polym. Sci., Polym. Phys. Ed.* **1987**, *25*, 2441–2469.
157. Patel, S. K.; Malone, S.; Cohen, C.; Gillmor, J. R.; Colby, R. H., Elastic Modulus and Equilibrium Swelling of Poly(dimethylsiloxane) Networks. *Macromolecules* **1992**, *25* (20), 5241–5251.
158. Venkatraman, S., Deformation-Behavior of Poly(dimethyl Siloxane) Networks 0.1. Applicability of Various Theories to Modulus Prediction. *J. Appl. Polym. Sci.* **1993**, *48* (8), 1383–1393.
159. Gent, A. N.; Liu, G. L.; Mazurek, M., Experimental-Study of Molecular Entanglement in Polymer Networks. *J. Polym. Sci., Part B: Polym. Phys.* **1994**, *32* (2), 271–279.
160. Graessley, W. W., *Polymeric Liquids and Networks. Structure and Properties.* Garland Science: New York, 2004.
161. Vilgis, T. A.; Erman, B., Comparison of the Constrained Junction and the Slip-Link Models of Rubber Elasticity. *Macromolecules* **1993**, *26*(24), 6657–6659.
162. Ferry, J. D., *Viscoelastic Properties of Polymers*, 3rd ed. Wiley: New York, 1980.
163. Oppermann, W.; Rennar, N., Stress-Strain Behaviour of Model Networks in Uniaxial Tension and Compression. Progr. Coll. & Polym. Sci **1992**, *75*(1) 49–54.
164. Rennar, N.; Oppermann, W., Swelling Behavior and Mechanical-Properties of Endlinked Poly(dimethylsiloxane) Networks and Randomly Cross-Linked Polyisoprene Networks. *Coll. Polym. Sci.* **1992**, *270*(6), 527–536.
165. Mark, J. E.; Eisenberg, A.; Graessley, W. W.; Mandelkern, L.; Samulski, E. T.; Koenig, J. L.; Wignall, G. D., *Physical Properties of Polymers*, 2nd ed. American Chemical Society: Washington, DC, 1993.
166. Andrady, A. L.; Llorente, M. A.; Mark, J. E., Model Networks of End-Linked Polydimethylsiloxane Chains. VII. Networks Designed to Demonstrate Non-Gaussian Effects Related to Limited Chain Extensibility. *J. Chem. Phys.* **1980**, *72*, 2282–2290.
167. Mandelkern, L., Crystallization and Melting. In *Comprehensive Polymer Science*, Allen, G., Ed. Pergamon Press: Oxford, 1989; p 363.
168. Mandelkern, L., *Crystallization of Polymers.* Cambridge University Press: Cambridge, 2003.
169. Mark, J. E., Elastomeric Networks with Bimodal Chain-Length Distributions. *Acc. Chem. Res.* **1994**, *27*, 271–278.
170. Mark, J. E.; Eisenberg, A.; Graessley, W. W.; Mandelkern, L.; Koenig, J. L., *Physical Properties of Polymers*, 1st ed. American Chemical Society: Washington, DC, 1984.
171. Andrady, A. L.; Llorente, M. A.; Sharaf, M. A.; Rahalkar, R. R.; Mark, J. E.; Sullivan, J. L.; Yu, C. U.; Falender, J. R., Model Networks of End-Linked Polydimethylsiloxane Chains. XII. Dependence of Ultimate Properties on Dangling Chain Irregularities. *J. Appl. Polym. Sci.* **1981**, *26*, 1829–1836.

172. Mark, J. E., Silicon-Containing Polymers. In *Silicon-Based Polymer Science. A Comprehensive Resource*, Zeigler, J. M.; Fearon, F. W. G., Eds. American Chemical Society: Washington, DC, 1990; Vol. 224, pp 47–68.
173. Landherr, L. J. T.; Cohen, C.; Archer, L. A., Effect of Pendent Chins on the Interfacial Properties of Thin Polydimethylsiloxane (PDMS) Networks. *Langmuir* **2011**, *27*, 5944–5952.
174. Sperling, L. H., *Interpenetrating Polymer Networks and Related Materials.* Plenum Press: New York, 1981.
175. Mark, J. E.; Ning, Y.-P., Interpenetrating Bimodal Networks. *Polym. Eng. Sci.* **1985**, *25*, 824–827.
176. Fichet, O.; Vidal, F.; Laskar, J.; Teyssie, D., Polydimethylsiloxane-Cellulose Acetate Butyrate Interpenetrating Polymer Networks Synthesis and Kinetic Study. Part I. *Polymer* **2005**, *46*, 37–47.
177. Vidal, F.; Fichet, O.; Laskar, J.; Teyssie, D., Polydimethylsiloxane-Cellulose Acetate Butyrate Interpenetrating Polymer Networks Close to True IPNs on a Large Composition Range. Part I. *Polymer* **2006**, *47*, 3747–3753.
178. Darras, V.; Fichet, O.; Perrot, F.; Boileau, S.; Teyssie, D., Polysiloxane-Poly(fluorinated acrylate) Interpenetrating Polymer Networks: Synthesis and Characterization. *Polymer* **2007**, *48*, 687–695.
179. Boileau, S.; Bouteiller, L.; Khalifa, R. B.; Liang, Y.; Teyssie, D., Polycarbonate-Polysiloxane-Based Interpenetrating Networks. In *Silicones and Silicone-Modified Materials*, Clarson, S. J.; Fitzgerald, J. J.; Owen, M. J.; Smith, S. D., Eds. American Chemical Society: Washington, DC, 2000; Vol. 729, pp 383–394.
180. Jia, L.-Y.; Zhang, C.; Du, Z.-J.; Li, C.-J.; Li, H.-Q., Preparation of Interpenetrating Polymer Networks of Epoxy/Polydimethylsiloxane in a Common Solvent of the Precursors. *Polym. J.* **2007**, *39*, 593–597.
181. Jia, L.-Y.; Zhang, C.; Du, Z.-J.; Li, C.-J.; Li, H.-Q., A Novel Approach to Interpenetrating Networks of Epoxy Resin and Polydimethylsiloxane. *J. Appl. Polym. Sci.* **2007**, *105*, 2663–2669.
182. Marques, R. S.; Leod, T. C. O. M.; Yoshida, I. V. P.; Mano, V.; Assis, M. D.; Schiavon, M. A., Synthesis and Characterization of Semi-Interpenetrating Networks Based on Poly(dimethylsiloxane) and Poly(vinyl alcohol). *J. Appl. Polym. Sci.* **2010**, *115*, 158–166.
183. Hillerstrom, A.; Kronberg, B., A Two-Step Method for the Synthesis of a Hydrophilic PDMS Interpenetrating Polymer Network. *J. Appl. Polym. Sci.* **2008**, *110*, 3059–3067.
184. Mark, J. E.; Zhang, Z.-M., The Use of Model Networks with Reptating Chains to Study Extraction Efficiencies. *J. Polym. Sci., Polym. Phys. Ed.* **1983**, *21*, 1971–1979.
185. Garrido, L.; Mark, J. E., Extraction and Sorption Studies Using Linear Polymer Chains and Model Networks. *J. Polym. Sci., Polym. Phys. Ed.* **1985**, *23*, 1933–1940.
186. Mark, J. E., Sorption, Extraction, and Trapping of Linear and Cyclic Molecules in Model Elastomeric Networks. *J. Appl. Polym. Sci. Symp.* **1989**, *44*, 209–216.
187. Mazan, J.; Leclerc, B.; Galandrin, N.; Couarraze, G., Diffusion of Free Polydimethylsiloxane Chains in Polydimethylsiloxane Elastomer Networks. *Eur. Polym. J.* **1995**, *31*(8), 803–807.
188. Dole, M., *The Radiation Chemistry of Macromolecules*. Academic Press: New York, 1973.
189. Peppas, N. A.; Langer, R., New Challenges in Biomaterials. *Science* **1994**, *263*(5154), 1715–1720.

190. Graessley, W. W., Viscoelasticity and Flow in Polymer Melts and Concentrated Solutions. In *Physical Properties of Polymers*, 2nd ed.; Mark, J. E.; Eisenberg, A.; Graessley, W. W.; Mandelkern, L.; Samulski, E. T.; Koenig, J. L.; Wignall, G. D., Eds. American Chemical Society: Washington, DC, 1993, pp 97–144.
191. Frechet, J. M. J., Functional Polymers and Dendrimers—Reactivity, Molecular Architecture, and Interfacial Energy. *Science* **1994**, *263*(5154), 1710–1715.
192. Tomalia, D. A., Dendrimer Molecules. *Sci. Am.* **1995**, *272*(5), 62–66.
193. Grayson, S. M.; Frechet, J. M. J., Convergent Dendrons and Dendrimers: From Synthesis to Applications. *Chem. Rev.* **2001**, *101*, 3819–3867.
194. Tomalia, D. A.; Frechet, J. M. J., Discovery of Dendrimers and Dendritic Polymers: A Brief Historical Perspective. *J. Polym. Sci., Part A: Polym. Chem.* **2002**, *40*, 2719–2728.
195. Yu, C. U.; Mark, J. E., Specific Solvent Effects in Swollen Polymer Networks. *Macromolecules* **1974**, *7*, 229–232.
196. Hoeve, C. A. J.; O'Brien, M. K., Specific diluent effects on polymer chain dimensions. *J. Polym. Sci. Part A: General Papers* **1963**, *1*(6), 1947–1954.
197. Dondos, A.; Benoit, H., The Influence of Solvents on Unperturbed Dimensions of Polymer in Solution. *Macromolecules* **1971**, *4*(3), 279–283.
198. Bueche, F., *Physical Properties of Polymers*. Wiley-Interscience: New York, 1962.
199. Mark, J. E., Recent Studies of Rubberlike Elasticity. *Makromol. Chem., Suppl.* **1979**, *2*, 87–97.
200. Mark, J. E., Recent Studies of Rubberlike Elasticity. *Polym. J.* **1985**, *17*, 265–269.
201. Mark, J. E., Recent Studies of Rubberlike Elasticity. In *Frontiers of Macromolecular Science*, Saegusa, T.; Higashimura, T.; Abe, A., Eds. Blackwell Scientific: Oxford, 1989; pp 289–294.
202. Mark, J. E., Improved Elastomers Through Control of Network Chain-Length Distributions. In *Science and Technology of Polymers and Advanced Materials*, Prasad, P. N.; Mark, J. E.; Kandil, S. H.; Kafafi, Z. H., Eds. Plenum Press: New York, 1998; pp 85–98.
203. Mark, J. E., Improved Elastomers Through Control of Network Chain-Length Distributions. *Rubber Chem. Technol.* **1999**, *72*, 465–483.
204. Mark, J. E., Elastomers with Multimodal Distributions of Network Chain Lengths. *Macromol. Symp., St. Petersburg issue* **2003**, *191*, 121–130.
205. Jeram, E. M.; Striker, R. A. High Strength Organopolysiloxane Compositions. 1976, US patent US3884866 A.
206. Lee, C. L.; Maxson, M. T.; Stebleton, L. F. High Strength, Extrudable Silicone Elastomer Compositions. 1979, US patent US4162243 A.
207. Galiatsatos, V.; Subramanian, P. R. Polymeric Networks of Polysiloxane with Increased Modulus. 1994, US patent US5376738 A.
208. Kilian, H.-G., Thermomechanics of Bimodal End-Linked Polydimethylsiloxane Networks with One of the Chains Much Shorter. *Coll. Polym. Sci.* **1981**, *259*, 1151–1160.
209. Menduina, C.; Freire, J. J.; Llorente, M. A.; Vilgis, T., Correctly Averaged Non-Gaussian Theory of Rubberlike Elasticity. Application to the Description of the Behavior of Poly(dimethylsiloxane) Bimodal Networks. *Macromolecules* **1986**, *19*, 1212–1217.
210. Llorente, M. A.; Rubio, A. M.; Freire, J. J., Moments and Distribution Functions of the End-to-End Distance of Short Poly(dimethylsiloxane) and Poly(oxyethylene) Chains. Application to the Study of Elasticity in Model Networks. *Macromolecules* **1984**, *17*, 2307–2315.

211. Erman, B.; Mark, J. E., Use of the Fixman-Alben Distribution Function in the Analysis of Non-Gaussian Rubberlike Elasticity. *J. Chem. Phys.* **1988**, *89*, 3314–3316.
212. Termonia, Y., Mechanical Model for the Mechanical Properties of Elastomers. 2. Synergistic Effects in Bimodal Cross-Linked Networks. *Macromolecules* **1990**, *23*, 1481–1483.
213. Kloczkowski, A.; Mark, J. E.; Erman, B., Fluctuations, Correlations and Small Angle Neutron Scattering from Endlinked Gaussian Chains in Regular Bimodal Networks. *Macromolecules* **1991**, *24*, 3266–3275.
214. Sakrak, G.; Bahar, I.; Erman, B., Computer Simulations of Two-Dimensional Trifunctional Bimodal Networks. *Macromol. Theory Simul.* **1994**, *3*, 151–161.
215. Vasilevskaya, V. V.; Semenov, A. N., On the Theory of Elasticity and Swelling of Polydisperse Polymer Networks. *Vysokomol. Soedin.* **1994**, *36*, 1535–1540.
216. Bahar, I.; Erman, B.; Bokobza, L.; Monnerie, L., Molecular Orientation in Deformed Bimodal Networks. 1. Theory. *Macromolecules* **1995**, *28*, 225–230.
217. Hagn, C.; Wittkop, M.; Kreitmeier, S.; Trautenberg, H. L.; Holzl, T.; Goritz, D., The Creation and Spatial Structure of End-Linked Bimodal Polymer Networks: A Monte Carlo Study. *Polym. Gels Networks* **1997**, *5*, 327–337.
218. Sotta, P., Local Order in Monomodal and Bimodal Polymer Networks: Monte Carlo Simulations. *Macromolecules* **1998**, *31*, 8417–8422.
219. Von Lockette, P. R.; Arruda, E. M., Topological Studies of Bimodal Networks. *Macromolecules* **1999**, *32*, 1990–1999.
220. Von Lockette, P. R.; Arruda, E. M., Computational Annealing of Simulated Unimodal and Bimodal Networks. *Comput. Theor. Polym. Sci.* **2001**, *11*, 415–428.
221. Von Lockette, P. R.; Arruda, E. M.; Wang, Y., Mesoscale Modeling of Bimodal Elastomeric Networks: Constitutive and Optical Theories and Results. *Macromolecules* **2002**, *35*, 7100–7109.
222. Kloczkowski, A.; Erman, B.; Mark, J. E., Effect of Non-Gaussian Chains on Fluctuations of Junctions in Bimodal Networks. *Polymer* **2002**, *43*, 2569–2574.
223. Von Lockette, P. R., Examination of the Effects of Computationally Determined Network Topology on an Analytical Constitutive Model for Bimodal Elastomers. *Polymer* **2008**, *49*, 5158–5168.
224. Clarson, S. J.; Mark, J. E., Siloxane Elastomers. In *Siloxane Polymers*, Clarson, S. J.; Semlyen, J. A., Eds. Prentice Hall: Englewood Cliffs, 1993; pp 616–648.
225. Xu, P.; Mark, J. E., Elasticity Measurements on Bimodal Networks in Elongation and Compression. Networks Cross-linked in Solution and Studied Unswollen, and Networks Cross-linked in the Undiluted State and Studied Swollen. *Polymer* **1992**, *33*, 1843–1848.
226. Goering, H.; Kruger, H.; Bauer, M., Multimodal Polymer Networks: Design and Characterisation of Nanoheterogeneous PU Elastomers. *Macromol. Mater. Eng.* **2000**, *278*, 23–25.
227. Hamurcu, E. E.; Baysal, B. M., Interpenetrating Polymer Networks of Poly(dimethylsiloxane): 1. Preparation and Characterization. *Polymer* **1993**, *34*, 5163–5168.
228. Sur, G. S.; Mark, J. E., A Novel Method for Preparing Bimodal Elastomeric Networks. *Polym. Bull.* **1985**, *13*, 505–509.
229. Falender, J. R.; Yeh, G. S. Y.; Mark, J. E., The Effect of Chain Length Distribution on Elastomeric Properties. 1. Comparisons Between Random Networks and Highly Nonrandom Networks. *J. Am. Chem. Soc.* **1979**, *101*, 7353–7356.

230. Viers, B. D.; Mark, J. E., Large-Scale Structures in Bimodal Poly(dimethylsiloxane) Elastomers. *J. Inorg. Organomet. Polym.* **2006**, *15*, 477–483.
231. Viers, B. D.; Mark, J. E., Elastomeric Properties of Polysiloxane Networks. Bimodal Elastomers That Are Spatially Inhomogeneous and Others That Are Very Broadly Multimodal. *J. Macro. Sci., Pure Appl. Chem.* **2007**, *44*, 131–138.
232. Viers, B. D.; Mark, J. E., Large-Scale Structures in Some End-Linked Polysiloxane Networks. In *Science and Technology of Silicones and Silicone-Modified Materials*, Clarson, S. J.; Fitzgerald, J. J.; Owen, M. J.; Smith, S. D.; Van Dyke, M. E., Eds. American Chemical Society: Washington, DC, 2007; Vol. 964, pp 136–164.
233. Viers, B. D.; Mark, J. E., Elastomeric Properties of Polysiloxane Networks: Birefringence Measurements on Bimodal Elastomers That Are Presumed to Be Spatially Inhomogeneous *J. Inorg. Organomet. Polym. Mater.* **2007**, *17*, 283–288.
234. Wu, W.-L.; Coyne, L. D.; Jong, L.; Hanyu, A.; Stein, R. S., Molecular Structure of Bimodal Polymer Networks. *Macromolecules* **1990**, *23*, 351–353.
235. Llorente, M. A.; Andrady, A. L.; Mark, J. E., Model Networks of End-Linked Polydimethylsiloxane Chains. XIII. The Effects of Junction Functionality on the Elastic Properties of the Bimodal Networks. *Coll. Polym. Sci.* **1981**, *259*, 1056–1061.
236. Llorente, M. A.; Andrady, A. L.; Mark, J. E., Model Networks of End-Linked Polydimethylsiloxane Chains. XI. The Use of Very Short Chains to Improve Ultimate Properties. *J. Polym. Sci., Polym. Phys. Ed.* **1981**, *19*(4), 621–630.
237. Tang, M.-Y.; Garrido, L.; Mark, J. E., The Effect of Crosslink Functionality on the Elastomeric Properties of Bimodal Networks. *Polymer* **1984**, *25*, 347–350.
238. Mark, J. E.; Tang, M.-Y., Dependence of the Elastomeric Properties of Bimodal Networks on the Length and Amount of the Short Chains. *J. Polym. Sci., Polym. Phys. Ed.* **1984**, *22* (11), 1849–1855.
239. Tang, M.-Y.; Mark, J. E., Effect of Composition and Cross-Link Functionality on the Elastomeric Properties of Bimodal Networks. *Macromolecules* **1984**, *17*, 2616–2619.
240. Kim, C. S. Y.; Ahmad, J.; Bottaro, J.; Farzan, M., Improvements in the Stress-Strain Behavior of Urethane Rubbers by Bimodal Network Formation. *J. Appl. Polym. Sci.* **1986**, *32*, 3027–3038.
241. Mark, J. E.; Andrady, A. L., Model Networks of End-Linked Polydimethylsiloxane Chains. X. Bimodal Networks Prepared in Two-Stage Reactions Designed to Give High Spatial Heterogeneity. *Rubber Chem. Technol.* **1981**, *54*, 366–373.
242. Pan, S.-J.; Mark, J. E., Model Networks of End-Linked Polydimethylsiloxane Chains. XV. Spatially Heterogeneous Networks Containing Domains of Very High Crosslink Density. *Polym. Bull.* **1982**, *7*, 553–559.
243. Tang, M.-Y.; Mark, J. E., Comparisons Between Poly(dimethylsiloxane) Networks Filled with Particles of Poly(dimethylsiloxane) Thermosets and Unfilled Bimodal Networks Prepared from the Same Components. *Polym. Bull.* **1984**, *11*, 573–578.
244. Mark, J. E., Effect of Swelling on the Short-Chain Reinforcement Observed in Bimodal Elastomeric Networks. *Macromolecules* **1984**, *17* (12), 2924–2925.
245. Jiang, C.-Y.; Mark, J. E.; Stebleton, L., Ultimate Properties of Filled Bimodal Networks Containing a Non-Volatile Diluent. *J. Appl. Polym. Sci.* **1984**, *29* (12), 4411–4414.

246. Clarson, S. J.; Galiatsatos, V.; Mark, J. E., An Investigation of the Properties of Non-Gaussian Poly(dimethylsiloxane) Model Networks in the Swollen State. *Macromolecules* **1990**, *23* (5), 1504–1507.
247. Galiatsatos, V.; Clarson, S. J., An Investigation of the Properties of Bimodal Poly(dimethylsiloxane) Elastomers Upon Swelling with a Linear Oligomeric Polymethylsiloxane. *Polym. Comm.* **1986**, *27*, 260–261.
248. Silva, L. K.; Mark, J. E.; Boerio, F. J., Some Spectroscopic Evidence on Deformation Mechanisms in Elastomeric Networks at Very High Elongations. *Makromol. Chem.* **1991**, *192*, 499–506.
249. Mark, J. E., The Rubber Elastic State. In *Physical Properties of Polymers,* 3rd ed.; Mark, J. E.; Ngai, K. L.; Graessley, W. W.; Mandelkern, L.; Samulski, E. T.; Koenig, J. L.; Wignall, G. D., Eds. Cambridge University Press: Cambridge, 2004; pp 3–71.
250. Fixman, M.; Alben, R., Polymer Conformational Statistics 0.1. Probability Distribution. *J. Chem. Phys.* **1973**, *58*(4), 1553–1558.
251. Kilian, H.-G., Bimodal Theory. *Proceedings, Network Group Meeting, Jerusalem* 1990.
252. Curro, J. G.; Mark, J. E., A Non-Gaussian Theory of Rubberlike Elasticity Based on Rotational Isomeric State Simulations of Network Chain Configurations. II. Bimodal Poly(dimethylsiloxane) Networks. *J. Chem. Phys.* **1984**, *80,* 4521–4525.
253. Mark, J. E., *Physical Properties of Polymers Handbook*. Springer-Verlag: New York, 1996.
254. Brandrup, J.; Immergut, E. H.; Grulke, E. A., *Polymer Handbook*, 4th ed. Wiley: New York, 1999.
255. Mark, J. E., *Polymer Data Handbook*. Oxford University Press: New York, 1999.
256. Sun, C.-C.; Mark, J. E., The Effect of Network Chain Length Distribution, Specifically Bimodality, on Strain-Induced Crystallization. *J. Polym. Sci., Polym. Phys. Ed.* **1987**, *25*, 2073–2083.
257. Clarson, S. J.; Mark, J. E.; Sun, C.-C.; Dodgson, K., Effect of Bimodal Chain Length Distribution on the Thermal Characteristics of Poly(oxyethylene) Model Networks. *Eur. Polym. J.* **1992**, *28* (7), 823–825.
258. Hanyu, A.; Stein, R. S., Segmental Orientation and Infrared Dichroism of Model Bimodal Elastomeric Networks. *Macromol. Symp.* **1991**, *45*, 189–203.
259. Roland, C. M.; Buckley, G. S., Thermal Crystallization of Polytetrahydrofuran Networks. *Rubber Chem. Technol.* **1991**, *64*, 74–82.
260. Tang, M.-Y.; Letton, A.; Mark, J. E., Impact Resistance of Unfilled and Filled Bimodal Thermosets of Poly(dimethylsiloxane). *Coll. Polym. Sci.* **1984**, *262,* 990–992.
261. Mark, J. E.; Erman, B., Elastomers and Rubber-like Elasticity. In *Performance of Plastics*, Brostow, W., Ed. Hanser: Cincinnati, 2001; pp 401–428.
262. Xu, P.; Mark, J. E., Biaxial Extension Measurements on Bimodal Elastomeric Networks. *J. Polym. Sci., Polym. Phys. Ed.* **1991**, *29*, 355–358.
263. Wang, S.; Mark, J. E., Unimodal and Bimodal Networks of Poly(dimethylsiloxane) in Shear. *J. Polym. Sci., Polym. Phys. Ed.* **1992**, *30*, 801–807.
264. Mark, J. E., Ceramic-Reinforced Polymers and Polymer-Modified Ceramics. *Polym. Eng. Sci.* **1996**, *36*, 2905–2920.
265. Wen, J.; Mark, J. E., Torsion Studies of Thermoelasticity and Stress-Strain Isotherms of Unimodal, Bimodal, and Filled Networks of Poly(dimethylsiloxane). *Polym. J.* **1994**, *26*, 151–157.

266. Yanyo, L. C.; Kelley, F. N., Effect of Chain Length Distribution on the Tearing Energy of Silicone Elastomers. *Rubber Chem. Technol.* **1987**, *60*, 78–88.
267. Smith, T. L.; Haidar, B.; Hedrick, J. L., Origin of the Self-Reinforcement in PDMS Bimodal Networks. *Rubber Chem. Technol.* **1990**, *63*, 256–264.
268. Shah, G. B., Effect of Bimodality on Tear Properties of Silicone Networks. *Macromol. Chem. Phys.* **1996**, *197*, 2201–2208.
269. Shah, G. B., Effect of Bimodality on the Tear Properties of Filled Silicone Networks. *J. Appl. Polym. Sci.* **2004**, *94*, 1719–1722.
270. Andrady, A. L.; Llorente, M. A.; Mark, J. E., Some Dynamic Mechanical Properties of Unimodal and Bimodal Networks of Poly(dimethylsiloxane). *Polym. Bull.* **1991**, *26* (3), 357–362.
271. Wen, J.; Mark, J. E.; Fitzgerald, J. J., Unimodal, Bimodal, and Filled Poly(dimethylsiloxane) Elastomers Under Cyclic Stress at Elevated Temperatures. *J. Macromol. Sci., Macromol. Rep.* **1994**, *31*, 429–438.
272. Kaneko, Y.; Watanabe, Y.; Okamoto, T.; Iseda, Y.; Matsunaga, T., Effect of Molecular Weight Distribution of Poly(oxytetramethylene)glycol on Cut Growth Resistance of Polyurethane. *J. Appl. Polym. Sci.* **1980**, *25*, 2467–2478.
273. Yuzhelevskii, Y. A.; Sarafanova, L. A.; Ramsh, A. S., Inhomogeneity of Networks of Siloxane Vulcanizates. *Polym. Sci. U.S.S.R.* **1993**, *35*, 1135–1138.
274. Zhang, Z.-M.; Mark, J. E., Model Networks of End-Linked Polydimethylsiloxane Chains. XIV. Stress-Strain, Thermoelastic, and Birefringence Measurements on the Bimodal Networks at Very Low Temperatures. *J. Polym. Sci., Polym. Phys. Ed.* **1982**, *20*, 473–480.
275. Galiatsatos, V.; Mark, J. E., Non-Gaussian Effects and Intermolecular Correlations in Bimodal Networks of Poly(dimethylsiloxane). In *Advances in Silicon-Based Polymer Science. A Comprehensive Resource*, Zeigler, J. M.; Fearon, F. W. G., Eds. American Chemical Society: Washington, DC, **1990**; Vol. 224, pp 201–206.
276. Riande, E.; Saiz, E., *Dipole Moments and Birefringence of Polymers*. Prentice Hall: Englewood Cliffs, NJ, 1992.
277. Subramanian, P. R.; Galiatsatos, V., Stress-Optical Properties of Bimodal Polymer Networks. *Macromol. Symp.* **1993**, *76*, 233–238.
278. Galiatsatos, V.; Neaffer, R. O.; Sen, S.; Sherman, B. J., Refractive Index, Stress Optical Coefficient, and Optical Configuration Parameter of Polymers. In *Physical Properties of Polymers Handbook*, Mark, J. E., Ed. Springer-Verlag: New York, 1996; pp 535–543.
279. Aguilera-Mercado, B. M.; Genesky, G. D.; Duncan, T. M.; Cohen, C.; Escobedo, F. A., ^{2}H NMR and Simulation Studies of Chain Segment Orientation in PDMS Bimodal Networks. *Macromolecules* **2010**, *43*, 7173–7184.
280. Jackson, C. L.; McKenna, G. B., The Glass Transition of Organic Liquids Confined to Small Pores. *J. Non-Cryst. Solids* **1991**, *131–133*, 221–224.
281. Goldstein, A. N.; Echer, C. M.; Alivisatos, A. P., Melting in Semiconductor Nanocrystals. *Science* **1992**, *256*, 1425–1427.
282. Grobler, J. H. A.; McGill, W. J., Anomalous Freezing-Point Depression of Solvents in a Swollen Homogeneous Rubber Network. *J. Polym. Sci., Part B: Polym. Phys.* **1993**, *31*(5), 575–577.
283. Baba, M.; Nedelec, J.-M.; Billamboz, N.; Lacoste, J., Porous Volume of Inorganic Materials and Degree of Swelling of Elastomers Monitored by DSC Measurements. *J. Phys. Chem. B* **2003**, *107*, 12884–12890.

284. Hoei, Y.; Ikeda, Y.; Sasaki, M., A Frozen Tube Model of Melting Point Depression of Solvents in Entangled Polymer Systems. *J. Phys. Chem. B* **2003**, *107*, 1483–1490.
285. Billamboz, N.; Nedelec, J.-M.; Grivet, M.; Baba, M., Cross-Linking of Polyolefins: A Study by Thermoporosimetry with Benzene Derivatives as Swelling Solvents. *ChemPhysChem* **2005**, *4*, 1126–1132.
286. Oodo, K.; Masuda, T.; Fujimori, H., Estimation of the Pore Diameter of Organosilane-Modified Silica Based on the Fusion Temperature of Ice. *J. Non-Cryst. Solids* **2011**, *357*, 683–685.
287. Madkour, T.; Mark, J. E., Some Evidence on Pore Sizes in Poly(dimethylsiloxane) Elastomers Having Unimodal, Bimodal, or Trimodal Distributions of Network Chain Lengths. *Polym. Bull.* **1993**, *31*, 615–621.
288. Clarson, S. J.; Mark, J. E.; Semlyen, J. A., Studies of Cyclic and Linear Poly(dimethylsiloxanes): 24. Topological Trapping of Cyclic Polymers into Unimodal and Bimodal Model Network Structures. *Polym. Comm.* **1987**, *28*, 151–153.
289. Oikawa, H., Dynamic Light Scattering Study on a Bimodal End-Linked Polydimethylsiloxane Network Structure. *Polymer* **1992**, *33*, 1116–1119.
290. Saalwachter, K., ^{1}H Multiple-Quantum Nuclear Magnetic Resonance Investigations of Molecular Order in Polymer Networks: Evidence for a Linear Mixing Law in Bimodal Systems. *J. Chem. Phys.* **2003**, *119*, 3468–3482.
291. Saalwachter, K.; Kleinschmidt, F., Swelling Heterogeneities in End-Linked Model Networks: A Combined Proton Multiple-Quantum NMR and Computer Simulation Study. *Macromolecules* **2004**, *37*, 8556–8568.
292. Saalwachter, K., ^{1}H Multiple-Quantum Nuclear Magnetic Resonance Investigations of Molecular Order in Polymer Networks. II. Intensity Decay and Restricted Slow Dynamics. *J. Chem. Phys.* **2004**, *120*, 454–464.
293. Nasir, M.; Teh, G. K., The Effects of Various Types of Crosslinks on the Physical-Properties of Natural-Rubber. *Eur. Polym. J.* **1988**, *24*(8), 733–736.
294. Sharaf, M. A.; Mark, J. E., Elastomers with Polysulfidic Cross Links Viewed as Bimodal Networks. *J. Macromol. Sci., Chem.* **1991**, *A28*, 67–73.
295. Sharaf, M. A.; Mark, J. E.; Hosani, Z. Y. A., Regular Bimodal Polydimethylsiloxane Networks. Elastomeric Properties of the Tetrafunctional Networks. *Eur. Polym. J.* **1993**, *29*, 809–817.
296. Sharaf, M. A.; Mark, J. E.; Al-Ghazal, A. A.-R., Elastomeric Properties of Poly(dimethylsiloxane) Networks Having High-Functionality Crosslinks and Bimodal Chain-Length Distributions. *J. Appl. Polym. Sci. Symp.* **1994**, *55*, 139–152.
297. Sharaf, M. A.; Mark, J. E., The Modulus of Randomly-Crosslinked Poly(dimethylsiloxane) Networks. *Polym. Gels Networks* **1993**, *1*, 33–44.
298. Sharaf, M. A.; Mark, J. E., Dependence of the Moduli of Random Bimodal Networks on Chain-Length Distribution. *Makromol. Symp.* **1993**, *76*, 13–31.
299. Quan, X., Properties of Post-Cured Siloxane Networks. *Polym. Eng. Sci.* **1989**, *29*, 1419–1425.
300. Madkour, T.; Hamdi, M. S., Bimodals from Two XL systems. *J. Appl. Polym. Sci.* **1996**, *61*(8), 1239–1244.
301. Isayev, A. I.; Kim, S. H.; Levin, V. Y., Superior Mechanical Properties of Reclaimed SBR with Bimodal Network. *Rubber Chem. Technol.* **1997**, *70*, 194–201.
302. Yun, J.; Isayev, A. I., Superior Mechanical Properties of Ultrasonically Recycled EPDM Rubber. *Rubber Chem. Technol.* **2003**, *76*, 253–270.

303. Holmes, G. A.; Letton, A., Bisphenol-A Bimodal Epoxy Resins. Part I: The Dynamic Mechanical Characterization of a 6300 (340/22,500) Weight Average Molecular Weight System. *Polym. Eng. Sci.* **1994**, *34*, 1635–1642.
304. Wu, W. L.; Hu, J. T.; Hunston, D. L., Structural Heterogeneity in Epoxies. *Polym. Eng. Sci.* **1990**, *30*(14), 835–840.
305. Tan, N. C. B.; Bauer, B. J.; Plestil, J.; Barnes, J. D.; Liu, D.; Matejka, L.; Dusek, K.; Wu, W. L., Bimodal Thermosets. *Polymer* **1999**, *40*, 4603–4614.
306. Shen, H.-W.; Luan, T.; Xie, B.-H.; Yang, W.; Yang, M.-B., Rheological Behaviors and Molecular Weight Distribution Characteristics of Bimodal High-Density Polyethylene. *J. Appl. Polym. Sci.* **2011**, *121*, 1543–1549.
307. Balzano, L.; Rastogi, S.; Peters, G., Self-Nucleation of Polymers with Flow; The Case of Bimodal Polyethylene. *Macromolecules* **2011**, *44*, 2926–2933.
308. Hakim, S.; Moballegh, L., Tailoring MWD of Bimodal Polyethylene in the Presence of Zeigler-Natta Catalyst. *Macromol. Symp.* **2011**, *302*, 191–197.
309. Long, Y.-Y.; Liu, S.-R.; Cui, L.; Li, Y.-S., Effect of Branching on the Thermal and Crystallization Behavior of Bimodal Polyethylenes Synthesized with Binary Late-Transition-Metal Catalyst Combinations. *J. Appl. Polym. Sci.* **2010**, *115*, 3045–3055.
310. Zhou, Z.; Hiltner, A.; Baer, E., Predicting Long-Term Creep Failure of Bimodal Polyethylene Pipe from Short-Term Fatigue Tests. *J. Mater. Sci.* **2011**, *46*, 174–182.
311. Sun, X.; Shen, H.; Xie, B.; Yang, W.; Yang, M., Fracture Behavior of Bimodal Polyethylene: Effect of Molecular Weight Distribution Characteristics. *Polymer* **2011**, *52*, 564–570.
312. Chen, C. C.; Borden, M. A., Ligand Conjugation to Bimodal Poly(ethylene glycol) Brush Layers on Microbubbles. *Langmuir* **2010**, *26*, 13183–13194.
313. Bucknall, C. B., *Toughened Plastics*. Applied Science Press: London, 1977.
314. Donald, A. M.; Kramer, E. J., Toughening of thermoplastics with elastomers. *J. Appl. Polym. Sci.* **1982**, *27*, 3729.
315. Okamoto, Y.; Miyagi, H.; Kakugo, M.; Takahashi, K., Impact Improvement Mechanism of HIPS with Bimodal Distribution of Rubber Particle Size. *Macromolecules* **1991**, *24*, 5639–5644.
316. Chen, T. K.; Jan, Y. H., Fracture Mechanism of Toughened Epoxy Resin with Bimodal Rubber-Particle Size Distribution. *J. Mater. Sci.* **1992**, *27*, 111–121.
317. Demirors, M., Role of rubber particle size distribution in toughness of High Impact Polystyrene (HIPS). *Abstr. Pap. Am. Chem. Soc.* **1998**, *216*, U837–U837.
318. Takahashi, J.; Watanabe, H.; Nakamoto, J.; Arakawa, K.; Todo, M., In-Situ Polymerization and Properties of Methyl Methacrylate-Butadiene-Styrene Resin with Bimodal Rubber Particle Size Distributions. *Polym. J.* **2006**, *38*, 835–843.
319. Arakawa, K.; Mada, T.; Takahashi, J.; Todo, M.; Ooka, S., Effect of Rubber Particle Size on the Impact Tensile Fracture Behavior of MBS Resin with a Bimodal Particle Size Distribution. *J. Mater. Sci.* **2007**, *42*, 8700–8706.
320. Li, D.; Peng, J.; Zhai, M.; Qiao, J.; Zhang, X.; Wei, G., Novel Methods for Synthesis of High-Impact Polystyrene with Bimodal Distribution of Rubber Particle Size. *J. Appl. Polym. Sci.* **2008**, *109*, 2071–2075.
321. Mae, H.; Omiya, M.; Kishimoto, K., Tensile Behavior of Polypropylene Blended with Bimodal Distribution of Styrene-Ethylene-Butadiene-Styrene Particle Size. *J. Appl. Polym. Sci.* **2008**, *107*, 3520–3528.

322. Dai, R.; Gao, G.-H.; Zhang, H., Different Deformation Mechanisms of Two Modified-Polystyrene Bimodal Systems. *Polym. Int.* **2010**, *59*, 738–742.
323. Wen, J.; Mark, J. E., Synthesis, Structure, and Properties of Poly(dimethylsiloxane) Networks Reinforced by in Situ-Precipitated Silica-Titania, Silica-Zirconia, and Silica-Alumina Mixed Oxides. *J. Appl. Polym. Sci.* **1995**, *58*, 1135–1145.
324. Scherer, C.; Utech, S.; Scholz, S.; Noskov, S.; Kindervater, P.; Graf, R.; Thunemann, A. F.; Maskos, M., Synthesis, Characterization and Fine-Tuning of Bimodal Poly(organosiloxane) Nanoparticles. *Polymer* **2010**, *51*, 5432–5439.
325. Olariu, C. I.; Yin, H. H. P.; Bouffier, L.; Nedjadi, T.; Costello, E.; Williams, S. R.; Halloran, C. M.; Rosseinsky, M. J., Multifunctional Fe_3O_4 Nanopartilcles for Targeted Bi-Modal Imaging of Pancreatic Cancer. *J. Mater. Chem.* **2011**, *21*, 12650–12659.
326. Shin, C.-K.; Kawthekar, R. B.; Kim, G.-J., Application of the Bimodal Meso/Macroporous Composite Synthesized from MCM-41 Sol. *J. Nanosci. Nanotech.* **2007**, *7*, 3876–3879.
327. Jiao, F.; Hill, A. H.; Harrison, A.; Berko, A.; Chadwick, A. V.; Bruce, P. G., Synthesis of Ordered Mesoporous NiO with Crystalline Walls and a Bimodal Pore Size Distribution. *J. Am. Chem. Soc.* **2008**, *130*, 5262–5266.
328. Ma, X.; Sun, H.; Yu, P., A Novel Way for Preparing High Surface Area Silica Monoliths with Bimodal Pore Structure. *J. Mater. Sci.* **2008**, *43*, 887–891.
329. Tao, S.; Yin, J.; Li, G., High-Performance TNT Chemosensory Materials Based on Nanocomposites with Bimodal Porous Structures. *J. Mater. Chem.* **2008**, *18*, 4872–4878.
330. Xing, R.; Lehmler, H.-J.; Knutson, B. L.; Rankin, S. E., Synthesis and Tuning of Bimodal Mesoporous Silica by Combined Hydrocarbon/Fluorocarbon Surfactant Templates. *Langmuir* **2009**, *25*, 6486–6492.
331. Carroll, N. J.; Pylypenko, S.; Atanassov, P. B.; Peter, D. N., Microparticles with Bimodal Nanoporosity Derived by Microemulsion Templating. *Langmuir* **2009**, *25*, 13540–13544.
332. Feng, F.; Paulson, S.; Young, J.; Birss, V., Synthesis of Bimodal Mesoporous Ni Oxide from Octylamine. *J. Sol-Gel Sci. Technol.* **2009**, *51*, 153–157.
333. Wang, W.; Pang, Y.; Hodgson, S. N. B., Design and Fabrication of Bimodal Meso-porous WO_3 Thin films and Their Electronic Properties. *J. Mater. Chem.* **2010**, *20*, 8591–8599.
334. Taki, K.; Okumura, S., Production of Porous Polymer Foams with Unimodal or Bimodal Pore-Size Distributions via Depressurization and Photopolymerization-Induced Bubble Nucleation in Low-Viscosity UV-Curable Monomer/High Pressure CO_2 Solutions. *Macromolecules* **2010**, *43*, 9899–9907.
335. Le Bideau, J.; Miah, M. Y.; Vioux, A.; Fajula, F.; Galarneau, A., Bimodal Porous Silica Monoliths Obtained by Phase Separation in Non-Aqueous Media. *J. Mater. Chem.* **2010**, *20*, 964–971.
336. Sun, Z.-Y.; Shen, M.-X.; Cao, G.-P.; Deng, J.; Liu, Y.; Liu, T.; Zhao, L.; Yuan, W.-K., Preparation of Bimodal Porous Copolymer Containing β-Cyclodextrin and Its Inclusion Adsorption Behavior. *J. Appl. Polym. Sci.* **2010**, *118*, 2176–2185.
337. Zhao, W.; Fierro, V.; Pizzi, A.; Celzard, A., Bimodal Cellular Activated Carbons Derived from Tannins. *J. Mater. Sci.* **2010**, *45*, 5778–5785.
338. De Jong, K. P.; Zecevic, J.; Friedrich, H.; De Jongh, P. E.; Bulut, M.; Van Donk, S.; Kenmogne, R.; Finiels, A.; Hulea, V.; Fajula, F., Zeolite Y Crystals with Trimodal Porosity as Ideal Hydrocracking Catalysts. *Angew. Chem. Int. Ed. Engl.* **2010**, *49*, 10074–10078.

339. Shen, W.; Ren, L.; Zhou, H.; Zhang, S.; Fan, W., Facile One-Pot Synthesis of Bimodal Mesoporous Carbon Nitride and Its Function as a Lipase Immobilization Support. *J. Mater. Chem.* **2011**, *21*, 3890–3894.
340. Vila, M.; Izquierdo-Bargba, I.; Bourgeois, A.; Vallet-Regi, M., Bimodal Meso/Macro Porous Hydroxyapatite Coatings. *J. Sol-Gel Sci. Technol.* **2011**, *57*, 109–113.
341. Zhao, J.; Jiang, Z.-T.; Li, J. T.-R., Sol-Gel Synthesis and Characterization of Titania Monolith with Bimodal Porosity. *J. Sol-Gel Sci. Technol.* **2011**, *58*, 436–441.
342. Moreno, J.; Van Grieken, R.; Carrero, A.; Paredes, B., Ethylene Polymerization by Metallocene Catalysts Supported over Materials with Bimodal Pore Size Distributions. *Macromol. Symp.* **2011**, *302*, 198–207.
343. Moreno, J.; Van Grieken, R.; Carrero, A.; Paredes, B., Development of Novel Chromium Oxide/Metallocene Hybrid Catalysts for Bimodal Polyethylene. *Polymer* **2011**, *52*, 1891–1899.
344. Suzuki, N.; Kiba, S.; Yamauchi, Y., Bimodal Filler System Consisting of Mesoporous Silica Particles and Silica Nanoparticles Toward Efficient Suppression of Thermal Expansion in Silica/Epoxy Composites. *J. Mater. Chem.* **2011**, *21*, 14941–14947.
345. Gao, L.; Sun, J.; Li, Y.; Zhang, L., Bimodal Mesoporous Silicas Functionalized with Different Level and Species of Amino Groups for Adsorption and Controlled Release of Aspirin. *J. Nanosci. Nanotech.* **2011**, *11*, 6692–6697.
346. Peterson, V. K.; Brown, C. M.; Liu, Y.; Kepert, C. J., Structural Study of D_2 within the Trimodal Pore System of a Metal Organic Framework. *J. Phys. Chem. C* **2011**, *115*, 8851–8857.
347. Gallagher, J. J.; Tekoriute, R.; O'Reilly, J.-A.; Kerskens, C.; Gun'ko, Y. K.; Lynch, M., Bimodal Magnetic-Fluorescent Nanostructures for Biomedical Applications. *J. Mater. Chem.* **2009**, *19*, 4081–4084.
348. Howes, P.; Green, M.; Bowers, A.; Parker, D.; Varma, G.; Kallumadil, M.; Hughes, M.; Warley, A.; Brain, A.; Botnar, R., Magnetic Conjugated Polymer Nanoparticles as Bimodal Imaging Agents. *J. Am. Chem. Soc.* **2010**, *132*, 9833–9842.
349. McCarthy, J. E.; Prina-Mello, A.; Rakovich, T.; Volkov, Y.; Gun'ko, Y. K., Fabrication and Characterization of Multimodal Magnetic-Fluorescent Polystyrene Nanowires as Selective Cell Imaging Probes. *J. Mater. Chem.* **2011**, *21*, 14219–14225.
350. Nair, B. P.; Pavithran, C., Bifunctionalized Hybrid Silica Spheres by Hydrolytic Cocondensation of 3-Aminopropyltriethoxysilane and Vinyltriethoxysilane. *Langmuir* **2010**, *26*, 730–735.
351. Kumar, S.; Singh, V.; Aggarwal, S.; Mandal, U. K.; Kotnala, R. K., Bimodal $Co_{0.5}Zn_{0.5}Fe_2O_4$/PANI Nanocomposites: Synthesis, Formation Mechanism and Magnetic Properties. *Comp. Sci. Tech.* **2010**, *70*, 249–254.
352. Yoo, S. I.; Yun, S. H.; Kim, H. K.; Sohn, B.-H., Highly Ordered Hexagonal Arrays of Hybridized Micelles from Bimodal Self-Assemblies of Diblock Copolymer Micelles. *Macromol. Rapid Commun.* **2010**, *31*, 645–650.
353. Chang, J.-J.; Kwon, J.-H.; Yoo, S. I.; Park, C.; Sohn, B.-H., Bimodal Arrays of Two Types of Nanoparticles by Mixtures of Diblock Copolymer Micelles. *J. Mater. Chem.* **2009**, *19*, 1621–1625.
354. Salerno, A.; Zeppetelli, S.; Di Maio, E.; Iannace, S.; Netti, P. A., Design of Bimodal PCL and PCL-HA Nanocomposite Scaffolds by Two Step

Depressurization During Solid-State Supercritical CO_2 Foaming. *Macromol. Rapid Commun.* **2011**, *32*, 1050–1156.
355. Van Berkel, K. Y.; Piekarski, A. M.; Kierstead, P. H.; Pressly, E. D.; Ray, P. C.; Hawker, C. J., A Simple Route to Multimodal Composite Nanoparticles. *Macromolecules* **2009**, *42*, 1425–1427.
356. Hwang, G. S.; Kaviany, M.; Gostick, J. T.; Kientiz, B.; Weber, A. Z.; Kim, M. H., Role of Water States on Water Uptake and Proton Transport in Nafion Using Molecular Simulations and Bimodal Networks. *Polymer* **2011**, *52*, 2584–2593.
357. Leung, K.; Nielsen, I. M. B.; Criscenti, L. J., Elucidating the Bimodal Acid-Base Behavior of the Water-Silica Interface from First Principles. *J. Am. Chem. Soc.* **2009**, *131*, 18358–18365.
358. Willis, S. A.; Dennis, G. R.; Zheng, G.; Price, W. S., Averaging Effects in PGSE NMR Attenuations Observed in Bimodal Molecular Weight PMMA Solutions. *Macromolecules* **2010**, *43*, 7351–7356.
359. Gentsch, R.; Boysen, B.; Lankenau, A.; Borner, H. G., Single-Step Electrospinning of Bimodal Fiber Meshes for Ease of Cellular Infiltration. *Macromol. Rapid Commun.* **2010**, *31*, 59–64.
360. May, A.; Stebe, M. J.; Gutierrez, J. M.; Blin, J. L., Coexistence of Two Kinds of Fluorinated Hydrogenated Micelles as Building Blocks for the Design of Bimodal Mesoporous Silica with Two Ordered Mesopore Networks. *Langmuir* **2011**, *27*, 1400–1404.
361. Bui, Q. H., Heterogeneous Plastic Deformation in Bimodal Bulk Ultrafine-Grained Nickel. *J. Mater. Sci.* **2012**, *47*, 1903–1908.
362. Peng, S.; Sun, Y., Ripening of Bimodally Distributed AgCl Nanoparticles. *J. Mater. Chem.* **2011**, *21*, 11644–11650.
363. Madkour, T.; Mark, J. E., Elastomeric Properties of Poly(dimethylsiloxane) Networks Having Bimodal and Trimodal Distributions of Network Chain Lengths. *J. Macromol. Sci., Macromol. Rep.* **1994**, *A31*, 153–160.
364. Genesky, G.; Cohen, C., Toughness and Fracture Energy of PDMS Bimodal and Trimodal Networks with Widely Separated Precursor Molar Masses. *Polymer* **2010**, *51*, 4152–4159.
365. Giuliani, J. R.; Gjersing, E. L.; Chinn, S. C.; Jones, T. V.; Wilson, T. S.; Alviso, C. T.; Herberg, J. L.; Pearson, M. A.; Maxwell, R. S., Thermal Degradation in a Trimodal Poly(dimethylsiloxane) Network Studied by ^{1}H Multiple Quantum NMR. *J. Phys. Chem. B* **2007**, *111*, 12977–12984.
366. Gjersing, E.; Chinn, S.; Giuliani, J. R.; Herberg, J.; Maxwell, R. S.; Eastwood, E.; Bowen, D.; Stephens, T., Investigation of Network Heterogeneities in Filled, Trimodal, Highly Functional PDMS Networks by ^{1}H Multiple Quantum NMR. *Macromolecules* **2007**, *40*, 4953–4962.
367. Buddenhagen, D. A.; Legge, N. R.; Zscheuschler, G. Elastomeric Triblock Copolymer Compositions and Articles Made Therewith. **1992**, US patent US5112900 A.
368. Buddenhagen, D. A.; Legge, N. R.; Zscheuschler, G. Elastomeric Triblock Copolymer Compositions and Articles Made Therewith. 1990, EU patent EP 0488021 A1.
369. Erman, B.; Mark, J. E., Calculations on Trimodal Elastomeric Networks. Effects of Chain Length and Composition on Ultimate Properties. *Macromolecules* **1998**, *31*, 3099–3103.
370. Hofmeister, C.; Yao, B.; Sohn, Y. H.; Delahanty, T.; Van Den Bergh, M.; Cho, K., Composition and Structure of Nitrogen-Containing Dispersoids in Trimodal Aluminum Metal-Matrix Composites. *J. Mater. Sci.* **2010**, *45*, 4871–4876.

371. Garrido, L.; Mark, J. E.; Clarson, S. J.; Semlyen, J. A., Studies of Cyclic and Linear Poly(dimethylsiloxanes): 16. Trapping of Cyclics Present During the End Linking of Linear Chains into Network Structures. *Polym. Comm.* **1985**, *26*, 53–55.
372. Garrido, L.; Mark, J. E.; Clarson, S. J.; Semlyen, J. A., Studies of Cyclic and Linear Poly(dimethylsiloxanes): 17. Elastomeric Properties of Model PDMS Networks Containing Either Cyclics or Linear Chains as Diluent. *Polym. Comm.* **1985**, *26*(2), 55–57.
373. Clarson, S. J.; Mark, J. E.; Semlyen, J. A., Studies of Cyclic and Linear Poly(dimethylsiloxanes): Effect of Ring Size on the Topological Trapping of Cyclic Polymers into Model Network Structures. *Polym. Comm.* **1986**, *27*, 243–245.
374. DeBolt, L. C.; Mark, J. E., Models for the Trapping of Cyclic Poly(dimethylsiloxane) (PDMS) Chains in PDMS Networks. *Macromolecules* **1987**, *20*, 2369–2374.
375. Fyvie, T. J.; Frisch, H. L.; Semlyen, J. A.; Clarson, S. J.; Mark, J. E., Polymeric Catenanes from Cross-Linked Poly(2,6-Dimethyl-1,4-Phenylene oxide) and Cyclic Poly(dimethylsiloxane). *J. Polym. Sci., Polym. Chem. Ed.* **1987**, *25*(9), 2503–2509.
376. Huang, W.; Frisch, H. L.; Hua, Y.; Semlyen, J. A., A Study of the Properties and Topological Trapping of Cyclic Poly(dimethylsiloxane) in Poly(2,6-dimethyl-1,4-phenylene oxide) Networks. *J. Polym. Sci., Part A: Polym. Chem.* **1990**, *28*(7), 1807–1812.
377. Flory, P. J., *Statistical Mechanics of Chain Molecules*. Interscience: New York, 1969.
378. Mattice, W. L.; Suter, U. W., *Conformational Theory of Large Molecules. The Rotational Isomeric State Model in Macromolecular Systems*. Wiley: New York, 1994.
379. Rehahn, M.; Mattice, W. L.; Suter, U. W., Rotational Isomeric State Models in Macromolecular Systems. *Adv. Polym. Sci.* **1997**, *131*, 1–18.
380. Mark, J. E.; Curro, J. G., A Non-Gaussian Theory of Rubberlike Elasticity Based on Rotational Isomeric State Simulations of Network Chain Configurations. I. Polyethylene and Polydimethylsiloxane Short-Chain Unimodal Networks. *J. Chem. Phys.* **1983**, *79*, 5705–5709.
381. Galiatsatos, V.; Eichinger, B. E., An Interpretation of the Topological Trapping of Cyclic Poly(dimethylsiloxane) in PDMS Network Structures. *Polym. Commun.* **1987**, *28*, 182–184.
382. Rigbi, Z.; Mark, J. E., Concurrent Chain Extension and Crosslinking of Hydroxyl-Terminated Poly(dimethylsiloxane): Possible Formation of Catenate Structures. *J. Polym. Sci., Polym. Phys. Ed.* **1986**, *24*, 443–449.
383. De Gennes, P. G., *Scaling Concepts in Polymer Physics*. Cornell University Press: Ithaca, New York, 1979.
384. Frisch, H. L.; Wasserman, E., Chemical Topology1. *J. Am. Chem. Soc.* **1961**, *83*(18), 3789–3795.
385. Gibson, H. W.; Bheda, M. C.; Engen, P. T., Rotaxanes, Catenanes, Polyrotaxanes, Polycatenanes and Related Materials. *Prog. Polym. Sci.* **1994**, *19*, 843–945.
386. Gong, C.; Gibson, H. W., Controlling Polymeric Topology by Polymerization Conditions: Mechanically Linked Network and Branched Poly(urethane rotaxane)s with Controllable Polydispersity. *J. Am. Chem. Soc.* **1997**, *119*, 8585–8591.

387. Iwata, K.; Ohtsuki, T., On Catenate Network Formation by End-Linking Reactions. *J. Polym. Sci., Part B: Polym. Phys.* **1993**, *31*(4), 441–446.
388. Bokobza, L.; Erman, B., A Theoretical and Experimental Study of Filler Effect on Stress-Deformation-Segmental Orientation Relations for Poly(dimethylsiloxane) Networks. *Macromolecules* **2000**, *33*, 8858–8864.
389. Bokobza, L., Filled Elastomers: A New Approach Based on Measurements of Chain Orientation. *Polymer* **2001**, *42*, 5415–5423.
390. Bokobza, L., The Reinforcement of Elastomeric Networks by Fillers. *Macromol. Mater. Eng.* **2004**, *289*, 607–621.
391. Bokobza, L., Infrared Analysis of Elastomeric Composites Under Uniaxial Extension. *Macromol. Symp.* **2005**, *220*, 45–59.
392. Klimov, E.; Hoffmann, G. G.; Gumenny, A.; Seisler, H. W., Low-Temperature FT-NIR Spectroscopy of Strain-Induced Orientation and Crystallization in a Poly(dimethylsiloxane) Network. *Macromol. Rapid Commun.* **2005**, *26*, 1093–1098.
393. Aguilera-Mercado, B. M.; Cohen, C.; Escobedo, F. A., Extraction of Segment Orientation Distributions in Polymer Networks by Inversion of ^{2}H NMR Spectra through the Maximum-Entropy Method. *Macromolecules* **2009**, *42*, 8889–8898.
394. Lorthioir, C.; Deloche, B.; Courtieu, J., Experimental Evidence for a Negative Orientational Order Parameter in a Uniaxially Compressed Poly(dimethylsiloxane) Network. *Macromolecules* **2001**, *34*, 8835–8837.
395. Rahalkar, R. R.; Lamb, J.; Harrison, G.; Barlow, A. J.; Hawthorn, W.; Semlyen, J. A.; North, A. M.; Pethrick, R. A., Viscoelastic Studies of Polydimethylsiloxanes. *Proc. R. Soc. London, Ser. A* **1984**, *394*, 207–222.
396. Bell, W.; Daly, J.; North, A. M.; Pethrick, R. A.; Teik, P. B., Chain Entanglement and Normal Mode Motions in Poly(dimethylsiloxane) + Toluene Mixtures. *J. Chem. Soc., Faraday Trans. II* **1979**, *75*, 1452–1464.
397. Eastwood, A. E.; North, A. M.; Pethrick, R. A., Ultrasonic Relaxation Studies of Poly(methylphenylsiloxane) and Its Solutions in Toluene. *J. Chem. Soc., Faraday Trans. II* **1982**, *78*, 95–102.
398. Smith, D. A.; Thomas, J. K., Dynamic and Equilibrium Studies of Silicone Rubber Vulcanizates. *Eur. Polym. J.* **1971**, *7*, 433–443.

CHAPTER 8

Copolymers and Interpenetrating Networks

8.1 RANDOM COPOLYMERS

Random copolymers are prepared by the copolymerization of a mixture of cyclic oligomers.[1–5] Although the resulting polymer can be quite blocky (figure 8.1), taking the reaction to equilibrium can give a polymer that is essentially random in its chemical sequencing.[4, 6] One reason for preparing copolymers is to introduce functional species, such as hydrogen or vinyl side groups, along the chain backbone to facilitate cross linking. Another reason is the introduction of sufficient chain irregularity to make the polymer inherently noncrystallizable.

Specific examples of comonomers include imides,[7–12] perylenediimide,[13] urethane-ureas,[14, 15] epoxies,[16–19] other siloxanes,[20–23] amides,[24] styrene,[25] divinylbenzene,[26] acrylics,[27, 28] silsesquioxanes,[29] polythiophenes,[30] and poly(lactic acid).[31] One novel combination is the preparation of polysiloxane-based episulfide resins.[32] An unusual application is the use of monomethylitaconate-grafted polymethylsiloxane to induce crystal growth of $CaCO_3$.[33] Polysiloxanes containing thermally curable brenzoxazine moieties in the main chain are also in the category.[34]

These and other copolymers have been extensively characterized by nuclear magnetic resonance (NMR) spectroscopy.[35]

8.2 BLOCK COPOLYMERS

The sequential coupling of functionally terminated chains of different chemical structure can be used to make block copolymers,[36, 37] including

Random

AABBBABBAABAAABBABAAA

Diblock

AAAAAABBBBB

Triblock

AAAAAABBBBBAAAAA

AAAAABBBBBCCCCC

Figure 8.1:
Types of copolymers.

those in which one or more of the blocks is a polysiloxane.[6, 38, 39] If the blocks are relatively long, separation into a two-phase system invariably occurs. Frequently, one block will be in a continuous phase and the other will be dispersed in domains having an average size the order of a few hundred angstroms. Such materials can have unique mechanical properties not available from homopolymer species. Sometimes similar properties can be obtained by the simple blending of two or more polymers.[40]

Examples of blocks used with polydimethylsiloxane (PDMS) include imides,[41–48] epoxies,[49] butadienes,[50–54] ε-caprolactones,[55] amides having trichlorogermyl pendant groups,[56] urethanes,[57–60] ureas,[61–65] poly(ethylene glycols),[66] polystyrene,[67–74] vinyl acetates,[75, 76] acrylates or methacrylates,[77–84] 2-vinylpyridine,[85] and even other polysiloxanes.[86–88] Some results have also been reported for polyesters,[89, 90] polyethers,[91] hydroxyethers of bisphenol A,[92] bisphenol A arylene ether sulfones,[93] vinylpyridinebenzoxazines,[94] methyloxazolines,[95] terpyridines,[96] polysulfones,[97, 98] γ-benzyl-L-glutamate,[99] and carboranes.[100] Two other examples are foamed polypropylene[101] and melamine resins.[102] Even ABA,[103, 104] ABC triblock copolymers,[105] and ABCBA pentablock copolymers involving PDMS have been reported.[106]

In a number of cases, the block used with the PDMS is sufficiently polar to give an amphiphilic block copolymer. Such materials form interesting structures in polar or nonpolar solvents. In the first case, the polar chains act like a corona around the nonpolar core, and in the latter, the nonpolar chains are a corona around the polar core. Examples include blocks of poly(ethylene oxide),[107–112] acrylamides,[113–115] sugars,[116] gluconolactone,[117] hydrolysable siloxanes,[118] and maleic anhydride—vinyl ethyl ethers.[119]

In some cases, grafting is used to form the copolymers. Examples of chains combined with PDMS in this way include poly(ethylene oxide),[120–123] fluorinated chains,[124] alkyl acrylates[125] and alkyl methacrylates,[126] poly(vinyl alcohol),[127] poly(ether sulfones),[128] and polyurethanes.[129] Grafting has also been used to introduce *t*-butylamine and *t*-butylammonium biocidal functionalities,[130] and to improve the adhesion of silicone rubber to polyurethane.[131] In other cases, the siloxane groups are simply placed on the ends of chains, such as onto some polybenzoxazines.[132]

In related work, NMR studies have been carried out on PDMS chains onto which octyl groups had been grafted.[133] Other examples of grafted systems include poly(ethylene oxide),[123] polyurethanes,[131] poly(butyl methacrylate),[126] and gelatin.[134]

8.3 INTERPENETRATING NETWORKS

In this type of material, two networks are formed, either simultaneously or sequentially, in such a way as to interpenetrate one another. The networks thus "communicate" with one another through interchain physical forces and entanglements, rather than through covalent bonds. A particularly simple example is the simultaneous formation of two PDMS networks, one by a condensation end-linking reaction and the other by an addition end-linking reaction, with the two types of chains mixed at the molecular level.[135, 136]

A more complex example, with more novel properties, is the preparation of interpenetrating networks (IPNs) between PDMS and an organic thermoplastic polymer such as a nylon, polyurethane, or polyester. The preparation of this material, called Rimplast® by its developers,[137] involves a chemical reaction between a vinyl-functional polysiloxane blended into the thermoplastic, and a hydride-functional PDMS blended into more of the same thermoplastic. Small amounts of platinum catalyst are also present in both types of components. The polysiloxane content is typically around 10%. Pellets of both components are placed in an extruder or other high-temperature processor where they melt into a uniform mass at approximately 300°C. The reaction between the two complementary types of PDMS results in a network evenly distributed throughout the thermoplastic as a "semi-IPN" in the sense that there is only one IPN (PDMS). The resulting composite has the desirable properties of both the PDMS (good lubricity, abrasion resistance, and dielectric properties), and the thermoplastic (good mechanical strength and molding characteristics).

Other examples involving PDMS as one of the networks involve epoxies,[138, 139] cellulose acetate butyrate,[140, 141] polycarbonate,[142] poly(vinyl alcohol),[143] acrylates,[144] fluorinated acrylate,[145] *N*-vinyl-2-pyrrolidone,[146] and methacrylate/acrylamides.[147]

REFERENCES

1. Warrick, E. L.; Pierce, O. R.; Polmanteer, K. E.; Saam, J. C., Silicone Elastomer Developments 1967–1977. *Rubber Chem. Technol.* **1979**, *52*, 437–525.
2. McGrath, J. E.; Riffle, J. S.; Banthia, A.K.; Yilgor, I.; Wilkes, G. L., An Overview of the Polymerization of Cyclosiloxanes. In *Initiation of Polymerization*, Bailey, F. E., Jr., Ed. American Chemical Society: Washington, DC, 1983; pp 145–172.
3. Babu, G. N.; Christopher, S. S.; Newmark, R. A., Poly(dimethylsiloxane-co-diphenylsiloxanes): Synthesis, Characterization, and Sequence Analysis. *Macromolecules* **1987**, *20*, 2654–2659.
4. Ziemelis, M. J.; Saam, J. C., Sequence Distribution in Poly(dimethylsiloxane-*co*-methylvinylsiloxanes). *Macromolecules* **1989**, *22*, 2111–2116.
5. Pouget, E.; Tonnar, J.; Lucas, P.; Lacroix-Desmazes, P.; Ganachaud, F.; Boutevin, B., Well-Architectured Poly(dimethylsiloxane)-Containing Copolymers Obtained by Radical Chemistry. *Chem. Rev.* **2010**, *110*, 1233–1277.
6. McGrath, J. E.; Sormani, P. M.; Elsbernd, C. S.; Kilic, S., Kinetics, Mechanisms, and Synthesis Studies of Difunctional Aminopropyl Terminated Polydimethylsiloxane Oligomers. *Makromol. Chem., Macromol. Symp.* **1986**, *6*, 67–80.
7. Hamciuc, E.; Hamciuc, C.; Cazacu, M.; Lisa, G.; Okrasa, L., Polyimide-Polysiloxane Copolymers. Evaluation of the Thermal and Electrical Properties. *J. Macromol. Sci., Pure Appl. Chem.* **2007**, *44*, 1069–1078.
8. Hamciuc, E.; Hamciuc, C.; Cazacu, M., Poly(1,3,4-Oxadiazole-ether-imide)s and Their Polydimethylxiloxane-Containing Copolymers. *Eur. Polym. J.* **2007**, *43*, 4739–4749.
9. Hamciuc, E.; Hamciuc, C.; Cazacu, M.; Ignat, M.; Zarnescu, G., Polyimide-Polydimethylsiloxane Copolymers Containing Nitrile Groups. *Eur. Polym. J.* **2009**, *45*, 182–190.
10. Xi, K.; Meng, Z.; Heng, L.; Ge, R.; He, H.; Yu, X.; Jia, X., Polyimide-Polydimethylsiloxane Copolymers for Low-Dielectric-Constant and Moisture-Resistant Applications. *J. Appl. Polym. Sci.* **2009**, *113*, 1633–1641.
11. Damaceanu, M.-D.; Musteata, V.-E.; Cristea, M.; Bruma, M., Viscoelastic and Dielectric Behavior of Thin Films Made from Siloxane-Containing Poly(oxadiazole-imide)s. *Eur. Polym. J.* **2010**, *46*, 1049–1062.
12. Adhikari, R.; Dao, B.; Hodgkin, J.; Mardel, J., Synthesis, Structure and Membrane Properties of Siloxane-Imide Co-polymers Produced by Aqueous Polymerization. *Eur. Polym. J.* **2011**, *47*, 1328–1337.
13. Fu, W.; He, C.; Jiang, S.; Chen, Z.; Zhang, J.; Li, Z.; Yan, S.; Zhang, R., Synthesis of Polymeric Electron Acceptor Based on Perylenediimide-Bridged Ladder Polysiloxane. *Macromolecules* **2011**, *44*, 203–207.
14. Lee, Y.-H.; Kim, E.-J.; Kim, H.-D., Synthesis and Properties of Waterborne Poly(urethane urea)s Containing Polydimethylsiloxane. *J. Appl. Polym. Sci.* **2011**, *120*, 212–219.

15. Alexandru, M.; Cazacu, M.; Cristea, M.; Nistor, A.; Grigoras, C.; Simionescu, B. C., Poly(siloxane-urethane) Crosslinked Structures Obtained by Sol-Gel Techniques. *J. Polym. Sci., Part A: Polym. Chem.* **2011**, *49*, 1708–1718.
16. Huang, W.; Zhang, Y.; Yu, Y.; Yuan, Y., Studies on UV-Stable Silicone-Epoxy Resins. *J. Appl. Polym. Sci.* **2007**, *104*, 3954–3959.
17. Herr, D.; Chaplinsky, S.; Romanelli, A. L.; Zadjura, R., Synthesis and Study of New Radical Organic/Inorganic Hybrid Epoxides. *J. Appl. Polym. Sci.* **2008**, *107*, 3244–3257.
18. Yang, S.; Kim, J. H.; Jin, J. H.; Bae, B.-S., Synthesis and Characterization of Nano-Sized Epoxy Oligosiloxanes for Fabrication of Transparent Nano Hybrid Materials. *J. Polym. Sci., Part B: Polym. Phys.* **2009**, *47*, 756–763.
19. Gunji, T.; Itagaki, S.; Kajiwara, T.; Abe, Y.; Hatakeyama, T.; Aoki, R., Preparation and Properties of Siloxane/Epoxy Organic-Inorganic Hybrid Thin Films, Self-Standing Films, and Bulk Bodies. *Polym. J.* **2009**, *41*, 541–546.
20. Curteanu, S.; Cazacu, M., Neural Networks and Genetic Algorithms Used for Modeling and Optimization of the Siloxane-Siloxane Copolymers Synthesis. *J. Macromol. Sci., Pure Appl. Chem.* **2008**, *45*, 21–34.
21. Qu, L.; Xie, Z.; Huang, G.; Tang, Z., Effect of Diphenylsiloxane Unit Content on Relaxation Behavior of Poly(dimethylsiloxane-co-diphenyisiloxane). *J. Polym. Sci., Part B: Polym. Phys.* **2008**, *46*, 1652–1659.
22. Cypryk, M.; Delczk, B.; Juhari, A.; Koynov, K., Controlled Synthesis of Trifluoropropylmethylsiloxane-Dimethylsiloxane Gradient Copolymers by Anionic ROP of Cyclotrisiloxanes. *J. Polym. Sci., Part A: Polym. Chem.* **2009**, *47*, 1204–1216.
23. Xu, Y.; Yin, H.; Zheng, H.; Yuan, S.; Chen, Z., Application Performance and Surface Morphologies of Amino Polysiloxanes with Different Amino Values and Amino Types. *J. Appl. Polym. Sci.* **2011**, *119*, 2326–2333.
24. Mosurkal, R.; Samuelson, L. A.; Smith, K. D.; Westmoreland, P. R.; Parmar, V. S.; Yan, F.; Kumar, J.; Watterson, A. C., Nanocomposites of TiO_2 and Siloxane Copolymers as Environmentally Safe Flame-Retardant Materials. *J. Macromol. Sci., Pure Appl. Chem.* **2008**, *45*, 942–946.
25. Mukbaniani, O.; Tatrishvili, T.; Titvinidze, G.; Mukbaniani, N., Hydrosilylation Reactions of Methylhydridesiloxane to Styrene and α-Methylstyrene. *J. Appl. Polym. Sci.* **2006**, *101*, 388–394.
26. Pinho, R. O.; Radovanovic, E.; Torriani, I. L.; Yoshida, I. V. P., Hybrid Materials Derived from Divinylbenzene and Cyclic Siloxane. *Eur. Polym. J.* **2004**, *40*, 615–622.
27. Yamamoto, K.; Ito, E.; Fukaya, S.; Takagi, H., Phase-Separated Conetwork Structure Induced by Radical Copolymerization of Poly(dimethylsiloxane)-α,ω-diacrylate and N,N-Dimethylacrylamide. *Macromolecules* **2009**, *42*, 9561–9567.
28. Manoli, K.; Goustouridas, D.; Raptis, J.; Valamontes, E.; Sanopoulou, M., Vapor-Induced Swelling of Supported Methacrylic and Siloxane Polymer Films: Determination of Interaction Parameters. *J. Appl. Polym. Sci.* **2010**, *116*, 184–190.
29. Hoque, M. A.; Kakihana, Y.; Shinke, S.; Kawakami, Y., Polysiloxanes with Periodically Distributed Isomeric Double-Decker Silsesquixane in the Main Chain. *Macromolecules* **2009**, *42*, 3309–3315.
30. Baba, A.; Xia, C.; Knoll, W.; Advincula, R. C., Electrochemical Surface Plasmon Resonance and Field-Enhanced Light-Scattering: Monomer Copolymerization

with a Polysiloxane-Conjugated Polythiophene Precursor. *Macromol. Chem. Phys.* **2010**, *211*, 2624–2635.

31. Wakita, T.; Obata, A.; Poologasundarampillai, G.; Jones, J. R.; Kasuga, T., Preparation of Electrospun Siloxane-poly(lactic acid)-Valerite Hybrid Fibrous Membranes for Guided Bone Regeneration. *Comput. Sci. Technol.* **2010**, *70*, 1889–1893.
32. Lan, X.; Huang, W.; Yu, Y., Synthesis, Characterization, and Properties of the Polysiloxane-Based Episulfide Resin. *Eur. Polym. J.* **2010**, *46*, 1545–1556.
33. Neira-Carrillo, A.; Vasquez-Quitral, P.; Yazdani-Pedram, M., Crystal Growth of $CaCO_3$ Induced by Monomethylitaconate Grafted Polymethylsiloxane. *Eur. Polym. J.* **2010**, *46*, 1184–1193.
34. Aydogan, B.; Sureka, D.; Kiskan, B.; Yagci, Y., Polysiloxane-Containing Benzoxazine Moieties in the Main Chain. *J. Polym. Sci., Part A: Polym. Chem.* **2010**, *48*, 5156–5162.
35. Kahlig, H.; Mayer-Helm, B. X., Characterization of Siloxane Copolymers by Solution ^{17}O NMR Spectroscopy. *Polymer* **2005**, *46*, 6447–6454.
36. Keohan, F. L.; Hallgren, J. E., Novel Poly (imide-siloxane) Polymers and Copolymers. In *Silicon-Based Polymer Science. A Comprehensive Resource*, Zeigler, J. M.; Fearon, F. W. G., Eds. American Chemical Society: Washington, DC, 1990; Vol. 224, pp 165–179.
37. Odian, G., *Principles of Polymerization*, 4th ed. Wiley-Interscience: New York, 2004.
38. Ibemesi, J.; Gvozdic, N.; Keumin, M.; Tarshiani, Y.; Meier, D. J., Synthesis and Properties of Copolymers of Diphenylsiloxane with Other Organosiloxanes. In *Polymer-Based Molecular Composites*, Schaefer, D. W.; Mark, J. E., Eds. Materials Research Society: Pittsburgh, PA, 1990; Vol. 171, pp 105–115.
39. Schmidt, U.; Zehetmaier, P. C.; Rieger, B., Direct Synthesis of Poly(dimethylsiloxane) Copolymers with TPE-Properties via CuAAC (Click Chemistry). *Macromol. Rapid Commun.* **2010**, *31*, 545–548.
40. Manson, J. A.; Sperling, L. H., *Polymer Blends and Composites*. Plenum Press: New York, 1976.
41. Ku, C.-K.; Lee, Y.-D., Microphase Separation in Amorphous Poly(imide siloxane) Segmented Copolymers. *Polymer* **2007**, *48*, 3565–3573.
42. Ghosh, A.; Banerjee, S., Synthesis, Characterization, and Comparison of Properties of Novel Fluorinated Poly(imide siloxane) Copolymers. *J. Appl. Polym. Sci.* **2008**, *107*, 1831–1841.
43. Ghosh, A.; Banerjee, S.; Komber, H.; Schneider, K.; Haussler, L.; Voit, B., Synthesis and Characterization of Fluorinated Poly(imide siloxane) Block Copolymers. *Eur. Polym. J.* **2009**, *45*, 1561–1569.
44. Novak, I.; Sysel, P.; Zemek, J.; Spirkova, M.; Velic, D.; Aranyosiova, M.; Florian, S.; Pollak, V.; Kleinova, A.; Lednicky, F.; Janigova, I., Synthesis and Adhesion Properties of (Polyimide-Siloxane) Block Copolymers. *Eur. Polym. J.* **2009**, *45*, 57–69.
45. Liaw, W.-C.; Cheng, Y.-L.; Huang, P.-C.; Chen, K.-P.; Fu, L.-W., SiO_2 Reinforcement and Morphology of a Novel Poly(imide-siloxane)-Based Thermal Plastic Elastomer Composite. *Polym. J.* **2009**, *41*, 822–834.
46. Hamciuc, E.; Hanciuc, C., Polyimide-Polydimethylsiloxane Copolymers. In *Recent Developments in Silicone-Based Materials*, Cazacu, M., Ed. Nova Science: New York, 2010; pp 77–106.

47. Jiang, X.; Gu, J.; Shen, Y.; Wang, S.; Tian, X., Formation of Honeycomb-Patterned Microporous Films Based on a Fluorinated Poly(siloxane imide) Segmented Copolymer. *J. Appl. Polym. Sci.* **2011**, *119*, 3329–3337.
48. Ghosh, A.; Banerjee, S.; Wang, D.-Y.; Komber, H.; Voit, B., Synthesis, Characterization, and Properties of New Siloxane Grafted Copolyimides. *J. Appl. Polym. Sci.* **2012**, *123*, 2959–2967.
49. Hu, D.; Zheng, S., Morphology and Thermomechanical Properties of Epoxy Thermosets Modified with Polysulfone-*block*-Polydimethylsiloxane Multiblock Copolymer. *J. Appl. Polym. Sci.* **2011**, *119*, 2933–2944.
50. Hetflejs, J.; Sabata, S.; Podesva, J.; Netopilik, M.; Latalova, P.; Spevacek, J., Syntheses of Triblock Polybutadiene-Polydimethylsiloxane Copolymers by Coupling Reaction. *J. Appl. Polym. Sci.* **2006**, *101*, 3233–3240.
51. Ciolino, A.; Sakellariou, G.; Pantazis, D.; Villar, M. A.; Valles, E.; Hadjichristidis, N., Synthesis and Characterization of Model Block Copolymers of Poly(dimethylsiloxane) with Poly(1,4-butadiene) or Poly(ethylene). *J. Polym. Sci., Part A: Polym. Chem.* **2006**, *44*, 1579–1590.
52. Ciolino, A.; Villar, M. A.; Valles, E.; Hadjichristidis, N., Synthesis and Characterization of Model Polybutadiene-1,4-b-Polydimethylsiloxane-b-Polybutadiene-1,4 Copolymers. *J. Polym. Sci., Part A: Polym. Chem.* **2007**, *45*, 1579–1590.
53. Ciolino, A. E.; Gomez, L. R.; Vega, D. A.; Villar, M. A.; Valle, E. R., Synthesis and Physicochemical Characterization of a Well-Defined Poly(butadiene 1,3)-*block*-Poly(dimethylsiloxane) Copolymer. *Polymer* **2008**, *49*, 5191–5194.
54. Schulte, L.; Grydgaard, A.; Jakobsen, M. R.; Szewczykowski, P. P.; Guo, F.; Vigild, M. E.; Berg, R. H.; Ndoni, S., Nanoporous Materials from Stable and Metastable Structures of 1,2-PB-*b*-PDMS Block Copolymers. *Polymer* **2011**, *52*, 422–429.
55. Zhang, D.; Giese, M. L.; Prukop, S. L.; Grunlan, M. A., Poly(ε-caprolactone)-Based Shape Memory Polymers with Variable Polydimethylsiloxane Soft Segment Lengths. *J. Polym. Sci. A, Polym. Chem.* **2011**, *49*, 754–761.
56. Gill, R.; Muzhar, M.; Siddiq, M., Structural Characterization and Thermal Behaviour of Block Copolymers of Polydimethylsiloxane and Polyamide Having Trichlorogermyl Pendant Groups. *Polym. Int.* **2010**, *59*, 1598–1605.
57. Hernandez, R.; Weksler, J.; Padsalgikar, A.; Runt, J., Microstructural Organization of Three-Phase Polydimethylsiloxane-Based Segmented Polyurethanes. *Macromolecules* **2007**, *40*, 5441–5449.
58. Jiang, H.; Zheng, Z.; Song, W.; Li, Z.; Wang, X., Alkoxysilane Functionalized Polyurethane/Polysiloxane Copolymers: Synthesis and the Effect of End-Capping Agents. *Polym. Bull.* **2007**, *59*, 53–63.
59. Choi, T.; Weksler, J.; Padsalgikar, A.; Runt, J., Influence of Soft Segment Composition on Phase-Separated Microstructure of Polydimethylsiloxane-Based Segmented Polyurethane Copolymers. *Polymer* **2009**, *50*, 2320–2327.
60. Joki-Korpela, F.; Pakkanen, T. T., Incorporation of Polydimethylsiloxane into Polyurethanes and Characterization of Copolymers. *Eur. Polym. J.* **2011**, *47*, 1694–1708.
61. Sheth, J. P.; Yilgor, E.; Erenturk, B.; Ozhalici, H.; Yilgor, I.; Wilkes, G. L., Structure-Property Behavior of Poly(dimethylsiloxane) Based Segmented Polyurea Copolymers Modified with Poly(propylene oxide). *Polymer* **2005**, *46*, 8185–8193.
62. Yilgor, I.; Yilgor, E., Silicone—Urea Copolymers Modified with Polyethers. In *Science and Technology of Silicones and Silicone-Modified Materials*, Clarson, S. J.;

Fitzgerald, J. J.; Owen, M. J.; Smith, S. D.; Van Dyke, M. E., Eds. American Chemical Society: Washington, DC, 2007; Vol. 964, pp 100–115.
63. Yilgor, I.; Eynur, T.; Yilgor, E.; Wilkes, G. L., Contribution of Soft Segment Entanglements on the Tensile Properties of Silicone-Urea Copolymers with Low Hard Segment Contents. *Polymer* **2009**, *50*, 4432–4437.
64. Yilgor, I.; Eynur, T.; Bilgin, S.; Yilgor, E.; Wilkes, G. L., Influence of Soft Segment Molecular Weight on the Mechanical Hysteresis and Set Behavior of Silicone-Urea Copolymers with Low Hard Segment Contents. *Polymer* **2011**, *52*, 266–274.
65. Yildirim, E.; Yurtsever, M.; Yurtsever, E.; Yilgor, I.; Yilgor, E., Multiscale Modeling of the Morphology and Properties of Segmented Silicone-Urea Copolymers. *J. Inorg. Organomet. Polym. Mater.* **2012**, *22*, 604–616.
66. Yang, M.; Mao, J.; Nie, W.; Dong, Z.; Wang, D.; Zhao, Z.; Ji, X., Facile Synthesis and Responsive Behavior of PDMS-*b*-PEG Diblock Copolymer Brushes via Photoinitiated "Thiol-ene" Click Reaction. *J. Polym. Sci., Part A: Polym. Chem.* **2012**, *50*, 2075–2083.
67. Politakos, N.; Ntoukas, E.; Averopoulos, A.; Krikorian, V.; Pate, B. D.; Thomas, E. L.; Hill, R. M., Strongly Segregated Cubic Microdomain Morphology Consistent with the Double Gyroid Phase in High Molecular Weight Diblock Copolymers of Polystyrene and Poly(dimethylsiloxane). *J. Polym. Sci., Part B: Polym. Phys.* **2009**, *47*, 2419–2427.
68. Wadley, M. L.; Cavicchi, K. A., Synthesis of Polydimethylsiloxane-Containing Block Copolymers via Reversible Addition Fragmentation Chain Transfer (RAFT) Polymerization. *J. Appl. Polym. Sci.* **2010**, *115*, 635–640.
69. Guan, C.-M.; Luo, Z.-H.; Tang, P.-P., Poly(dimethylsiloxane-*b*-styrene) Diblock Copolymers Prepared by Reversible Addition-Fragmentation Chain-Transfer Polymerization: Synthesis and Characterization. *J. Appl. Polym. Sci.* **2010**, *116*, 3283–3290.
70. Guan, C.-M.; Luo, Z.-H.; Qiu, J.-J.; Tang, P.-P., Novel Fluorosilicone Triblock Copolymers Prepared by Two-Step RAFT Polymerization: Synthesis, Characterization, and Surface Properties. *Eur. Polym. J.* **2010**, *46*, 1582–1593.
71. Ninago, M. D.; Satti, A. J.; Ciolino, A. E.; Valles, E. M.; Villar, M. A.; Vega, D. A.; Sanz, A.; Nogales, A.; Rueda, D. R., Synthesis and Morphology of Model PS-*b*-PDMS Copolymers. *J. Polym. Sci. A, Polym. Chem.* **2010**, *48*, 3119–3127.
72. Giacomelli, F. C.; Riegel, I. C.; Stepanek, P.; Petzhold, C. L.; Ninago, M. D.; Satti, A. J.; Ciolino, A. E.; Villar, M. A.; Schmidt, V.; Giacomelli, C., Micelles Formed by Highly Asymmetric Polystyrene-*b*-Polydimethylsiloxane and Polystyrene-b-poly[5-(N,N-diethylamino)isoprene] Diblock Copolymers. *Langmuir* **2010**, *26*, 14494–14501.
73. Malik, M. I.; Sinha, P.; Bayley, G.M.; Mallon, P. E.; Pasch, H., Characterization of Polydimethylsiloxane-Block-Polystyrene (PDMS-B-Ps) Copolymers by Liquid Chromatography at Critical Conditions. *Macromol. Chem. Phys.* **2011**, *212*, 1221–1228.
74. Del Valle-Carrandi, L.; Alegria, A.; Arbe, A.; Colmenero, J., Unexpected PDMS Behavior in Segregated Cylindrical and Spherical Nanophases of PS-PDMS Asymmetric Diblock Copolymers. *Macromolecules* **2012**, *45*, 491–502.
75. Tonnar, J.; Pougeta, E.; Lacroix-Desmazes, P.; Boutevin, B., Synthesis of Poly(vinyl acetate)-b-Poly(dimethylsiloxane)-b-Poly(vinyl acetate) Triblock Copolymers by Iodine Transfer Polymerization. *Eur. Polym. J.* **2008**, *44*, 318–328.

76. Tonnar, J.; Pouget, E.; Lacroix-Desmazes, P.; Boutevin, B., Synthesis of Poly(vinyl acetate)-block-Poly(dimethylsiloxane)-Poly(vinyl acetate) Copolymers by Iodine Transfer Photopolymerization in Microemulsion. *Macromol. Symp.* **2009**, *281*, 20–30.
77. Fragouli, P.; Iatrou, H.; Lohse, D.J.; Hadjichristidis, N., Linear Pentablock Quintopolymers (I-SIDMV) with Five Incompatible Blocks: Polystyrene, Polyisoprene-1,4-, Poly(dimethylsiloxane), Poly(tert-butyl methacrylate, and Poly(2-vinylpyridine). *J. Polym. Sci., Part A: Polym. Chem.* **2008**, *46*, 3938–3946.
78. Luo, Z.-H.; Yu, H.-J.; He, T.-Y., Synthesis and Characterization of the Hydrophobic Diblock Copolymers of Poly(dimethylxiloxane)-block-poly(ethyl methyl acrylate) Through Atom Transfer Radical Polymerization. *J. Appl. Polym. Sci.* **2008**, *107*, 1201–1208.
79. Zhou, Y.-N.; Guan, C.-M.; Luo, Z.-H., Kinetic Modeling of Two-Step RAFT Process for the Production of Novel Fluorosilicone Triblock Copolymers. *Eur. Polym. J.* **2010**, *46*, 2164–2173.
80. Vargun, E.; Usanmaz, A., Synthesis and Surface Properties of Polydimethylsiloxane-Based Block Copolymers: Poly[dimethylsiloxane-*block*-ethyl methacrylate)] and Poly[dimethylsiloxane-block-hydroxyethyl methacrylate)] *Polym. Int.* **2010**, *59*, 1586–1597.
81. Lian, K.-J.; Chen, C.-Q.; Liu, H.; Wang, N.-X.; Yu, H.-J.; Luo, Z.-H., Surface Microphase Separation in PDMS-*b*-PMMA-*b*-PHFBMA Triblock Copolymer Films. *J. Appl. Polym. Sci.* **2011**, *120*, 156–164.
82. Martinelli, E.; Galli, G.; Krishnan, S.; Paik, M. Y.; Ober, C. K.; Fischer, D. A., New Poly (dimethylsiloxane)/Poly(perfluroocty acrylate) Block Copolymers: Structure and Order Across Multiple Length Scales in Thin films. *J. Mater. Chem.* **2011**, *21*, 15357–15368.
83. Muppalla, R.; Jewrajka, S. K., Synthesis, Morphology and Properties of Poly(dimethylsiloxane)/Poly(*n*-butyl acrylate) mixed Soft Block-Based Copolymers: A New Class of Thermoplastic Elastomer. *Polymer* **2012**, *53*, 1453–1464.
84. Sugimoto, H.; Nishino, G.; Koyama, H.; Daimatsu, K.; Inomata, K.; Nakanishi, E., Preparation and Morphology of Transparent Poly(methyl Methacrylate) Poly(dimethylsiloxane) Hybrid Materials Using Multifunctional Silicone Macromonomer. *J. Appl. Polym. Sci.* **2012**, *124*, 1316–1322.
85. Jeong, J. W.; Park, W. I.; Kim, M.-J.; Ross, C. A.; Jung, Y. S., Highly Tunable Self-Assembled Nanostructures from a Poly(2-vinyl;pyridine-*b*-dimethylsiloxane) Block Copolymer. *Nano Lett.* **2011**, *11*, 4095–4101.
86. Gadda, T. M.; Webber, W. P., Copolymers Based on Dimethylsiloxane and Diphenylsiloxane. In *Science and Technology of Silicones and Silicone-Modified Materials*, Clarson, S. J.; Fitzgerald, J. J.; Owen, M. J.; Smith, S. D.; Van Dyke, M. E., Eds. American Chemical Society: Washington, DC, 2007; Vol. 964, pp 82–99.
87. Su, T.; Wang, G. Y.; Wang, S. L.; Hu, C. P., Fluorinated Siloxane-Containing Waterborne Polyurethane Ureas with Excellent Hemocompatibility, Waterproof and Mechanical Properties. *Eur. Polym. J.* **2010**, *46*, 472–483.
88. Wang, R.; Fang, L.; Xu, C., Synthesis, Characterization, and Thermal Properties of New Silarylene-Siloxane-Acetylene Polymers. *Eur. Polym. J.* **2010**, *46*, 465–471.

89. Abduallah, A. B. E.; Mallon, P. E., Microscopic Surface and Bulk Morphology of Semicrystalline Poly(dimethylsiloxane)-Polyester Copolymers. *J. Appl. Polym. Sci.* **2010**, *115*, 1518–1533.
90. Vuckovic, M. V.; Antic, V. V.; Govedarica, M. N.; Djonlagic, J., Synthesis and Characterization of Copolymers Based on Poly(butylene terephthalate) and Ethylene Oxide-Poly(dimethylsiloxane)-Ethylene Oxide. *J. Appl. Polym. Sci.* **2010**, *115*, 3205–3216.
91. An, Q.; Yang, G.; Wang, Q.; Huang, L., Synthesis and Morphology of Carboxylated Polyether-*block*-Polydimethylsiloxane and the Supermolecule Self-Assembled from It. *J. Appl. Polym. Sci.* **2008**, *110*, 2595–2600.
92. Gong, W.; Zeng, K.; Wang, L.; Zheng, S., Poly(hydroxy ether of bisphenol A)-block-Polydimethylsiloxane Alternating Block Copolymer and Its Nanostructured Blends with Epoxy Resins. *Polymer* **2008**, *49*, 3318–3326.
93. Cureton, L. T.; Turner, S. R., Synthesis and Properties of Bisphenol A-Terphenyl Poly(aryl ether sulfone)-Polydimethylsiloxane Segmented Block Copolymers. *Eur. Polym. J.* **2011**, *47*, 2303–2310.
94. Wang, L.; Zheng, S., Morphology and Thermomechanical Properties of Main-Chain Polybenzoxazine-block-Polydimethylsiloxane Multiblock Polymers. *Polymer* **2010**, *51*, 1124–1132.
95. Theogarajan, L.; Li, H.; Busse, K.; Desai, S.; Kressler, J.; Scholz, C., Self-Assembly of ABA Triblock Copolymers Based on Functionalized Polydimethylsiloxane and Polymethyloxazoline. *Polym. Int.* **2010**, *59*, 1204–1211.
96. Landsmann, S.; Winter, A.; Chiper, M.; Fuston, C.-A.; Hoeppener, S.; Wouters, D.; Gohy, J.-F.; Schubert, U. S., Poly(dimethylsiloxane)-Substituted 2,2':6,2"-Terpyridines: Synthesis and Characterization of New Amphiphilic Supramolecular Diblock Copolymers. *Macromol. Chem. Phys.* **2008**, *209*, 1666–1672.
97. Cureton, L. T.; Beyer, F. L.; Turner, S.R., Synthesis and Characterization of Hexafluoroisopropylidene Bisphenol Poly(arylene ether sulfone) and Polydimethylsiloxane Segmented Block Copolymers. Polymer **2010**, *51*, 1679–1686.
98. Rusu, G. I.; Arinei, A.; Hamciuc, V.; Rusu, G. G.; Rambu, P.; Diciu, M.; Garoi, P.; Rusu, M., Electronic and Optical Properties of Some Polysulfone-Polydimethylsiloxane Copolymers in Thin Films. *J. Macromol. Sci., Part B: Phys.* **2009**, *48*, 238–253.
99. Ibarboure, E.; Rodriguez-Hernandez, J., Supramolecular Structures from Self-Assembled Poly(γ-benzyl-L-glutamate)-Polydimethylsiloxane-Poly(γ-benzyl-L-glutamate) Triblock Copolypeptides in Thin Films. *Eur. Polym. J.* **2010**, *46*, 891–899.
100. Zhang, X.; Kong, L.; Dai, L.; Zhang, X.; Wang, Q.; Tan, Y.; Zhang, Z., Synthesis, Characterization, and Thermal Properties of Poly(siloxane-carbone)s. *Polymer* **2011**, *52*, 4777–4784.
101. Wang, M.; Ma, J.; Chu, R.; Park, C. B.; Nanqiao, Z., Effect of the Introduction of Polydimethylsiloxane on the Foaming Behavior of Block-Copolymerized Polypropylene. *J. Appl. Polym. Sci.* **2012**, *123*, 2726–2732.
102. Nishino, G.; Kanda, S.; Sugimoto, H.; Inomata, K.; Nakanishi, E., Preparation and Coating Properties of an Acrylic Melamine Resin Containing Silicone Segments. *Polym. Bull.* **2012**, *68*, 2049–2060.
103. Gall, B. T.; Thomann, R.; Mulhaupt, R., Synthesis and Characterization of Semicrystalline Triblockcopolymers of Isotactic Polystyrene and Polydimethylsiloxane. *J. Polym. Sci., Part A: Polym. Chem.* **2011**, *49*, 2339–2345.

104. Sun, Y.; Liu, W., Synthesis and Properties of Triblock Copolymers Containing PDMS via AGET ATRP. *Polym. Bull.* **2012**, *68*, 1815–1829.
105. Qiu, J.-J.; Luo, Z.-H., Self-Assembly of ABC Nonamphiphilic Fluorosilicone Triblock Copolymers in Dilute Solution: The First Example. *J. Polym. Sci., Part A: Polym. Chem.* **2011**, *49*, 2513–2519.
106. Pavlovic, D.; Linhardt, J. G.; Kunzler, J. F.; Shipp, D. A., Synthesis and Characterization of PDMS-, PVP, and PS-Containing ABCBA Pentablock Copolymers. *Macromol. Chem. Phys.* **2010**, *211*, 1482–1487.
107. Da Rocha, S. R. P.; Dickson, J.; Cho, D.; Rossky, P. J.; Johnston, K. P., Stubby Surfactants for Stabilization of Water and CO_2 Emulsions: Trisiloxanes. *Langmuir* **2003**, *19*, 3114–3120.
108. Dong, J.; Mao, G.; Hill, R.M., Nanoscale Aggregate Structures of Trisiloxane Surfactants at the Solid-Liquid Interface. *Langmuir* **2004**, *20*, 2695–2700.
109. Kaneko, M.; Matsuzawa, K.; Uddin, M. H.; Lopez-Quintela, M. A.; Kunieda, H., Effect of Hydrophilic Chain Length of Amphiphilic Silicone Oil (Copolymer) on the Nonionic Surfactant-Layer Curvature. *J. Phys. Chem. B* **2004**, *108*, 12736–12743.
110. Liang, W.-J.; Wu, C.-P.; Hsu,C.-Y.; Kuo, P.-L., Synthesis, Characterization, and Proton-Conducting Properties of Organic-Inorganic Hybrid Membranes Based on Polysiloxane Zwitterionomer. *J. Polym. Sci. A, Polym. Chem.* **2006**, *44*, 3444–3453.
111. Srividhya, M.; Chandrasekar, K.; Baskar, G.; Reddy, B. S. R., Physico-Chemical Properties of Siloxane Surfactants in Water and Their Surface Energy Characteristics. *Polymer* **2007**, *48*, 1261–1268.
112. Klasner, S. A.; Metto, E. C.; Roman, G. T.; Culbertson, C. T., Synthesis and Characterization of a Poly(dimethylsiloxane)-Poly(ethylene oxide) Block Copolymer for Fabrication of Amphiphilic Surfaces on Microfluidic Devices. *Langmuir* **2009**, *25*, 10390–10396.
113. Erdodi, G.; Kennedy, J. P., Third-Generation Amphiphilic Conetworks. I. Synthesis and Swelling Behavior of Poly(N,N-dimethyl acrylamide)/Poly(dimethylsiloxane) Conetworks. *J. Polym. Sci. A, Polym. Chem.* **2007**, *45*, 295–307.
114. Pavlovic, D.; Linhardt, J. G.; Kunzier, J. F.; Shipp, D. A., Synthesis of Amphiphilic Multiblock and Triblock Copolymers of Polydimethylsiloxane and Poly(*N,N*-dimethylacrylamide). *J. Polym. Sci., Part A: Polym. Chem.* **2008**, *46*, 7033–7048.
115. Kang, J.; Erdodi, G.; Kennedy, J. P., Thermoplastic Amphiphilic Conetworks. *J. Polym. Sci. Part A: Polym. Chem.* **2009**, *47*, 682–691.
116. Beppu, K.; Kaneko, Y.; Kadokawa, J.-I.; Mori, H.; Nishiksawa, T., Synthesis of Sugar-Polysiloxane Hybrids Having Rigid Main-Chains and Formation of Their Nano Aggregates. *Polym. J.* **2007**, *39*, 1065–1070.
117. Iwakiri, N.; Nishikawa, T.; Kaneko, Y.; Kadokawa, J.-I., Synthesis of Amphiphilic Polysiloxanes and Their Properties for Formation of Nano-Aggregates. *Coll. Polym. Sci.* **2009**, *287*, 577–582.
118. Sakamoto, S.; Shimojima, A.; Miyasaka, K.; Ruan, J.; Terasaki, O.; Kurada, K., Formation of Two- and Three-Dimensional Hybrid Mesostructures from Branched Siloxane Molecules. *J. Am. Chem. Soc.* **2009**, *131*, 9634–9635.
119. Mecham, S.; Sentman, A.; Sambasivam, M., Amphiphilic Silicone Copolymers for Pressure Sensitive Adhesive Applications. *J. Appl. Polym. Sci.* **2010**, *116*, 3265–3270.

120. Neugebauer, D.; Zhang, Y.; Pakula, T.; Matyjaszowski, K., PDMS-PEO Densely Grafted Copolymers. *Macromolecules* **2005**, *38*, 8687–8693.
121. Hu, A.; Cui, Y.; Wei, X.; Lu, Z.; Ngai, T., Hydrogen-Bonding-Induced Complexation of Polydimethylsiloxane-*graft*-poly(ethylene oxide) and Poly(acrylic acid)-*block*-Polyacrylonitrile Micelles in Water. *Langmuir* **2010**, *26*, 14502–14508.
122. Murthy, R.; Bailey, B. M.; Valentin-Rodriguez, G.; Ivanisevic, A.; Grunlan, M. A., Amphiphilic Silicones Prepared from Branched PEO-Silanes with Siloxane Tethers. *J. Polym. Sci. A, Polym. Chem.* **2010**, *48*, 4108–4119.
123. Benkova, Z.; Szefczyk, B.; Cordeiro, M. N. D. S., Molecular Dynamics Study of Hydrated Poly(ethylene oxide) Chains Grafted on Siloxane Surface. *Macromolecules* **2011**, *44*, 3639–3648.
124. Cheng, C.; Powell, K. T.; Khoshdel, E.; Wooley, K. L., Polydimethylsiloxane-(PDMS-) Grafted Fluorocopolymers by a "Grafting through" Strategy Based on Atom Transfer Radical (Co)polymerization. *Macromolecules* **2007**, *40*, 7195–7207.
125. Yang, X.-F.; Yao, C., Synthesis and Comparative Properties of Poly(dimethylsiloxane) Grafted Alkyl Acrylate. *J. Appl. Polym. Sci.* **2007**, *106*, 3600–3604.
126. Li, J.; Yi, L.; Lin, H.; Hou, R., Synthesis of Poly(*tert*-butyl methacrylate)-*graft*-Poly(dimethylsiloxane) Graft Copolymers via Reversible Addition-Fragmentation Chain Transfer Polymerization. *J. Polym. Sci. A, Polym. Chem.* **2011**, *49*, 1483–1493.
127. Subramanyam, U.; Kennedy, J. P., PVA Networks Grafted with PDMS Branches. *J. Polym. Sci., Part A: Polym. Chem.* **2009**, *47*, 5272–5277.
128. Chen, Z.; Yang, J.; Guan, S.; Li, X.; Rong, C.; Jiang, Z., Synthesis and Characterization of Poly(ether sulfone)-*Graft*-Polydimethylsiloxane Copolymers. *J. Appl. Polym. Sci.* **2010**, *118*, 2434–2441.
129. Zong, J.; Zhang, Q.; Sun, H.; Yu, Y.; Wang, S.; Liu, Y., Characterization of Polydimethylsiloxane-Polyurethanes Synthesized by Graft or Block Copolymerization. *Polym. Bull.* **2010**, *65*, 477–493.
130. Mizerska, U.; Fortuniak, W.; Chojnowski, J.; Turecka, K.; Konopacka, A.; Werel, W., Antimicrobial Siloxane Statistical and Graft Copolymers Substituted with *t*-Butylamine and t-Butylammonium Biocidal Functions. *J. Inorg. Organomet. Polym.* **2010**, *20*, 554–563.
131. Wang, J.; Xia, W.; Liu, K.; Tuo, X., Improved Adhesion of Silicone Rubber to Polyurethane by Surface Grafting. *J. Appl. Polym. Sci.* **2011**, *121*, 1245–1253.
132. Ardhyananta, H.; Kawauchi, T.; Ismail, H.; Takeichi, T., Effect of Pendant Group of Polysiloxanes on the Thermal and Mechanical Properties of Polybenzoxazine Hybrida. *Polymer* **2009**, *50*, 5959–5969.
133. Jancelewicz, M.; Fojud, Z.; Waszkowiak, W.; Jurga, S., Local Mobility in Grafted Polydimethylsiloxane Melts. *Eur. Polym. J.* **2011**, *47*, 48–51.
134. Zhang, M.; Wu, Y.; Zhang, Q.; Xia, Y.; Li, T., Synthesis and Characterization of Gelatin-Polydimethylsiloxane Graft Copolymers. *J. Appl. Polym. Sci.* **2011**, *120*, 2130–2137.
135. Mark, J. E.; Ning, Y.-P., Interpenetrating Bimodal Networks. *Polym. Eng. Sci.* **1985**, *25*, 824–827.
136. Yoo, S. H.; Cohen, C.; Hui, C.-Y., Mechanical and Swelling Properties of PDMS Interpenetrating Polymer Networks. *Polymer* **2006**, *47*, 6226–6235.
137. Arkles, B.; Rediger, P., *Silicones in Biomedical Applications*. Technomic Publishing: Lancaster, PA, 1983.

138. Jia, L.-Y.; Zhang, C.; Du, Z.-J.; Li, C.-J.; Li, H.-Q., Preparation of Interpenetrating Polymer Networks of Epoxy/Polydimethylsiloxane in a Common Solvent of the Precursors. *Polym. J.* **2007**, *39*, 593–597.
139. Jia, L.-Y.; Zhang, C.; Du, Z.-J.; Li, C.-J.; Li, H.-Q., A Novel Approach to Interpenetrating Networks of Epoxy Resin and Polydimethylsiloxane. *J. Appl. Polym. Sci.* **2007**, *105*, 2663–2669.
140. Fichet, O.; Vidal, F.; Laskar, J.; Teyssie, D., Polydimethylsiloxane-Cellulose Acetate Butyrate Interpenetrating Polymer Networks Synthesis and Kinetic Study. Part I. *Polymer* **2005**, *46*, 37–47.
141. Vidal, F.; Fichet, O.; Laskar, J.; Teyssie, D., Polydimethylsiloxane-Cellulose Acetate Butyrate Interpenetrating Polymer Networks Close to True IPNs on a Large Composition Range. Part I. *Polymer* **2006**, *47*, 3747–3753.
142. Boileau, S.; Bouteiller, L.; Khalifa, R. B.; Liang, Y.; Teyssie, D., Polycarbonate-Polysiloxane-Based Interpenetrating Networks. In *Silicones and Silicone-Modified Materials*, Clarson, S. J.; Fitzgerald, J. J.; Owen, M. J.; Smith, S. D., Eds. American Chemical Society: Washington, DC, 2000; Vol. 729, pp 383–394.
143. Marques, R. S.; Leod, T. C. O. M.; Yoshida, I. V. P.; Mano, V.; Assis, M. D.; Schiavon, M. A., Synthesis and Characterization of Semi-Interpenetrating Networks Based on Poly(dimethylsiloxane) and Poly(vinyl alcohol). *J. Appl. Polym. Sci.* **2010**, *115*, 158–166.
144. Alizadeh, M.; Abbasi, F.; Farahi, M.; Jalili, K., Silicone-Based Hydrogels Prepared by Interpenetrating Polymer Network Synthesis: Swelling Properties and Confinements Effects on the Formation Kinetics. *J. Appl. Polym. Sci.* **2012**, *124*, 985–992.
145. Darras, V.; Fichet, O.; Perrot, F.; Boileau, S.; Teyssie, D., Polysiloxane-Poly(fluorinated acrylate) Interpenetrating Polymer Networks: Synthesis and Characterization. *Polymer* **2007**, *48*, 687–695.
146. Hillerstrom, A.; Kronberg, B., A Two-Step Method for the Synthesis of a Hydrophilic PDMS Interpenetrating Polymer Network. *J. Appl. Polym. Sci.* **2008**, *110*, 3059–3067.
147. Tang, Q.; Yu, J.-R.; Chen, L.; Zhu, J.; Hu, Z.-M., Porous Silicone Hydrogel Interpenetrating Polymer Networks Prepared Using a Template Method for Biomedical Use. *Polym. Int.* **2010**, *60*, 1136–1141.

CHAPTER 9

Composites

9.1 SOL-GEL CERAMICS

A relatively new area that involves silicon-containing materials is the synthesis of "ultrastructure" materials (i.e., materials in which structure can be controlled at the level of 100 Å). An example is the "sol-gel" hydrolysis of alkoxysilanes (organosilicates) to give silica, SiO_2.[1–20] The reaction is complicated,[7, 21–26] involving polymerization and branching, but the overall reaction may be written

$$Si(OR)_4 + 2H_2O \rightarrow SiO_2 + 4ROH \tag{9.1}$$

where the $Si(OR)_4$ organometallic species is typically tetraethoxysilane such as tetraethylorthosilicate (TEOS, with R being C_2H_5). In this application, the precursor compound is hydrolyzed and then condensed to yield branched polymers. Eventually a continuous swollen gel is formed. The gel is dried at moderately low temperatures to remove volatile species, and then it is fired into a porous ceramic object that can then be densified and machined into a final ceramic part.

The production of ceramics by this novel route triggered interest in the ceramics community because of advantages over the conventional powder-processing approach to ceramics. Advantages include (i) the higher purity of the starting materials, (ii) the relatively low temperatures required, (iii) the possibility of controlling the ultrastructure to reduce the microscopic flaws that lead to failure, (iv) the ease with which ceramic coatings can be formed, and (v) the ease with which ceramic *alloys* can be prepared (e.g., by hydrolyzing solutions of both silicates and titanates).

The sol-gel approach has been used to form ceramic-like phases in a variety of polymers. Poly(dimethylsiloxane) (PDMS) is the most popular.

PDMS is relatively weak and stands to benefit most from reinforcement. PDMS is easily absorbs the precursor materials generally used in the sol-gel process. Nearly monodisperse silica microparticles can be obtained using siloxane elastomer mixtures.[27] In some cases, the PDMS has been part of a copolymer, with ureas,[28] imides,[29] amideimides,[30, 31] and dianilines.[20] In other approaches, the particle surface is modified,[32] for example, with a polysiloxane.[33] Siloxane/silica nanocomposites have also been used as "culture-stone-protective materials."[34]

9.2 FILLERS IN ELASTOMERS

9.2.1 Approximately Spherical Particles

Sol-gel hydrolysis and condensation can be carried out within a polymeric matrix to generate *particles* of the ceramic material, typically with an average diameter of a few hundred angstroms.[13–15, 35–37] Silica has been of particular interest in this regard[38] because of its ecodesign, sustainability, and recyclability.[39] The use of hard, polysilicate particles has been explored in considerable detail.[40, 41] The polymer typically has end groups, such as hydroxyls, that can participate in the hydrolysis-condensation reactions.[42, 43] In some cases, amphiphilic PDMS networks have been used.[44]

Polysiloxanes have been particularly useful as templates and building blocks for nanostructured materials.[45, 46] End groups improve bonding between the two rather disparate phases, but bonding agents may also be introduced for this purpose.[47] Considerable reinforcement of elastomers, including those prepared from PDMS, can be achieved. Some PDMS samples have even been electrospun into fibers and fiber mats.[48] In another novel approach, an emulsion technique was used to prepare particles in which the PDMS is entrapped in a silica-like matrix.[49] It is also possible to use simultaneous vapor-phase polymerization to make hybrid organic/inorganic thin films.[50]

Many of these nanocomposites have been characterized by nuclear magnetic resonance (NMR) spectroscopy,[51–54] single-molecule spectroscopy,[55] and small-angle scattering.[37, 56–58] Some applications require other measurements, such as permeability[59] or ionic conductivity.[60] In the case of permeability, the presence of the filler can substantially reduce flux by blocking low-resistance pathways and by modification of the elastomer in the vicinity of the filler (figure 9.1).

Other properties of PDMS-silica nanocomposites that have been reported include the glass transition temperatures, segmental dynamics,[61–63]

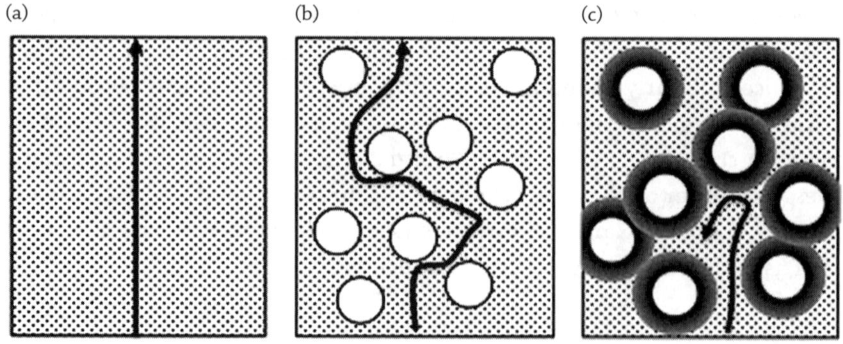

Figure 9.1:
Permeability of a small molecule through PDMS as illustrated for the unfilled polymer (a), a nonpermeable filler decreasing the permeability of the PDMS by requiring a more tortuous route around the filler particles (b), and a filler that interacts so strongly with the PDMS that the PDMS segments close to the filler surface are so restricted that they are frequently called "bound rubber" (c).[406–409]

surface properties,[64] thermomechanical behavior,[65] optical properties,[66] and degree of cross linking from the silica-based nanoparticles.[67] Fluorescence analysis is also possible in silica-PDMS nanocomposites in which the polymer is labeled with dansyl chromophores[68] or lanthanide complexes.[69–71]

A variety of other nanoparticles have been formed by such in situ sol-gel reactions. Examples are the oxides of titanium,[72–79] aluminum,[72, 73] tantalum,[72, 80] zirconium,[73, 77] niobium,[80] and vanadium.[81] Some nanocomposites of this type have also included barium titanate,[82] calcium oxide,[83] calcium salts,[84] borates,[85] $HTiNbO_5$,[86] and Eu^{3+} dopants.[87]

The sol-gel has a number of advantages over the conventional approach in which separately prepared filler particles are blended into the un-cross-linked elastomer before vulcanization.[88–90] The time-honored ex situ technique is difficult to control because the filler particles are generally agglomerated[91] and the polymer is typically of high enough molecular weight to make the viscosity of the mixture exceedingly high. Thus, the blending technique is energy intensive and time consuming, and frequently not entirely successful.

Because of the nature of the in situ precipitation, the particles are essentially unagglomerated (as demonstrated by electron microscopy). The mechanism for their growth seems to involve simple homogeneous nucleation. Since the particles are separated by polymer, they do not have the opportunity to coalesce. Figure 9.2 shows a typical transmission electron micrograph of such a silica-filled material.[92] The particles are relatively monodisperse, most having diameters in the range of 100–200 Å. Similar results have been obtained with other particles formed by sol-gel

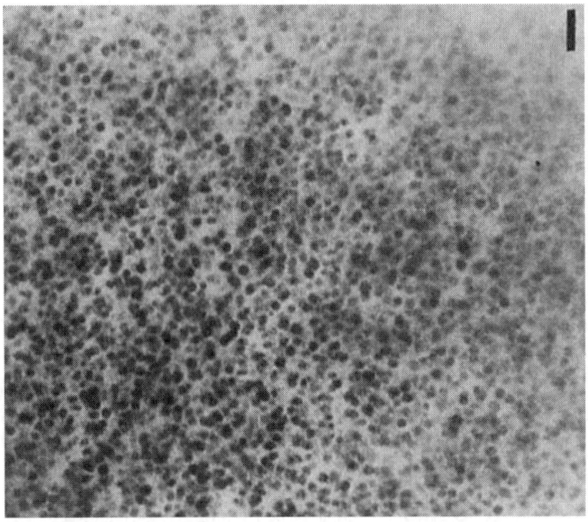

Figure 9.2:
Transmission electron micrograph of a PDMS network containing 34.4 wt % in situ precipitated silica particles.[92] The length of the bar corresponds to 1,000 Å.
Reproduced by permission of John Wiley and Sons.

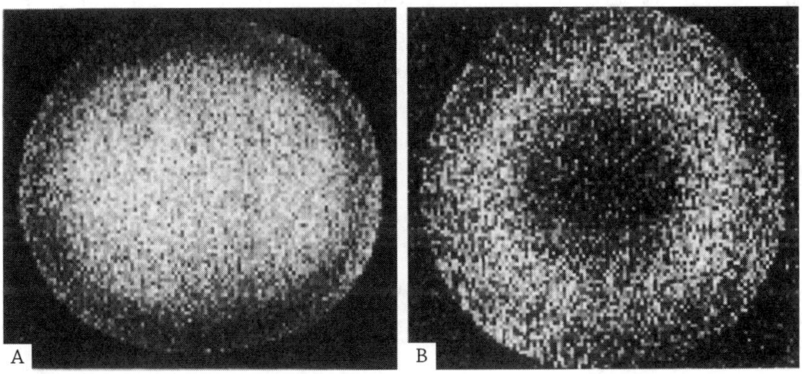

Figure 9.3:
^1H NMR images of a SiO_2-PDMS elastomeric sample obtained with a two-dimensional spin echo sequence having an echo time of 3.3 ms (portion A) and 22.7 ms (portion B).[99] The view is down the axis, and the resolution is 128 X by 128 pixels of 211 and 236 mm, respectively (in A), and 211 mm in both axes (in B).

reactions, including titania.[42, 93–97] In some cases, the particles are formed in a separate step, and then blended into the polymer.[98]

In the in situ approach, the growth of the particles from the surface of a PDMS sample can be followed by NMR (figure 9.3) utilizing ^1H and ^{29}Si magic-angle spinning, with two-dimensional Fourier transform spin-echo techniques.[99] The ^1H spin-spin (T_2) relaxation time of the protons in the PDMS is monitored as the chains become constrained by the growing silica-like material. For testing purposes, this composite was intentionally made to be inhomogeneous, with much larger amounts of silica on the surface. The dark rim at the edge of the sample indicates a reduced mobility of the network chains due to the presence of the silica. The evolution of the dark rim reveals the movement of the reaction front into the sample. This technique is nondestructive, but if the sample can be sacrificed, then slices can be further studied in a gradient column with regard to density, by electron microscopy[100] or by x-ray or neutron scattering.

Sol-gel methods are quite general[36, 42] in that a variety of precursor materials can be hydrolyzed, photolyzed, or thermolyzed to give reinforcing, ceramic-type particles. Titanates, for example, can be hydrolyzed to titania, aluminates hydrolyzed to alumina, and metal carbonyls photolyzed or thermolyzed to metals or metal oxides. Magnetic particles would be particularly interesting if they can be manipulated with an external magnetic field during the curing process (section 9.2.5). It is also possible to polymerize conducting polymers such as polyaniline within polysiloxane matrices.[101, 102] The method can be used in a variety of polymers (organic as well as inorganic, nonelastomeric as well as elastomeric). Even nonpolar, purely hydrocarbon polymers can be reinforced, provided the sol-gel precursor has sufficient miscibility in the matrix.

A variety of catalysts work well in hydrolyses reactions, including acids, bases, and salts.[103] Basic catalysts give precipitated phases that are generally well-defined particles, whereas the acidic catalysts give more poorly defined, diffuse particles.[4, 104] The solvent can be of considerable importance.[105] In some cases, particles are not formed at all, and bicontinuous (interpenetrating) phases result.[37, 106, 107] Another approach uses poly(allylamine hydrochloride) in a buffered solution to obtain what was called "bioinspired silica."[108]

These sol-gel polymer reactions can be carried out in three ways.[36, 42] In the first, the polymer is cross linked and swelled with the organometallic reagent, which is then hydrolyzed in situ. In the second, hydroxyl-terminated chains are blended with enough of the organometallic compound (TEOS) to both end link the chains and generate silica by the hydrolysis reaction. Thus, curing and filling take place simultaneously, in a one-step procedure. In the third technique, TEOS is blended into a polymer that has end groups (e.g., vinyl units) that are unreactive under hydrolysis conditions. The silica is formed in the usual manner (Equation 9.1), and the mixture is dried. The resultant slurry of polymer and silica is stable and can be cross linked at a later time using any of the standard techniques, including peroxide cures, vinyl-silane coupling, or irradiation. It is also possible to generate the catalyst and other reactants in situ, to give composites of unusually high transparency.[109]

Interesting "aging" effects are frequently observed in these systems. If the precipitated particles are left in contact with the hydrolysis catalyst and water they appear to reorganize, so that their surfaces become better defined and their sizes become more uniform.[35] The process seems analogous to the "Ostwald ripening."[110]

The reinforcing ability of such in situ generated particles has been amply demonstrated for a variety of deformations, including uniaxial extension (simple elongation), biaxial extension (compression), shear, and torsion.[36, 42, 111, 112] In the case of uniaxial extension, the reduced stress $[f^*]$ frequently increases by more than an order of magnitude, with the isotherms generally showing the upturns at high elongation, which is the signature of good reinforcement.[113, 114] The left portion of figure 9.4 shows typical results, where α is the extension.[115] As is generally the case in filled elastomers, there is irreversibility in the isotherms, which is thought to be due to irrecoverable sliding of the chains over the surfaces of the filler particles. The right portion of the figure documents the reinforcement observed in biaxial extension. The maxima and minima exhibited by such results will be a challenge to those seeking a better molecular understanding of filler reinforcement.

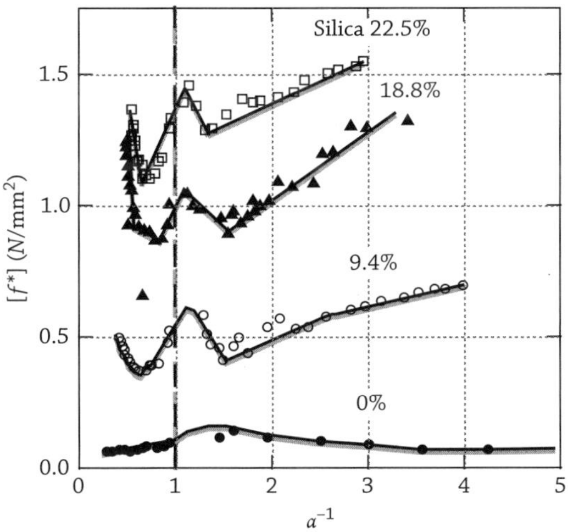

Figure 9.4:
Stress-strain isotherms for PDMS-silica in situ–reinforced elastomers in elongation (region to the left of the vertical dashed line, with $\alpha^{-1} < 1$), and in biaxial extension (compression, to the right, with $\alpha^{-1} > 1$).[115] The filled symbols represent the data obtained out of sequence to test for reversibility. $[f^*]$ is the reduced stress, ($[f^*] = f^*/(\alpha - \alpha^{-2})$ where $f^* = f/A^*$, f = elastic force, A^* = undeformed area and α = elongation.

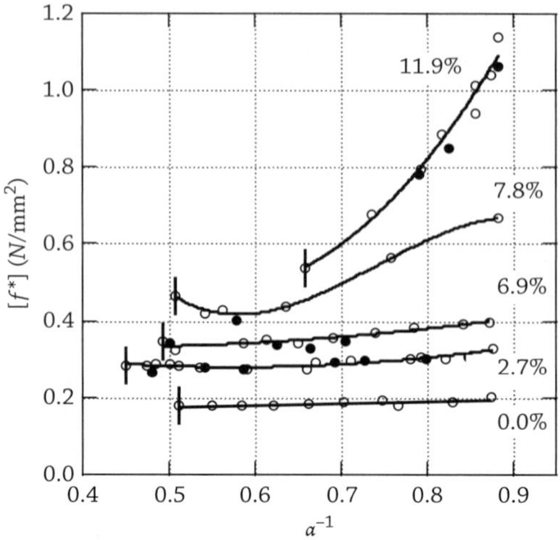

Figure 9.5:
Stress-strain isotherms for PDMS networks reinforced with in situ–generated titania particles.[116] Each curve is labeled with the wt % of filler. The filled circles locate results used to test for reversibility.

Some fillers (e.g., TiO_2) do give stress-strain isotherms that are reversible, indicating interesting differences in surface chemistry, including increased ability of the chains to slide along the particle surfaces.[116] Figure 9.5 illustrates such results.[116] Nevertheless, the bonding of PDMS to silica, titania, or silica-titania mixed oxide particles is strong enough to suppress *swelling* of the polymer (figure 9.6) These results involve equilibrium

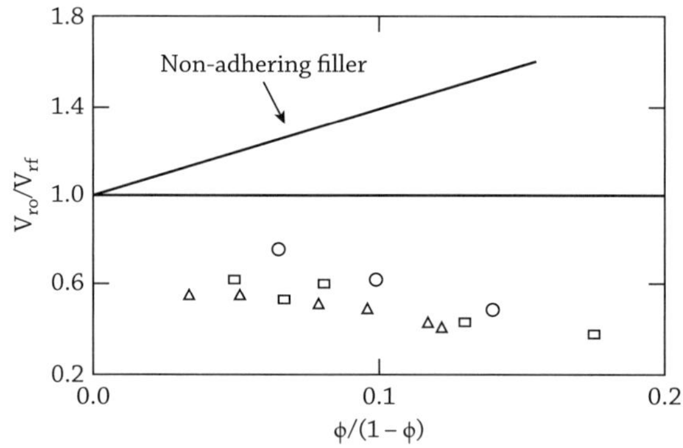

Figure 9.6:
Plot of volume fraction ratio V_{ro}/V_{rf} characterizing the swelling of an unfilled PDMS network relative to that of a filled PDMS network, against filler loading. ϕ is the volume fraction of filler.[118] Bonding of PDMS to the oxide particles is strong enough to suppress swelling of the polymer. Types of filler were silica-titania mixed oxides (□), silica (O), and titania (Δ).

swelling measurements obtained on unfilled and filled PDMS elastomers to estimate the degree of adhesion between elastomer and filler particles.[117–119] The results differ greatly from those for nonadhering fillers, indicating good bonding between the two phases. Resistance to separation from the surface in such swelling tests does not contradict the chains having considerable mobility *along* the surface.

In situ–generated silica fillers improve creep resistance and compression set in cyclic deformation.[120] The filled samples show very little compression set (figure 9.7). Thermal stability also improves as reflected by higher decomposition temperatures. (figure 9.8)[121] A possible mechanism

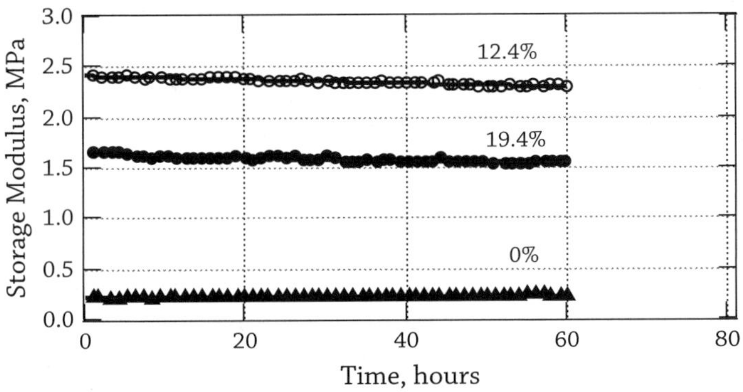

Figure 9.7:
Effect of cyclic stress on unfilled PDMS and PDMS elastomers filled with in situ precipitated silica: (Δ) unfilled, (●) 12.0 wt % silica, (O) 19.4 wt % silica.[120]

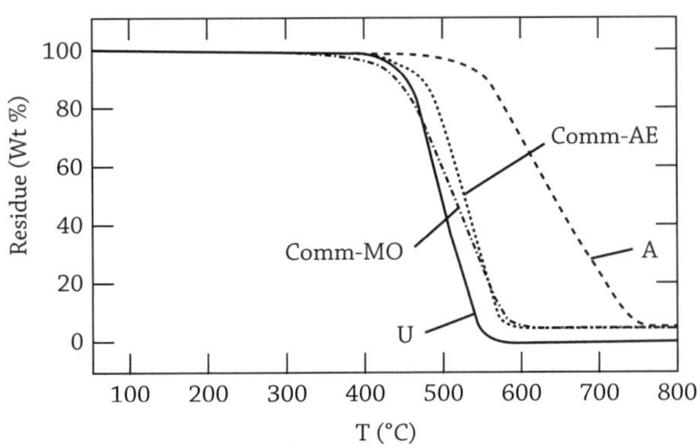

Figure 9.8:
Comparison of thermogravometric plots for PDMS networks that were unfilled (U), or contained either in situ precipitated silica (A) or commercial fume silica (COMM-MO and COMM-AE).[121] The heating was under nitrogen.

for this improvement is increased capability of the in situ–produced silica to tie up hydroxyl chain ends that participate in the degradation reaction. Attempts at further improvements of silica fillers are continuing.[122]

A variety of techniques have been used to further characterize *in situ*–filled elastomers.[36, 42] Density measurements, for example, yield information on the nature of the particles. Specifically, the densities of the ceramic-type particles are significantly less than that of silica itself, which implies the particles contain some unhydrolyzed alkoxy groups or voids, or both.

The low-temperature properties of some in situ–filled materials have been studied by the calorimetry techniques mentioned in chapter 1. Of particular interest is the way in which reinforcing particles affect the crystallization of a polysiloxane, both in the undeformed state and at high elongations.[123]

Electron microscopy (both transmission and scanning) has been used to reveal (i) the nature of the precipitated phase (particulate or nonparticulate), (ii) the average particle size, if particulate, (iii) the distribution of particle sizes, (iv) the integrity of the particles, and (v) the degree of agglomeration of the particles.[42]

A number of studies using x-ray and neutron scattering[124–127] have also been carried out on filled PDMS elastomers.[36, 37] Although the results are generally consistent with those obtained by electron microscopy, there are some intriguing differences. Some fillers that appear to be particulate in electron microscopy appear to consist of a continuously interpenetrating phase by scattering measurements. Additional experiments will certainly be forthcoming. Determining whether particle-induced network formation occurs is obviously a question of considerable importance.[128]

9.2.2 Glassy Particles Deformable into Ellipsoidal Shapes

It is possible to obtain reinforcement of a PDMS elastomer by polymerizing a monomer such as styrene to yield hard glassy domains within the elastomer.[129, 130] Low concentrations of styrene give a low molecular weight polymer that acts more like a plasticizer than a reinforcing filler. At higher styrene concentrations, however, roughly spherical polystyrene (PS) particles are formed, leading to good reinforcement. The particles are relatively easy to extract from the elastomeric matrix, which implies minimal bonding exists between the components. A trifunctional $R'Si(OC_2H_5)_3$ where R' is an unsaturated group can be used to couple the particles to

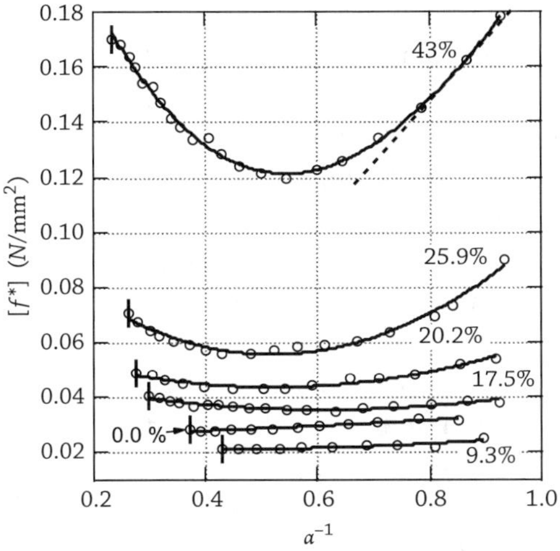

Figure 9.9:
Stress-strain isotherms for PDMS-polystyrene (PS) composites.[129] Each curve is labeled with the wt % PS present in the composite. The dashed line locates the linear portion of the curve useful for quantitative interpretation.[384]

the matrix. The R' groups on the particle surfaces participate in the polymerization, thereby bonding the elastomer chains to the reinforcing particles. Alternatively, the $R'Si(OC_2H_5)_3$ can be used as one of the end-linking agents, to place unsaturated groups at the cross links. In any case, good reinforcement is observed for PS domains in the roughly spherical state. (figure 9.9)[129]

Ex situ techniques can also be used. The effects of particle dispersion on the melt state have been documented using nanocomposites containing PS homopolymers or PS-grafted silica.[131] PS has also been grafted onto the cross links in a siloxane polymer[132] and onto filler particles used to reinforce an elastomer.[133]

The PS domains have the disadvantage of having a relatively low glass transition temperature ($T_g \approx 100^\circ C$)[134–136] and in being totally amorphous. Above T_g they soften and presumably lose their reinforcing ability. For this reason, studies have been carried out using crystalline poly(diphenylsiloxane) as the reinforcing phase.[137] Measurements on copolymers containing diphenylsiloxane blocks indicate it has a melting point (and thus a softening temperature) as high as 550°C.[138]

It is possible to convert the essentially spherical PS particles just described into ellipsoids.[42, 139–143] First, the PS-PDMS composite is raised to a temperature well above the T_g of PS; it is then deformed and cooled while in the stretched state. The particles are thereby deformed into ellipsoids, and they retain this shape when cooled. Uniaxial deformation of the composite give prolate (needle-shaped) ellipsoids, and biaxial deformations give oblate (disc-shaped) ellipsoids.[140, 144] Prolate particles can be thought of as a conceptual bridge between the roughly spherical particles used to reinforce elastomers and the long fibers frequently used for reinforcement in thermoplastics and thermosets. Similarly, oblate particles can be considered analogues of the clay platelets used to reinforce a variety of polymers,[145–150] but with dimensions and compositions that are controllable.

Ellipsoidal particles have been characterized using both scanning and transmission electron microscopy to determine their axial ratios and measure of the extent to which their axes were aligned in the stretch direction. In these anisotropic materials, elongation moduli in the direction of the stretching are significantly larger than those of the untreated PS-PDMS elastomer, whereas in the perpendicular direction they are significantly lower (figure 9.10).[140] Such differences are expected from the anisotropic nature of the filler particles.

In the case of nonspherical particles degree of orientations is also of considerable importance. The effects of orientation can be removed by dissolving away the host polymer and redispersing the particles isotropically in another, elastomer matrix. There have been simulations to better understand the mechanical properties of such composites.[151–154]

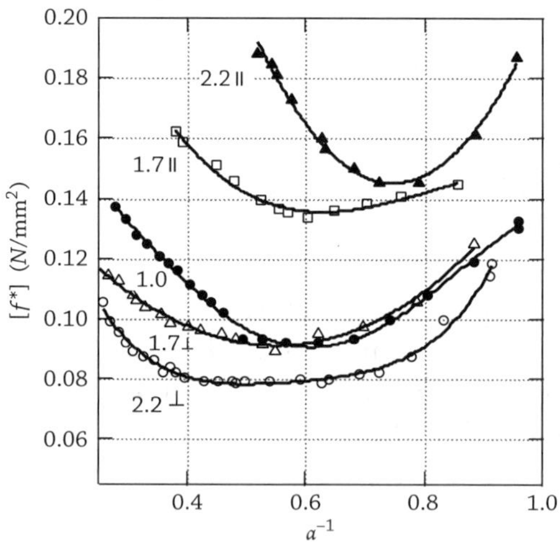

Figure 9.10:
Stress-strain isotherms of the PDMS-PS composites.[140] Values of the draw ratio and testing directions are indicated on each curve. The PS introduced during the in situ polymerizations was 35.0 wt %.

There are endless variations of the in situ filler generation method. Polysiloxane can be made rod-like by cationic charges from ammonium groups.[155] Epoxies have also been studied for coating applications.[156] In some cases, the particles can give an elastomer “self-healing” properties.[157] Particles have been introduced in emulsions,[158] or in the curing agent to provide reinforcement as well as cross linking.[159] In one set of studies, hyperbranched poly(ethoxysiloxane) was used to reinforce organic polymers, including poly(ethylene terephthalate), isotactic polypropylene, and high-density polyethylene.[160]

9.2.3 Ex Situ and Modified Silicas

Separately prepared silica has long been used to reinforce polysiloxanes by blending them into the polymer prior to its cross linking. The technology of this ex situ method is well developed, including the characterization of the resulting composites.[128, 161–169] Mesoporous silica particles have also been used to improve reinforcement of polysiloxane elastomers.[170] Theory and simulations have also been carried out on these materials.[63, 171–173]

One goal is to modify the surfaces of silica particles to improve bonding with PDMS, for example, with vinylethoxysilane[174] or by silanization.[175] Similarly, tetraphenyl-modified fumed silica has been used to increase PDMS radiation resistance.[176] Such materials can be difficult to characterize quantitatively. For example, in some cases fumed silica particles in PDMS formed secondary domain structures that made it difficult to characterize nanoparticle formation by tapping-mode atomic force microscopy.[177] Ultra small angle x-ray[91] and neutron scattering[178] are useful for characterization of such complex morphologies.

More novel structures have also been produced. Biogenic opaline silica short fibers modified with vinyltrimethoxysilane, for example, are good alternative reinforcing fillers for PDMS.[179] Other materials in this category include mesoporous silica,[180] composites with polyindole,[181] and self-assembled structures such as molecularly ordered phenylene-bridged mesoporous organosilica nanofilaments.[182] Also of interest are polysiloxanes with quaternary ammonium salt biocidal functionality,[183] conversion of hyperbranched polyethoxysiloxane,[184] and hybrid core-shell systems based on molecular silicasols.[185] Perhaps the most novel structures are some recently developed enzyme-responsive snap-top covered silica nanocontainers.[186]

NMR[187–189] and positron annihilation lifetime spectroscopy[188] have been used to characterize nanocomposites of this type.

9.2.4 Layered Fillers

Exfoliated layered particles such as the clays, mica, graphite, or graphite oxide give effective reinforcement of polymers at loading levels much smaller than in the case of spherical particles such as carbon black and silica.[190–193] It should be recognized, however, that reinforcement is typically well below that expected for idealized sheet-like fillers.[57] Graphite-oxide reinforced PDMS, for example, gives a 71% enhancement of young's modulus at 3 wt % loading.[194] This value, however, is just 8% that expected for flat sheets. The shortfall is attributed to crumpling of the sheets.[57]Other properties can also be substantially improved, including processability, resistance to solvents, and reduced permeability and flammability.[190–192, 195–202] The viscoelastic properties of graphite oxide suspensions in PDMS have also been studied.[203]

The layered silicate most often used to improve the properties of PDMS is montmorillonite.[119, 204–225] Improvements have also been reported upon the introduction of graphite,[226] graphite oxide,[194] mica,[227, 228] fluoromica,[210] hectorite,[207] fluorohectorite,[210] laponite,[210] wollastonite,[229] perovskite,[230] sepiolite,[231] and titanium-niobium oxide.[232, 233] Kaolin[234] and exfoliated layered double hydroxides[235] have also been used in PDMS nanocomposites. In some cases, diphenylsiloxane-clay hybrids have been used to reinforce polyethylene.[236]

Segmental motions of poly(methylphenylsiloxane) chains confined in intercalated nanocomposites have been studied by quasielastic neutron scattering and, surprisingly, were found to be *faster* than in the bulk polymer.[237]

Computations have been carried out to better understand the structures and properties of these types of nanocomposites.[238–243] One subject of particular interest is the observed *enhanced* gas transport performance of PDMS nanocomposite membranes containing layered silicates,[244] which is not expected based on studies of magnetically alligned particles disucssed in the next section. Clays have also been used to improve the properties of silsesquioxane polymers.[245, 246]

9.2.5 Magnetic and Metallic Particles

Incorporating reinforcing particles that respond to a magnetic field is important with regard to aligning particles to improve mechanical

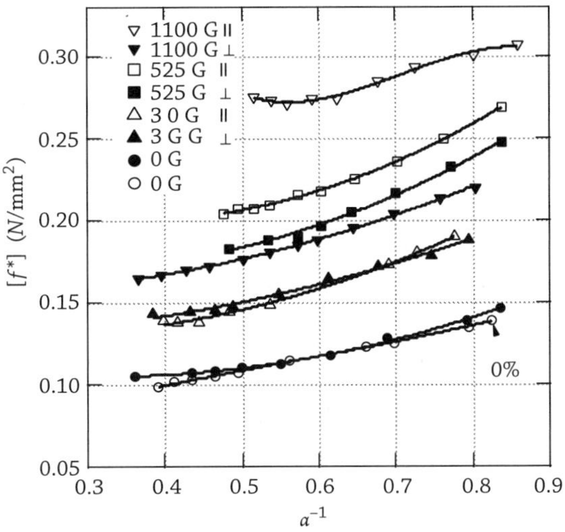

Figure 9.11:
Stress-strain isotherms for PDMS composites containing magnetically responsive particles.[250] Each curve is identified by the strength of the magnetic field (in Gauss) applied during cross linking, and the orientation (parallel or perpendicular) of the sample strip relative to the lines of force of the magnetic field. Loading is 30 wt % silica except for the control, which has 0.0%.

properties anisotropically.[42, 102, 247–249] Considerable anisotropy in structure and mechanical properties can be obtained.[250] Specifically, the reinforcement of PDMS was found to be significantly higher in the direction parallel to the magnetic lines of force (figure 9.11). Sterically stabilized single-particle complexes with a specifically tailored size have been reported in magnetite-PDMS systems.[251]

Making a polysiloxane electrically conductive is best carried out using anisometric particles such as films and fibers, because of their low values of the percolation threshold for conducting pathways.[57] Nonetheless, roughly spherical particles of carbon black have also been used for this purpose.[252] The use of layer-like particles is illustrated by functionalized graphene sheets,[253] while needle shapes are illustrated by unmodified or modified carbon nanotubes or nanofibers.[254–263] In the case of the carbon nanotubes, CH-π interactions between the methyl groups in the PDMS and the nanotube surface can greatly improve dispersion.[264] The same is presumably true for carbon nanofiber interconnected porous network structures.[265]

Magnetic properties have also been induced into polysiloxanes by incorporation of particles of a ferrite,[250, 266, 267] iron,[268, 269] or magnetite.[270, 271] These materials can show large field-induced elongations,[268] and could also find applications as elastomeric bearings and vibration absorbers.[269]

The types of particles seem endless and new applications appear regularly. Iron particles have been used to modify the dielectric properties of polyoxides[272] and alumina particles to increase thermal conductivity.[273] PDMS composites with Al_2O_3-NbC have also been prepared.[274] Also, gold nanoparticles can serve as catalysts for the polymerization of polysiloxanes into nanowires, filaments, and tubes.[275] Another catalytic application involves gold nanoparticles mimicking the catalytic activity of a polysiloxane-synthesizing enzyme.[276] One novel aspect of such gold-polysiloxane nanocomposites is the ability to switch between several colors by swelling.[277] It is also possible to form striped patterns on a gold film deposited on a stretched polysiloxane film,[278] and silica-coated gold particles have also been used to make polysiloxane films with unusual optical properties.[279]

9.2.6 Polyhedral Oligomeric Silsesquioxanes

Polyhedral silsesquioxanes fillers are cage-like silicon-oxygen structures, and have been called the smallest possible silica particles.[280–285] The most common structure has eight silicon atoms, each carrying an organic group. The particles on which none of the groups are functionally reactive can be simply blended into elastomers such as PDMS using mixing or compounding techniques. In this case, the inert groups are chosen to improve miscibility with the elastomeric host matrix. Polyhedral oligomeric silsesquioxanes (POSS) molecules having one reactive functional group can be attached to a polymer as side chains.[286] Those with two reactive groups can be incorporated into polymer backbones by copolymerization, and those with more than two can be used for forming cross links. As opposed to particulate fillers where interfacial bonding is of secondary importance,[287, 288] bonding polyhedral silsesquioxanes to the elastomeric network is critical to effective reinforcement. Bonding is not the source of reinforcement, but rather it aids in dispersion of the cage structures.[285]

POSS particles have been extensively used to reinforce polysiloxanes.[285, 289–291] In some cases, the surfaces of the nanoparticles were made strongly hydrophobic by fluorination.[292] C_{60} fullerene particles have also been used to improve the mechanical properties of PDMS.[293] The polymer had been functionalized and the particles were either monofunctionalized or multifunctionalized with complementary groups.

9.2.7 Nanotubes

Carbon nanotubes are also of considerable interest with regard to both reinforcement and possible increases in electrical conductivity.[280, 281, 283, 284, 294–306] If the goal is to reduce the percolation threshold for electrical conductivity, aggregation can be helpful if it occurs in a random way to create fractal structures.[57] There is considerable interest in characterizing the flexibility of these nanotube structures, in managing their tendencies to aggregate, and in maximizing their miscibility with inorganic as well as organic polymers. Although an enormous effort has been expended on surface functionalization to improve coupling between the filler and the matrix, it is not clear that functionalization does more than improve dispersion.[57]

9.2.8 Dual Fillers

There can be a considerable advantage to using a combination of fillers of different types, such as particles and layered sheets. One frequently obtains a synergistic effect in that the improvements in properties obtained can be larger than expected from simple additivity. In addition, the first filler may have a solubilizing effect, making incorporation of the second one easier. Relatively little has been done in this regard.[307–310]

9.2.9 Porous Fillers

Some fillers such as zeolites are sufficiently porous to harbor monomers, which can then be polymerized. This approach threads the chains through the cavities, with unusually intimate interactions between the reinforcing phase and the host elastomeric matrix (figure 9.12).[295, 311, 312] Good reinforcement is observed. Because of the constraints imposed by the cavity walls, these confined polymers frequently show no glass transition

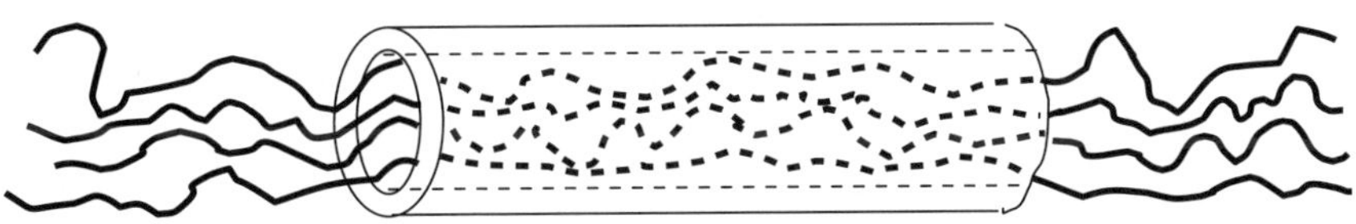

Figure 9.12:
Polymer chains being threaded through a porous inorganic material such as a zeolite by polymerizing monomer that had been absorbed into one of the channels or cavities.

temperature or melting point.[152] PDMS chains have also been threaded through cyclodextrins, to form *pseudo*-rotaxanes.[313]

9.2.10 Fillers with Controlled Interfaces

By choosing the appropriate chemical structures chains that span filler particles in a PDMS-based composite can be designed so that they are durable, irreversibly breakable, or breakable reversibly.[287, 288, 314, 315]

9.2.11 Silicification and Biosilicification

There has been some interest in generating silica-like particles using templates, as occurs naturally in biosilicification processes.[108, 316–325] Various particle shapes have been obtained. Platelet forms are of particular interest with regard to their ability to reinforce and decrease permeability.

9.2.12 Miscellaneous Fillers

There are a variety of miscellaneous fillers that are of interest for reinforcing elastomers such as PDMS. Examples are lignocellulose,[326] ground-up silica xerogels,[327] carbon-coated silica,[328] nickel-coated graphite,[329] and functionalized silica particles.[330–332] Even cross-linked PDMS particles ("silicone powders") have been used,[333–335] as well as functionalized polysiloxane nanospheres.[336] Polystyrene/PDMS core-shell particles have been used for the modification of mechanical properties.[337] Trimethylsiloxy silicates[338] and glass fibers[339] have also been used for this purpose. In some cases, nanoparticles have been generated by phase separation, for example, in PDMS containing epoxy resins and poly(methyl dmethacrylate),[340] and polyamides or polyamideimides.[341]

Additional novel particles, including nylon 66,[342] alumina,[343] gold,[344] garnet,[345] diamond,[346] graphite,[347] polyaniline,[348] and boron nitride nanosheets,[349] and nanoparticles of low molecular weight borate esters, have been introduced to increase ionic conductivities.[350] Boron-doped PDMS has been used as scintillators for thermal neutron detection.[351] Nanowires of ZnO have been inserted into PDMS to give tunable, flexible antireflection layers,[352] and ZnO has also been introduced into such polymers as quantum dots.[353, 354] Similarly, color-tunable luminescent materials have been based on functionalized PDMS and lanthanide ions,[355] or CdSe quantum dots.[356]

Other unusual additives include oxetanes,[357] vinylic macromonomers,[358] silicon carbide,[359] superconductive carbon blacks,[360] silver-coated fly ash,[361] metal oxides,[362] Tb^{3+} for green emission,[363] antibacterial agents,[364] and organic-inorganic hybrid copolymer fibers.[365] In an interesting reversal, transition element acetylacetonate salts were decomposed in a PDMS matrix to give membranes with catalytic activity.[366]

9.2.13 Unusual Shapes

Porous materials, including aerogels and xerogels,[30, 31] can be synthesized in cylindrical, hexagonal, or monoclinic[367] and hierarchically ordered structures.[368, 369] Methods have been developed to prepare hybrid silica-silicone PDMS nanofibers[370] and nanocapsules that are hollow and have controlled diameters.[371–373] Some polysiloxane capsules can be deformed by external electric fields.[374] PDMS sheets bombarded with gallium ions form skins that are under strains not experienced by the underlying polymer.[375] This strain causes a wrinkling into wavy or hierarchical patterns. Gold films can also be used for this purpose.[278] Such materials could be useful in microfluidic devices, for example, for mixing and stretching proteins and polynucleotides.

An origami type of structure can be obtained by placing a droplet of a liquid such as water on a sheet of PDMS placed on a superhydrophobic surface.[376] Capillary forces then wrap the PDMS sheet around the droplet. Square sheets of PDMS envelope the droplet in a cylindrical shape, while triangular sheets seal it into the shape of a tetrahedral pyramid.

Replication techniques are available for transforming complex silica shapes into the corresponding shapes of various polymers.[377] The process can also be extended in the opposite direction, by converting the silica into silicon. Specifically, a low-temperature reduction process has been developed to convert three-dimensional nanostructured silica micro-assemblies into microporous nanocrystalline silicon replicas.[378] Such materials could be useful in a variety of applications, including sensors and biomedical devices.

9.2.14 Simulations on Fillers

Monte Carlo computer simulations have been carried out on a variety of filled elastomers, including PDMS,[379–382] in an attempt to obtain a better molecular interpretation of how such dispersed phases reinforce elastomers. The approach taken enabled estimation of the effect of the excluded volume of the filler particles on the network chains and on the elastic properties of the networks. In the first step, distribution functions for the end-to-end vectors of the chains were obtained by applying Monte Carlo methods to rotational isomeric state representations of the chains.[383] Conformations of chains that overlapped with any filler particle during the simulation were rejected. The resulting perturbed distributions were then used in the three-chain elasticity model[384] to obtain the desired stress-strain isotherms in elongation.

In one application, a filled PDMS network was modeled as a composite of cross-linked polymer chains and spherical filler particles arranged on a cubic lattice.[385] The filler particles increase the non-Gaussian behavior of the chains and increase the moduli. It is interesting to note that composites with such structural regularity have actually been produced[386] and mechanical properties have been reported.[387, 388]

In a subsequent study, the reinforcing particles were randomly distributed within the PDMS matrix.[382] One effect of the filler was to increase the end-to-end separation of the chains. These results on the chain-length distributions are in agreement with some subsequent neutron scattering experiments on silicate-filled PDMS.[389] The polymers contained silica particles that were surface treated to make them inert to the polymer chains, as was implicitly assumed in the simulations. These experimental results also indicated chain extension when the particles were relatively small, and chain compression when they were relatively large.

Two issues need to be addressed by simulations regarding the molecular origin of reinfocement: increases in modulus with loading and upturns in the modulus with increasing deformation. Results are typically expressed as the reduced-nominal or engineering stress as a function of deformation. The area under such curves up to the rupture point then gives the energy of rupture, which is the standard measure of the toughness of a material.[42] The stress-strain isotherms in elongation[380] from simulated distributions showed substantial increases in modulus that increased with increase in filler loading, as expected. Additional increases would be expected by taking into account other mechanisms for reinforcement such as physisorption, chemisorption, and so on.

Simulations have also been carried out on ellipsoidal particles such as the polystyrene prolate and oblate ellipsoids. For example, oriented prolate particles[390] showed that the anisotropy causes the modulus in the longitudinal direction to be significantly higher in the transverse direction.

These simulated results are in at least qualitative agreement with the experimental differences in longitudinal and transverse moduli.[140]

The silica or carbon black particles used to reinforce commercial materials are seldom completely dispersed,[57, 88, 89, 391–393] as is assumed in the simulations described. The primary particles are generally aggregated. These aggregates are frequently clustered into less stable arrangements called agglomerates.[91]

Simulations are being carried out on more ordered structures.[394] The shapes of the aggregates include linear, globular, branched, star-shaped, and fractal. It is well known in the industry that such structures are important in maximizing the reinforcement, as evidenced by the fact that being too persistent in removing such aggregates and agglomerates in blending procedures gives materials with less than optimal mechanical properties.[88, 89, 391–393] Simulations should provide guidance on optimizing the degree of particle dispersion.

9.3 POLYMER-MODIFIED CERAMICS

If hydrolysis in silane precursor-polymer systems is carried out using relatively large amounts of the silane, the silica generated can become the continuous phase, with elastomeric polysiloxane dispersed in it.[13–15, 395–402] The resultant composite is a polymer-modified glass or ceramic, frequently of very good transparency. Although thermal stability is inferior to that of the neat ceramic component, there are many applications for ceramic-type materials where reduced thermal stability is not a concern.

As might be expected, the properties of polymer-toughened ceramic materials depend greatly on the relative amounts of the two phases. Properties of particular interest are modulus, impact resistance, ultimate strength, maximum extensibility, viscoelastic responses, and transparency. The hardness of such a composite, for example, can be varied by control of the molar ratio of alkyl R groups to Si atoms, as illustrated for PDMS in figure 9.13.[397] Low values of R/Si yield a brittle ceramic, and high values give a relatively hard elastomer. The most interesting range, $R/Si \cong 1$, can yield a tough ceramic of increased impact resistance.

Figure 9.14 illustrates some improvement in impact strength in such composites.[403] Specifically, impact strengths of some PDMS-SiO_2 samples were determined by the Charpy pendulum test and by the falling-weight test.[403] The samples investigated were PDMS-modified SiO_2 and SiO_2/TiO_2

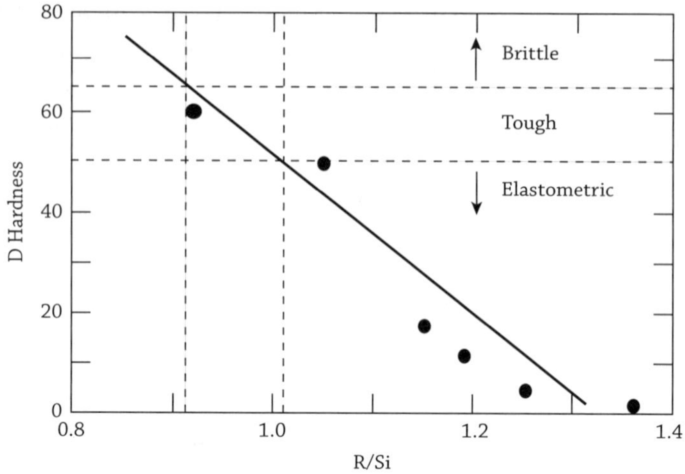

Figure 9.13:
The hardness of a silica-PDMS composite as a function of the molar ratio of alkyl groups to silicon atoms.[397]
Reproduced by permission of Springer Verlag.

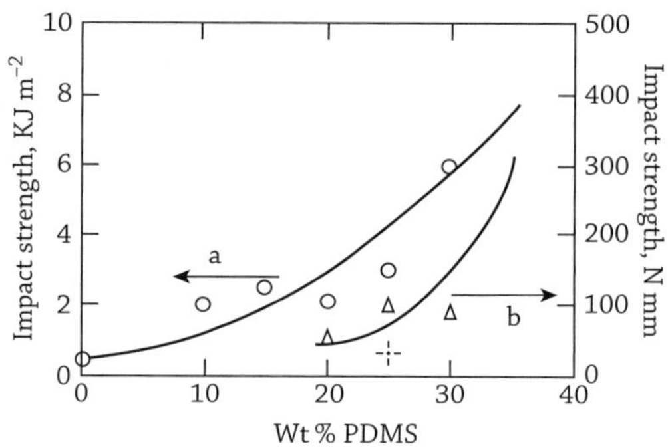

Figure 9.14:
Dependence of two estimates on the impact strength on the amount of PDMS in PDMS-modified SiO_2 glasses.[403] The impact strengths were obtained from: (a) the Charpy pendulum impact test and (b) the falling-weight impact test.

glasses with PDMS contents ranging from 0 to 65 wt %. Only samples with relatively high ceramic content were sufficiently brittle to be studied in this manner. The larger the amount of PDMS used, the higher the impact strength.

For PDMS-modified SiO_2 glasses, structural analysis shows that this hybrid material exhibits localized phase separation of the PDMS component, even though OH-terminated PDMS can be successfully incorporated into the SiO_2 network by chemical bonding. The PDMS component can behave as an elastomeric phase because the glass transition temperature of PDMS is far below room temperature.[134–136] When the material is subjected to an impact test, the PDMS component absorbs energy by motion of the PDMS chains, thus ameliorating the growth of cracks and fracture. Therefore, considerable toughening of the glass can be achieved by increasing the amount of PDMS. The impact resistance was also observed to increase with increasing PDMS molecular weight, possibly due to an increase in the phase separation that leads to the energy-absorbing domains[403] as was found by introducing PDMS domains into a polyamide.[404]

For composites having relatively low PDMS content, microscopy of fracture typically shows smooth fracture surfaces,[405] with little evidence of effective resistance to either initiation or propagation of cracks. In contrast, samples with high PDMS contents had fracture surfaces showing some degree of "whitening" or shearing,[403] implying a ductile, energy-absorbing response to the impacts, with increased resistance to crack propagation.

REFERENCES

1. Hench, L. L.; West, J. K., The Sol-Gel Process. *Chem. Rev.* **1990**, *90*, 33–72.
2. Schaefer, D. W.; Mark, J. E., *Polymer-Based Molecular Composites.* Materials Research Society: Pittsburgh, PA, 1990; Vol. 171.
3. Brinker, C. J.; Scherer, G. W., *Sol-Gel Science. The Physics and Chemistry of Sol-Gel Processing.* Academic Press: New York, 1990.
4. Keefer, K. D., Structure and Growth of Silica Condensation Polymers. In *Silicon-Based Polymer Science. A Comprehensive Resource*, Zeigler, J. M.; Fearon, F. W. G., Eds. American Chemical Society: Washington, DC, 1990; Vol. 224, pp 227–240.
5. Doughty, D. H.; Assink, R. A.; Kay, B. D., Hydrolysis and Condensation Kinetics of Dimeric Sol-Gel Species by 29Si NMR Spectroscopy. In *Silicon-Based Polymer Science. A Comprehensive Resource*, Zeigler, J. M.; Fearon, F. W. G., Eds. American Chemical Society: Washington, DC, 1990; Vol. 224, p 241.
6. Ulrich, D. R., Special Report—Chemical-Processing of Ceramics. *Chem. Eng. News* **1990**, *68*(1), 28–40.
7. Holmes, R. R., The Stereochemistry of Nucleophilic Substitution of Tetracoordinated Silicon. *Chem. Rev.* **1990**, *90*, 17–31.
8. Coltrain, B. K.; Sanchez, C.; Schaefer, D. W.; Wilkes, G. L., *Better Ceramics Through Chemistry VII: Organic/Inorganic Hybrid Materials.* Materials Research Society: Pittsburgh, PA, 1996; Vol. 435.
9. Brinker, C. J.; Giannelis, E. P.; Laine, R. M.; Sanchez, C., *Better Ceramics Through Chemistry VIII: Hybrid Materials.* Materials Research Society: Warrendale, PA, 1998; Vol. 519.

10. Klein, L. C.; Pope, E. J. A.; Sakka, S.; Woolfrey, J. L., *Sol-Gel Processing of Advanced Materials*. American Ceramic Society: Westerville, OH, 1998.
11. Beaucage, G.; Mark, J. E.; Burns, G. T.; Hua, D.-W., *Nanostructured Powders and Their Industrial Application*. Materials Research Society: Warrendale, PA, 1998; Vol. 520.
12. Klein, L. C.; Francis, L. F.; De Guire, M. R.; Mark, J. E., *Organic/Inorganic Hybrid Materials II*. Materials Research Society: Warrendale, PA, 1999; Vol. 576.
13. Laine, R. M.; Sanchez, C.; Giannelis, E.; Brinker, C. J., *Organic/Inorganic Hybrid Materials – 2000*. Materials Research Society: Warrendale, PA, 2001; Vol. 628.
14. Nakatani, A. I.; Hjelm, R. J.; Gerspacher, M.; Krishnamoorti, R., *Filled and Nanocomposite Polymer Materials*. Materials Research Society: Warrendale, PA, 2001; Vol. 661.
15. Sanchez, C.; Laine, R. M.; Yang, S.; Brinker, C. J., *Organic/Inorganic Hybrid Materials–2002*. Materials Research Society: Warrendale, PA, 2002; Vol. 726.
16. Rurack, K.; Martinez-Manez, R., *The Supramolecular Chemistry of Organic-Inorganic Hybrid Materials*. John Wiley and Sons: Hoboken, NJ, 2010.
17. Corriu, R. J. P.; Ahn, N. T., *Molecular Chemistry of Sol-Gel Derived Materials*. John Wiley and Sons: Hoboken, NJ, 2009.
18. Ma, Y.; Kanezashi, M.; Tsuru, T., Preparation of Organic/Inorganic Hybrid Silica Using Methyltriethoxysilane and Tetraethoxysilane as Co-Precursors. *J. Sol-Gel Sci. Technol.* **2010**, *53*, 93–99.
19. Morris, R. E., *The Sol-Gel Process: Uniformity, Polymers and Applications*. Nova Science: New York, 2011.
20. Alexandru, M.; Cazacu, M.; Vlad, A.; Macsim, A. M.; Hitruc, G.; Ionita, D., Polydimethylsiloxane Cross-linked by 4,4'-(1,3-phenylenefioxy)-Dianiline Within In Situ Generated Silica Network. *Polym-Plast. Tech. Eng.* **2011**, *50*, 539–546.
21. Bechtold, M. F.; Vest, R. D.; Plambeck, L., Jr., Silicic Acid from Tetraethyl Silicate Hydrolysis. Polymerization and Properties. *J. Am. Chem. Soc.* **1968**, *86*, 4590–4598.
22. Hench, L. L.; West, J. K., The Sol-Gel Process. *Chem. Rev.* **1990**, *90*, 33–72.
23. Ueda, N.; Gunji, T.; Abe, Y., Synthesis of Linear Ethoxysiloxanes by the Oxidative Condensation of Triethoxysilane. *J. Sol-Gel Sci. Technol.* **2008**, *48*, 163–167.
24. Ye, F.; Collinson, M. M.; Higgins, D. A., What Can Be Learned from Single Molecule Spectroscopy? Applications to Sol-Gel-Derived Silica Materials. *Phys. Chem. Chem. Phys.* **2009**, *11*, 66–82.
25. Kanamori, K.; Nakanishi, K.; Hanada, T., Spinodal Decomposition in Siloxane Sol-Gel Systems in Macroporous Media. *Soft Matter* **2009**, *5*, 3106–3113.
26. Malani, A.; Auerbach, S. M.; Monson, P. A., Probing tthe Mechanism of Silica Polymerization at Ambient Temperatures Using Monte Carlo Simulations. *J. Phys. Chem. Lett.* **2010**, *1*, 3219–3224.
27. Rajendra, V.; Gonzaga, F.; Brook, M. A., Nearly Monodisperse Silica Microparticles Form in Silicone (Pre)elastomer Mixtures. *Langmuir* **2011**, *27*, 1470–1477.
28. Fir, M.; Orel, B.; Vuk, A. S.; Vilcnik, A.; Jese, R.; Francetic, V., Corrosion Studies and Interfacial Bonding of Urea/Poly(dimethylsiloxane) Sol/Gel Hydrophobic Coatings on AA 2024 Aluminum Alloy. *Langmuir* **2007**, *23*, 5505–5514.

29. Wu, K. H.; Chang, T. C.; Yang, J. C.; Chen, H. B., Organic-Inorganic Hybrid Materials. II. Chain Mobility and Stability of Polysiloxaneimide-Silica Hybrids. *J. Appl. Polym. Sci.* **2001**, *79*, 965–973.
30. Park, Y.-W.; Lee, D.-S.; Kim, S.-H., Mechanical, Surface, and Thermal Properties of Polyamideimide-Polydimethylsiloxane Nanocomposites Fabricated by Sol-Gel Process. *J. Appl. Polym. Sci.* **2004**, *91*, 1774–1783.
31. Park, Y.-W.; Lee, D.-S., Surface and Electrical Characteristics of Polyamideimide-Poly(dimethylsiloxane) Nanocomposites. *J. Appl. Polym. Sci.* **2004**, *91*, 342–347.
32. Zhang, H.; Han, J.; Yang, B., Structural Fabrication and Functional Modulation of Nanoparticle-Polymer Composites. *Adv. Funct. Mater.* **2010**, *20*, 1533–1550.
33. Ogawa, T.; Watanabe, J.; Eguchi, K.; Oshima, Y., Synthesis of Polysiloxane-Modified Hybrid Particles by a High Temperature Water Technology. *Polymer* **2010**, *51*, 2836–2842.
34. Huang, Y.; Liu, W.; Zhou, X., Silicone/Silica Nanocomposites as Culture-Stone Protective Materials. *J. Appl. Polym. Sci.* **2012**, *125*, 282–290.
35. Xu, P.; Wang, S.; Mark, J. E., Particle Growth and Development During the In-Situ Precipitation of Silica in a Polymeric Matrix. In *Better Ceramics Through Chemistry IV*, Zelinski, B. J. J.; Brinker, C. J.; Clark, D. E.; Ulrich, D. R., Eds. Materials Research Society: Pittsburgh, PA, 1990; Vol. 180, pp 445–451.
36. Mark, J. E.; Schaefer, D. W., Reinforcement of Elastomers by the In-Situ Generation of Filler Particles. In *Polymer-Based Molecular Composites*, Schaefer, D. W.; Mark, J. E., Eds. Materials Research Society: Pittsburgh, PA, 1990; Vol. 171, pp 51–56.
37. Schaefer, D. W.; Mark, J. E.; McCarthy, D. W.; Jian, L.; Sun, C.-C.; Farago, B., Structure of Microphase-Separated Silica/Siloxane Molecular Composites. In *Polymer-Based Molecular Composites*, Schaefer, D. W.; Mark, J. E., Eds. Materials Research Society: Pittsburgh, PA, 1990; Vol. 171, pp 57–63.
38. Zou, H.; Wu, S.; Shen, J., Polymer/Silica Nanocomposites: Preparation, Characterization, Properties, and Applications. *Chem. Rev.* **2008**, *108*, 3893–3957.
39. Baccile, N.; Babonneau, F.; Thomas, B.; Coradin, T., Introducing Ecodesign in Silica Sol-Gel Materials. *J. Mater. Chem.* **2009**, *19*, 8537–8559.
40. Gordon, G. V.; Schmidt, R. G.; Quintero, M.; Benton, N. J.; Cosgrove, T.; Krukonis, V. J.; Williams, K.; Wetmore, P. M., Impact of Polymer Molecular Weight on the Dynamics of Poly(dimethylsiloxane)-Polysilicate Nanocomposites. *Macromolecules* **2010**, *43*, 10132–10142.
41. Schmidt, R. G.; Gordon, G. V.; Dreiss, C. A.; Cosgrove, T.; Krukonis, V. J.; Williams, K.; Wetmore, P. M., A Critical Size Ratio for Viscosity Reduction in Poly(dimethylsiloxane)-Polysilicate Nanocomposites. *Macromolecules* **2010**, *43*, 10143–10151.
42. Erman, B.; Mark, J. E., *Structures and Properties of Rubberlike Networks*. Oxford University Press: New York, 1997.
43. Mark, J. E., Some Unusual Elastomers and Experiments on Rubberlike Elasticity. *Prog. Polym. Sci.* **2003**, *28*, 1205–1221.
44. Alexandru, M.; Cazacu, M.; Racles, C.; Grigoras, C., Amphiphile Polydimethylsiloxane-Based Networks Reinforced with In Situ Generated Silica. *Polym. Eng. Sci.* **2011**, *51*, 78–86.

45. Kickelbick, G., Polysiloxanes as Templates and Building Blocks in Nanostructured Materials. In *Silicon-Based Inorganic Polymers*, De Jaeger, R.; Gleria, M., Eds. Nova Science: New York, 2008; pp 163–189.
46. Iji, M.; Morishita, N.; Kai, H., Self-Asembling Siloxane Nanoparticles with Three Phases That Increase Tenacity of Poly L-Lactic Acid. *Polym. J.* **2011**, *43*, 101–104.
47. Ahmad, Z.; Wang, S.; Mark, J. E., Polyamides. In *Better Ceramics Through Chemistry VI*, Cheetham, A. K.; Brinker, C. J.; Mecartney, M. L.; Sanchez, C., Eds. Materials Research Society: Pittsburgh, PA, 1994; Vol. 346, p 127.
48. Kim, Y. B.; Cho, D.; Park, W. H., Electrospinning of Poly(dimethyl siloxane) by Sol-Gel Method. *J. Appl. Polym. Sci.* **2009**, *114*, 3870–3874.
49. Sertchook, H.; Elimelech, H.; Avnir, D., Amino-Polysiloxane Hybrid Materials for Bone Reconstruction. *Chem. Mater.* **2005**, *17*, 4711–4716.
50. Han, Y.-H.; Travas-Sejdic, J.; Wright, B.; Yim, J.-H., Simultaneous Vapor-Phase Polymerization of PEDOT and a Siloxane into Organic/Inorganic Hybrid Thin films. *Macromol. Chem. Phys.* **2011**, *212*, 521–530.
51. Brus, J.; Dybal, J., Solid-State NMR Study of Structure, Size and Dynamics of Domains in Hybrid Siloxane Networks. *Polymer* **2000**, *41*, 5269–5282.
52. Brus, J.; Skrdlantova, M., ^{1}H MAS NMR Study of Structure of Hybrid Siloxane-Based Networks and the Interaction with Quartz Filler. *J. Non-Cryst. Solids* **2001**, *281*, 61–71.
53. Apperley, D.; Hay, J. N.; Raval, H. M., Silica-Dimethylsiloxane Hybrids - Non-Hydrolytic Sol-Gel Synthesis and Characterization by NMR Spectroscopy. *Chem. Mater.* **2002**, *14*, 983–988.
54. Sarmento, V. H. V.; Dahmouche, K.; Santilli, C. V.; Pulcinelli, S. H., Gelation of Siloxane–poly(oxypropylene) Composites. *J. Non-Cryst. Solids* **2002**, *304*, 134–142.
55. Higgins, D. A.; Collinson, M. M., Gaining Insight into the Nanoscale Properties of Sol-Gel Thin Films by Single-Molecule Spectroscopy. *Langmuir* **2005**, *21*, 9023–9031.
56. Hyeon-Lee, J.; Guo, L.; Beaucage, G.; Macip-Boulis, M. A., Morphological Development in PDMS/TEOS Hybrid Materials. *J. Polym. Sci., Part B: Polym. Phys.* **1996**, *34*, 3073–3080.
57. Schaefer, D. W.; Justice, R. S., How Nano Are Nanocomposites? *Macromolecules* **2007**, *40*, 8502–8517.
58. Dahmouche, K.; Santilli, C. V.; Pulcinelli, S. H.; Craievich, A. F., Small-Angle X-ray Scattering Study of Sol–Gel-Derived Siloxane–PEG and Siloxane–PPG Hybrid Materials. *J. Phys. Chem. B* **1999**, *103*, 4937–4942.
59. Jose, N. M.; Prado, L. A. S. A.; Yoshida, I. V. P., Synthesis, Characterization, and Permeability Evaluation of Hybrid Organic/Inorganic Films. *J. Polym. Sci., Part B: Polym. Phys.* **2004**, *42*, 4281–4292.
60. Chaker, J. A.; Dahmouche, K.; Santilli, C. V.; Pulcinelli, S. H.; Briois, V.; Flank, A.-M.; Judeinstein, P., Siloxane–Polypropyleneoxide Hybrid Ormolytes: Structure–Ionic Conductivity Relationships. *J. Non-Cryst. Solids* **2002**, *304*, 109–115.
61. Fragiadakis, D.; Pissis, P.; Bokobza, L., Glass Transition and Molecular Dynamics in Poly(dimethylsiloxane)/Silica Nanocomposites. *Polymer* **2005**, *46*, 6001–6008.
62. Fragiadakis, D.; Pissis, P., Glass Transition and Segmental Dynamics in Poly(dimethylsiloxane)/Silica Nanocomposites Studied by Various Techniques. *J. Non-Cryst. Solids* **2007**, *353*, 4344–4352.

63. Fragiadakis, D.; Pissis, P.; Bokobza, L., Modified Chain Dynamics in Poly(dimethylsiloxane)/Silica Nanocomposites. *J. Non-Cryst. Solids* **2006**, *352*, 4969–4972.
64. Li, Z.; Han, W.; Kozodaev, D.; Brokken-Zijp, J. C. M.; De With, G. P.; Thune, C., Surface Properties of Poly(dimethylsiloxane)-Based Inorganic/Organic Hybrid Materials. *Polymer* **2006**, *47*, 1150–1158.
65. Chevalier, P.; Ou, D.-L.; Lee, Y.; Robson, S.; Dupont, A., Thermo-Mechanical Properties of In-Situ Silica Filled Silicone Resin. *Macromol. Symp.* **2006**, *233*, 117–122.
66. Yamada, N.; Yoshinaga, I.; Katayama, S., Formation Behavior and Optical Properties of Transparent Inrganic-Organic Hybrids Prepared from Metal Alkoxides and Polydimethylsiloxane. *J. Sol-Gel Sci. Technol.* **2000**, *17*, 123–130.
67. Lebeau, R.; Maquet, J.; Sanchez, C.; Beaume, F.; Laupretre, F., Structural and Dynamical Studies of Hybrid Siloxane-Silica Materials. *J. Mater. Chem.* **1997**, *7*, 989–995.
68. Leezenberg, P. B.; Frank, C. W., In-Situ Precipation of Silica in Dansyl-Labeled Poly(dimethylsiloxane) Elastomers. *Chem. Mater.* **1995**, *7*, 1784–1792.
69. Curteanu, S.; Nistor, A.; Curteanu, N.; Airinei, A.; Cazacu, M., Applying Soft Computing Methods to Fluorescence Modeling of the Polydimethylsiloxane/Silica Composites Containing Lanthanum. *J. Appl. Polym. Sci.* **2010**, *117*, 3160–3169.
70. Lu, H.; Liu, L.; Feng, S., A New Way to Construct Luminescent Functionalized Polysiloxane Based on Ternary Lanthanide Complexes of 1,10-Phenanthroline. *J. Appl. Polym. Sci.* **2012**, *123*, 1884–1888.
71. Wang, X., Construction and Photoluminescence of Amorphous Silicone Resins Containing Lanthanide Ions as Luminescent Center. *J. Non-Cryst. Solids* **2012**, *358*, 765–770.
72. Yamada, N.; Yoshinaga, I.; Katayama, S., Synthesis and Dynamic Mechanical Behaviour of Inorganic-Organic Hybrids Containing Various Inorganic Components. *J. Mater. Chem.* **1997**, *7*, 1491–1495.
73. Dire, S., Sol-Gel Derived Polysiloxane-Oxide Hybrid Materials: Extent of Phase Interactions. *J. Sol-Gel Sci. Technol.* **2003**, *26*, 285–290.
74. Telez, L.; Rubio, J.; Rubio, F.; Morales, E.; Oteo, J. L., Synthesis of Inorganic-Organic Hybrid Materials from TEOS, TBT, and PDMS. *J. Mater. Sci.* **2003**, *38*, 1773–1780.
75. Nakade, M.; Kameyama, K.; Ogawa, M., Synthesis and Properties of Titanium Dioxide/Polydimethylsiloxane Hybrid Particles. *J. Mater. Sci.* **2004**, *39*, 4131–4137.
76. Pena-Alonso, R.; Tellez, L.; Rubio, F.; Rubio, J., Surface Chemical and Physical Properties of TEOS-TBOT-PDMS Hybrid Materials. *J. Sol-Gel Sci. Technol.* **2006**, *38*, 133–145.
77. Dworak, D. P.; Soucek, M. D., Effect of Mixed Sol-Gel Precursors on the Metal-Oxo Phase Within a UV-Curable Silicone Hybrid Material. *Macromol. Chem. Phys.* **2006**, *207*, 1230–1232.
78. Klonos, P.; Panagopoulou, A.; Bokobza, L.; Kyristis, A.; Peoglos, V.; Pissis, P., Comparative Studies on Effects of Silica and Titania Nanoparticles on Crystallization and Complex Segmental Dynamics in Poly(dimethylsiloxane). *Polymer* **2010**, *51*, 5490–5499.
79. Alexandru, M.; Cazacu, M.; Nistor, A.; Musteata, V. E.; Stoica, I.; Grigoras, C.; Simionescu, B. C., Polydimethylsiloxane/Silica/Titania Composites Prepared by Solvent-Free Sol-Gel Technique. *J. Sol-Gel Sci. Technol.* **2010**, *56*, 310–319.

80. Yoshinaga, I.; Yamada, N.; Katayama, S., Inorganic-Organic Hybrids Synthesized from Metal Alkoxides and Silanol-Terminated Poly(dimethylsiloxane). In *Sol-Gel Processing of Advanced Materials*, Klein, L. C.; Pope, E. J. A.; Sakka, S.; Woolfrey, J. L., Eds. American Ceramic Society: Westerville, OH, 1998; pp 317–322.
81. Alonso, B.; Sanchez, C., Structural Investigation of Polydimethylsiloxane-Vanadate Hybrid Materials. *J. Mater. Chem.* **2000**, *10*, 377–386.
82. Mimura, K.-I.; Nishimoto, Y.; Orihara, H.; Moriya, M.; Sakamoto, W.; Yogo, T., Synthesis of Transparent and Field-Responsive $BaTiO_3$ Particle/Organosiloxane Hybrid Fluid. *Angew. Chem. Int. Ed. Engl.* **2010**, *49*, 4902–4906.
83. Kamitakahara, M.; Kawashita, M.; Miyata, N.; Kokubo, T.; Nakamura, T., Bioactivity and Mechanical Properties of Polydimethylsiloxane (PDMS)-CaO-SiO_2 Hybrids with Different PDMS Contents. *J. Sol-Gel Sci. Technol.* **2001**, *21*, 75–81.
84. Colilla, M.; Sallinas, A. J.; Vallet-Regi, M., Amino-Polysiloxane Hybrid Materials for Bone Reconstruction. *Chem. Mater.* **2006**, *18*, 5676–5683.
85. Pena-Alonso, R.; Tamayo, A.; Rubio, F.; Rubio, J., Influence of Boron Concentration on the Surface Properties of TEOS-PDMS Hybrid Materials. *J. Sol-Gel Sci. Technol.* **2005**, *36*, 113–124.
86. Beigbeder, A.; Bruzaud, S.; Mederic, P.; Aubry, T.; Brohens, Y., Rheological Characterization of Poly(dimethylsiloxane)/$HTiNbO_5$ Nanocomposites Prepared from Different Routes. *Polymer* **2005**, *46*, 12279–12286.
87. Molina, C.; Dahmouche, K.; Santilli, C. V.; Craievich, A. F.; Ribeiro, S. J., Structure and Luminescence of Eu^{3+}-Doped Class I Siloxane-Poly(ethylene glycol) Hybrids. *Chem. Mater.* **2001**, *13*, 2818–2823.
88. Warrick, E. L.; Pierce, O. R.; Polmanteer, K. E.; Saam, J. C., Silicone Elastomer Developments 1967–1977. *Rubber Chem. Technol.* **1979**, *52*, 437–525.
89. Rigbi, Z., Reinforcement of Rubber by Carbon Black. *Adv. Polym. Sci.* **1980**, *36*, 21–68.
90. Medalia, A. I.; Kraus, G., Reinforcement of Elastomers by Particulate Fillers. In *Science and Technology of Rubber*, 2nd ed.; Mark, J. E.; Erman, B.; Eirich, F. R., Eds. Academic Press: San Diego, 1994; pp 387–418.
91. Schaefer, D. W.; Rieker, T.; Agamalian, M.; Lin, J. S.; Fischer, D.; Sukumaran, S.; Chen, C. Y.; Beaucage, G.; Herd, C.; Ivie, J. J., Multilevel Structure of Reinforcing Silica and Carbon. *J. Appl. Crystallog.* **2000**, *33*(1), 587–591.
92. Ning, Y.-P.; Tang, M.-Y.; Jiang, C.-Y.; Mark, J. E.; Roth, W. C., Particle Sizes of Reinforcing Silica Precipitated into Elastomeric Networks. *J. Appl. Polym. Sci.* **1984**, *29*, 3209–3212.
93. Mark, J. E.; Erman, B., *Rubberlike Elasticity: A Molecular Primer*, 2nd ed. Cambridge University Press: Cambridge, 2007.
94. Kim, Y. B.; Cho, D.; Park, W. H., Fabrication and Characterization of TiO_2/Poly(dimethylsiloxne) Composite Fibers with Thermal and Mechanical Stability. *J. Appl. Polym. Sci.* **2010**, *116*, 449–454.
95. Tamayo, A.; Tellez, L.; Rubio, J.; Rubio, F.; Oteo, J. L., Effect of Reaction Conditions on Surface Properties of TEOS-TBOT-PDMS Hybrid Materials. *J. Sol-Gel Sci. Technol.* **2010**, *55*, 94–104.
96. Klonos, P.; Panagopoulou, A.; Kyritsis, A.; Bokobza, L.; Pissis, P., Dielectric Studies of Segmental Dynamics in Poly(dimethylsiloxane)/Titania Nanocomposites. *J. Non-Cryst. Solids* **2011**, *357*, 610–614.

97. Xiang, H.; Ge, J.; Cheng, S.; Han, H.; Cui, S., Synthesis and Characterization of Titania/MQ Silicone Resin Hybrid Nanocomposite via Sol-Gel Process. *J. Sol-Gel Sci. Technol.* **2011**, *59*, 636–639.
98. Calvo, M. E.; Miguez, H., Flexible, Adhesive, and Biocompatible Bragg Mirrors Based on Polydimethylsiloxane Infuilitrated Nanoparticle Multilayers. *Chem. Mater.* **2010**, *22*, 3909–3915.
99. Garrido, L.; Mark, J. E.; Sun, C. C.; Ackerman, J. L.; Chang, C., NMR Characterization of Elastomers Reinforced with In-Situ Precipitated Silica. *Macromolecules* **1991**, *24*, 4067–4072.
100. Wang, S.; Mark, J. E., TEM Studies of the Distribution of Silica Generated In-Situ in an Elastomeric Matrix. *J. Macromol. Sci., Macromol. Rep.* **1994**, *A31*, 253–260.
101. Murugesan, S.; Sur, G. S.; Beaucage, G.; Mark, J. E., In-Situ Generation of Polyaniline in Poly(dimethylsiloxane) Networks. *Silicon Chem.* **2003**, *2*, 217–221.
102. Zhou, D.; Subramaniam, S.; Mark, J. E., In-Situ Synthesis of Polyaniline in Poly(dimethylsiloxane) Networks Using an Inverse Emulsion Route. *J. Macro. Sci.-Chem.* **2005**, *42*, 113–126.
103. Jiang, C.-Y.; Mark, J. E., The Effects of Various Catalysts in the In-Situ Precipitation of Reinforcing Silica in Polydimethylsiloxane Networks. *Makromol. Chem.* **1984**, *185*, 2609–2617.
104. Schaefer, D. W.; Keefer, K. D., Fractal Geometry of Silica Condensation Polymers. *Phys. Rev. Lett.* **1984**, *53*, 1383–1386.
105. Tamayo, A.; Rubio, J., Structural Modification by Solvent Addition into TEOS/PDMS Hybrid Materials. *J. Non-Cryst. Solids* **2010**, *356*, 1742–1748.
106. Schaefer, D. W.; Jian, L.; Sun, C.-C.; McCarthy, D. W.; Jiang, C.-Y.; Ning, Y.-P.; Mark, J. E.; Spooner, S., Structure-Property Relationships in Silica-Siloxane Molecular Composites. In *Ultrastructure Processing of Advanced Materials*, Uhlmann, D. R.; Ulrich, D. R., Eds. Wiley: New York, 1992; pp 361–375.
107. Ahmad, Z.; Sagheer, F. A.; Arbash, A. A.; Ali, A. A. M., Synthesis and Characterization of Chemically Cross-Linked Polyimide-Siloxane Hybrid Films. *J. Non-Cryst. Solids* **2009**, *355*, 507–517.
108. Patwardhan, S. V.; Taori, V. P.; Hassan, M.; Agashe, N. R.; Franklin, J. E.; Beaucage, G.; Mark, J. E.; Clarson, S. J., An Investigation of the Properties of Poly(dimethylsiloxane)-Bioinspired Silica Hybrids. *Eur. Polym. J.* **2006**, *42*, 167–178.
109. Rajan, G. S.; Sur, G. S.; Mark, J. E.; Schaefer, D. W.; Beaucage, G., Preparation and Characterization of Some Unusually Transparent Poly(dimethylsiloxane) Nanocomposites. *J. Polym Sci. B, Polym. Phys.* **2003**, *41*, 1897–1901.
110. Ostwald, W., Ostwald Ripening. *Z. Phys.Chem.* **1900**, *34*, 495.
111. Mark, J. E.; Erman, B., *Rubberlike Elasticity: A Molecular Primer.* Wiley-Interscience: New York, 1988.
112. Mark, J. E., The Sol-Gel Route to Inorganic-Organic Composites. *Heterogen. Chem. Rev.* **1996**, *3*, 307–326.
113. Mark, J. E.; Ning, Y.-P., Effects of Ethylamine Catalyst Concentration in the Precipitation of Reinforcing Silica Filler in an Elastomeric Network. *Polym. Bull.* **1984**, *12*, 413–417.
114. Mark, J. E.; Wang, S.; Xu, P.; Wen, J., Reinforcement from In-Situ Precipitated Silica in Polysiloxane Elastomers Under Various Types of Deformation. In *Submicron Multiphase Materials*, Baney, R. H.; Gilliom, L. R.; Hirano, S.-I.; Schmidt,

H. K., Eds. Materials Research Society: Pittsburgh, PA, 1992; Vol. 274, pp 77–84.
115. Wang, S.; Xu, P.; Mark, J. E., Shear and Biaxial Extension Measurements of Reinforcement from In-Situ Precipitated Silica. *Rubber Chem. Technol.* **1991**, *64*, 746–759.
116. Wang, S.-B.; Mark, J. E., In-Situ Precipitation of Reinforcing Titania Fillers. *Polym. Bull.* **1987**, *17*, 271–277.
117. Kraus, G., *Reinforcement of Elastomers*. Interscience: New York, 1965.
118. Wen, J.; Mark, J. E., Precipitation of Silica-Titania Mixed-Oxide Fillers into Poly(dimethylsiloxane) Networks. *Rubber Chem. Technol.* **1994**, *67*, 806–819.
119. Burnside, S. D.; Giannelis, E. P., Synthesis and Properties of New Poly(dimethylsiloxane) Nanocomposites. *Chem. Mater.* **1995**, *7*, 1597–1600.
120. Wen, J.; Mark, J. E.; Fitzgerald, J. J., Unimodal, Bimodal, and Filled Poly(dimethylsiloxane) Elastomers Under Cyclic Stress at Elevated Temperatures. *J. Macromol. Sci., Macromol. Rep.* **1994**, *31*, 429–438.
121. Sohoni, G. B.; Mark, J. E., Thermal Stability of In-Situ Filled Siloxane Elastomers. *J. Appl. Polym. Sci.* **1992**, *45*, 1763–1775.
122. Mora-Barrantes, I.; Rodriguez, A.; Ibarra, I.; Gonzales, L.; Valentin, J. L., Overcoming the Disadvantages of Fumed Silica as Filler in Composites. *J. Mater. Chem.* **2011**, *21*, 7381–7392.
123. Aranguren, M. I., Crystallization of Polydimethylsiloxane: Effect of Silica Filler and Curing. *Polymer* **1998**, *39*, 4897–4903.
124. Pethrick, R. A.; Dawkins, J. V., *Modern Techniques for Polymer Characterization*. John Wiley & Sons: New York, 1999.
125. Roe, R.-J., *Methods of X-Ray and Neutron Scattering in Polymer Science*. Oxford University Press: Oxford, 2000.
126. Sperling, L. H., *Introduction to Physical Polymer Science*, 3rd ed. Wiley Interscience: New York, 2001.
127. Wignall, G. D., Small-Angle-Neutron-Scattering Characterization of Polymers. In *Physical Properties of Polymers*, 3rd ed.; Mark, J. E.; Ngai, K. L.; Graessley, W. W.; Mandelkern, L.; Samulski, E. T.; Koenig, J. L.; Wignall, G. D., Eds. Cambridge University Press: Cambridge, 2004; pp 424–511.
128. Serbescu, A.; Saalwachter, K., Particle-Induced Network Formation in Linear PDMS Filled with Silica. *Polymer* **2009**, *50*, 5434–5442.
129. Fu, F.-S.; Mark, J. E., Elastomer Reinforcement from a Glassy Polymer Polymerized In-Situ. *J. Polym. Sci., Polym. Phys. Ed.* **1988**, *26*, 2229–2235.
130. Fu, F.-S.; Mark, J. E., Polystyrene-Polyisobutylene Network Composites from In-Situ Polymerizations. *J. Appl. Polym. Sci.* **1989**, *37*, 2757–2768.
131. Akcora, P.; Kumar, S. K.; Moll, J.; Lewis, S.; Schadler, L. S.; Li, Y.; Benicewicz, B. C.; Sandy, A.; Narayanan, S.; Ilavsky, J.; Thiyagarajan, P.; Colby, R. H.; Douglas, J. F., "Gel-Like" Mechanical Reinforcement in Polymer Nanocomposite Melts. *Macromolecules* **2010**, *43*, 1003–1010.
132. Sur, G. S.; Mark, J. E., Grafting of Polystyrene onto the Cross Links in a Siloxane Elastomer. *Eur. Polym. J.* **1988**, *24*, 913–916.
133. Sur, G. S.; Mark, J. E., Grafting of Polymer onto Filler Particles Used to Reinforce an Elastomer. *Polym. Bull.* **1988**, *20*, 131–136.
134. Mark, J. E., *Physical Properties of Polymers Handbook*. Springer-Verlag: New York, 1996.
135. Brandrup, J.; Immergut, E. H.; Grulke, E. A., *Polymer Handbook*, 4th ed. Wiley: New York, 1999.

136. Mark, J. E., *Polymer Data Handbook*. Oxford University Press: New York, 1999.
137. Wang, S.; Mark, J. E., Reinforcement of Elastomeric Poly(dimethylsiloxane) by Glassy Poly(diphenylsiloxane). *J. Mater. Sci.* **1990**, *25*, 65–68.
138. Ibemesi, J.; Gvozdic, N.; Keumin, M.; Tarshiani, Y.; Meier, D. J., Synthesis and Properties of Copolymers of Diphenylsiloxane with Other Organosiloxanes. In *Polymer-Based Molecular Composites*, Schaefer, D. W.; Mark, J. E., Eds. Materials Research Society: Pittsburgh, PA, 1990; Vol. 171, pp 105–115.
139. Nagy, M.; Keller, A., Ellipsoidal Polymer Particles with Predesigned Axial-Ratio. *Polymer Comm.* **1989**, *30*, 130–132.
140. Wang, S.; Mark, J. E., Generation of Glassy Ellipsoidal Particles within an Elastomer by In-Situ Polymerization, Elongation at an Elevated Temperature, and Finally Cooling Under Strain. *Macromolecules* **1990**, *23*, 4288–4291.
141. Ho, C. C.; Hill, M. J.; Odell, J. A., Morphology of Ellipsoidal Latex-Particles. *Polymer* **1993**, *34*, 2019–2023.
142. Ho, C. C.; Keller, A.; Odell, J. A.; Ottewill, R. H., Monodisperse Ellipsoidal Polystyrene Latex Particles: Preparation and Characterisation. *Polym. Int.* **1993**, *30*, 207–212.
143. Ho, C. C.; Keller, A.; Odell, J. A.; Ottewill, R. H., Preparation of Monodisperse Ellipsoidal Polystyrene Particles. *Coll. Polym. Sci.* **1993**, *271*, 469–479.
144. Wang, S.; Xu, P.; Mark, J. E., Method for Generating Oriented Oblate Ellipsoidal Particles in an Elastomer and Characterization of the Reinforcement They Provide. *Macromolecules* **1991**, *24*, 6037–6039.
145. Okada, A.; Kawasumi, M.; Usuki, A.; Kojima, Y.; Kurauchi, T.; Kamigaito, O., Nylon 6-Clay Hybrid. In *Polymer-Based Molecular Composites*, Schaefer, D. W.; Mark, J. E., Eds. Materials Research Society: Pittsburgh, PA,1990; Vol. 171, pp 45–50.
146. Pinnavaia, T. J.; Lan, T.; Wang, Z.; Shi, H.; Kaviratna, P. D., Clay-Reinforced Epoxy Nanocomposites: Synthesis, Properties, and Mechanism of Formation. In *Nanotechnology. Molecularly Designed Materials*, Chow, G.-M.; Gonsalves, K. E., Eds. American Chemical Society: Washington, DC, 1996; Vol. 622, pp 250–261.
147. Giannelis, E. P., Organoceramic Nanocomposites. In *Biomimetic Materials Chemistry*, Mann, S., Ed. VCH Publishers: New York, 1996; pp 337–359.
148. Vaia, R. A.; Giannelis, E. P., Liquid Crystal Polymer Nanocomposites: Direct Intercalation of Thermotropic Liquid Crystalline Polymers into Layered Silicates. *Polymer* **2001**, *42*, 1281–1285.
149. Vaslilakos, S. P.; Tarantili, P. A., Mechanical Properties and Nanostructure Correlation of Condensed-Type Poly(dimethylsiloxane)/Layered Silicate Hybrids. *J. Appl. Polym. Sci.* **2012**, *125*, 548–560.
150. Kim, F. S.; Kim, E. J.; Lee, T. H.; Yoon, J.-S., Clay Modification and Its Effect on the Physical Properties of Silicone Rubber/Clay Composites. *J. Appl. Polym. Sci.* **2012**, *125*, 298–304.
151. Sharaf, M. A.; Mark, J. E., Monte Carlo Simulations on Filler-Induced Network Chain Deformations and Elastomer Reinforcement from Oriented Oblate Particles. *Polymer* **2002**, *43*, 643–652.
152. Mark, J. E., New Developments and Directions in the Area of Elastomers and Rubberlike Elasticity. *Macromol. Symp., Kyoto issue* **2003**, *201*, 77–83.
153. Sharaf, M. A.; Kloczkowski, A.; Sen, T. Z.; Jacob, K. I.; Mark, J. E., Molecular Modeling of Matrix Chain Deformation in Nanofiber Filled Composites. *Coll. Poly. Sci.* **2006**, *284*, 700–709.

154. Mark, J. E., Some Novel Polymeric Nanocomposites. *Acc. Chem. Res.* **2006**, *39*, 881–888.
155. Kaneko, Y.; Iyi, N.; Matsumoto, T.; Kitamura, K., Preparation of Higher-Ordered Inorganic-Organic Nanocomposite Composed of Rodlike Cationic Polysioxane and Polyacrylate. *J. Mater. Chem.* **2005**, *15*, 1572–1575.
156. Pant, R. R.; Buckley, J. L.; Fulmer, P. A.; Wynne, J. H.; McCluskey, D. M.; Phillips, J. P., Hybrid Siloxane Epoxy Coatings Containing Quaternary Ammonium Moieties. *J. Appl. Polym. Sci.* **2008**, *110*, 3080–3086.
157. Cho, S. H.; Andersson, H. M.; White, S. R.; Sottos, N. R.; Braun, P. V., Polydimethylsiloxane-Based Self-Healing Materials. *Adv. Mater.* **2006**, *18*, 997–1000.
158. Hariri, K.; Akhrass, S. A.; Delaite, C.; Moireau, P.; Riess, G., Polymerizable Oil-in-Oil Emulsions: Poly(vinyl pyrolidone) Dispersions in Reactive PDMS Medium. *Polym. Int.* **2007**, *56*, 1200–1205.
159. Dai, L.; Zhang, Z.; Zhao, Y.; Xie, Z., Polymeric Curing Agent Reinforced Silicone Rubber Composites with Low Viscosity and Low Volume Shrinkage. *J. Appl. Polym. Sci.* **2008**, *110*, 1624–1631.
160. Trofimchuk, E. S.; Nikonorova, N. I.; Nesterova, E. A.; Eliseev, A. S.; Semenova, E. V.; Meshkov, I. R.; Kazakova, V. V.; Muzafarov, A. M.; Volynskii, A. L.; Bakeev, N. F., Crazing of Polymers in the Presence of Hyperbranched Poly(ethoxysiloxane). *Polym. Sci. U.S.S.R.* **2007**, *49*, 1107–1113.
161. Maiti, M.; Bhattacharya, M.; Bhowmick, A. K., Elastomer Nanocomposites. *Rubber Chem. Technol.* **2008**, *81*, 384–469.
162. Paul, D. R.; Robeson, L. M., Polymer Nanotechnology: Nanocomposites. *Polymer* **2008**, *49*, 3187–3204.
163. Nalwa, H. S., *Polymeric Nanostructures and Their Applications*. American Scientific: Stevenson Ranch, CA, 2007.
164. Mihara, S.; Datta, R. N.; Noordermeer, J. W. M., Flocculation in Silica Reinforced Rubber Compounds. *Rubber Chem. Technol.* **2009**, *82*, 524–540.
165. Cazacu, M.; Vlad, A.; Alexandru, M.; Budrugeac, P.; Racles, C.; Iacomi, F., Polydimethylsiloxane/Silica Interconnected Networks: Preparation and Properties Evaluation. *Polym. Bull.* **2010**, *64*, 421-434.
166. Camenzind, A.; Schweizer, T.; Sztucki, M.; Pratsinis, S. E., Structure & Strength of Silica-PDMS Nanocomposites. *Polymer* **2010**, *51*, 1796–1804.
167. Chakrabarty, S.; Zhang, X.; Bharti, P.; Chujo, Y.; Miyake, J.; Wynne, K. J.; Yadavalli, V. K., Processing Dependence of Surface Morphology in Condensation Cured PDMS Nanocomposites. *Polymer* **2010**, *51*, 5756–5763.
168. Yilgor, E.; Eynur, T.; Cosak, C.; Bilgin, S.; Yilgor, I.; Malay, O.; Menceloglu, Y.; Wilkes, G. L., Fumed Silica Filled Poly(dimethylsiloxane-Urea) Segmented Copolymers: Preparation and Properties. *Polymer* **2011**, *52*, 4189–4198.
169. Sumi, D.; Dhanabalan, A.; Thimmappa, B. H. S.; Krishnamurthy, S., Effect of Colloidal Silica Dispersions on the Properties of PDMS-Colloidal Silica Composites. *J. Appl. Polym. Sci.* **2012**, *125*, 515–522.
170. Suzuki, N.; Kiba, S.; Kamachi, Y.; Miyamoto, N.; Yamauchi, Y., Unusual Reinforcement of Silicone Rubber Compounds Containing Mesoporous Sillica Particles as Inorganic Fillers. *Phys. Chem. Chem. Phys.* **2012**, *14*, 3400–3407.
171. Nath, S. K.; Frischknecht, A. L.; Curro, J. G.; McCoy, J. D., Density Functional Theory and Molecular Dynamics Simulation of Poly(dimethylsiloxane) Melts Near Silica Surfaces. *Macromolecules* **2005**, *38*, 8562–8573.

172. Smith, J. S.; Borodin, O.; Smith, G. D.; Kober, E. M., A Molecular Dynamics Simulation and Quantum Chemistry Study of Poly(dimethylsiloxane)-Silica Nanoparticle Interactions. *J. Polym. Sci., Polym. Phys. Ed.* **2007**, *45*, 1599–1615.
173. Liu, J.; Wu, S.; Zhang, L.; Wang, W.; Cao, D., Molecular Dynamics Simulation for Insight into Microscopic Mechanism of Polymer Reinforcement. *Phys. Chem. Chem. Phys.* **2011**, *13*, 518–529.
174. Jia, L.; Du, Z.; Zhang, C.; Li, C.; Li, H., Reinforcement of Polydimethylsloxane Through Formation of Inorganic-Organic Hybrid Networks. *Polym. Eng. Sci.* **2008**, *48*, 74–79.
175. Akoum, R. A.; Vaulot, C.; Schwartz, D.; Hirn, M.-P.; Haidar, B., How Silanization of Silica Particles Affects the Adsorption of PDMS Chains on Its Surface. *J. Polym. Sci., Polym. Phys. Ed.* **2010**, *48*, 2371–2378.
176. Diao, S.; Zhang, S.; Yang, Z.; Feng, S.; Zhang, C.; Wang, Z.; Wang, G., Effect of Tetraphenylphenyl-Modified Fumed Silica on Silicone Rubber Radiation Resistance. *J. Appl. Polym. Sci.* **2011**, *120*, 2440–2447.
177. Inagi, S.; Ogoshi, T.; Miyake, J.; Bertolucci, M.; Fujiwara, T.; Galli, G.; Chiellini, E.; Chujo, Y.; Wynne, K. J., Appearing, Disappearing, and Reappearing Fumed Silica Nanoparticles: Tapping-Mode AFM Evidence in a Condensation Cured Polydimethylsiloxane Hybrid Elastomer. *Chem. Mater.* **2007**, *19*, 2141–2143.
178. Schaefer, D. W.; Agamalian, M., Ultra-Small-Angle Neutron Scattering: A New Tool For Materials Research. *Curr. Opin. Solid State Mater. Sci.* **2004**, *8*(1), 39–47.
179. Silva, V. P.; Goncalves, M. C.; Yoshida, I. V. P., Biogenic Silica Short Fibers as Alternative Reinforcing Fillers of Silicone Rubbers. *J. Appl. Polym. Sci.* **2006**, *101*, 290–299.
180. Suzuki, N.; Kiba, S.; Kamachi, Y.; Miyamoto, N.; Yamauchi, Y., Mesoporous Silica as Smart Inorganic Filler: Preparation of Robust Silicone Rubber with Low Thermal Expansion Property. *J. Mater. Chem.* **2011**, *21*, 5338–5344.
181. Urkmez, G.; Sari, B.; Unal, H. I., Synthesis and Characterization of Novel Poly(dimethylsiloxane)/Polyindole Composites. *J. Appl. Polym. Sci.* **2011**, *121*, 1600–1609.
182. Camarota, B.; Mann, S.; Onida, B.; Garrone, E., Heirarchical Self-Assembly in Molecularly Ordered Phenylene-Bridged Mesoporous Organosilica Nanofilaments. *ChemPhysChem* **2007**, *8*, 2363–2366.
183. Fortuniak, W.; Mazerska, U.; Chojnowski, J.; Basinska, T.; Slomkowski, S.; Chehimi, M. M.; Konopacka, A.; Turecka, K.; Werel, W., Polysiloxanes with Quaternary Ammonium Salt Biocidal Functions and Their Behavior When Incorporated into a Silicone Elastomer Network. *J. Inorg. Organomet. Polym.* **2011**, *21*, 576–589.
184. Kazakova, V. V.; Rebrov, E. A.; Myakushev, V. B.; Strelkova, T. V.; Ozerin, A. N.; Ozerina, L. A.; Chenskaya, T. B.; Sheiko, S. S.; Sharipov, E. Y.; Muzafarov, A. M., From a Hyperbranched Polyethoxysiloxane Toward Molecular Forms of Silica: A Polymer-Based Aproach to the Monitoring of Silica Properties. In *Silicones and Silicone-Modified Materials*, Clarson, S. J.; Fitzgerald, J. J.; Owen, M. J.; Smith, S. D., Eds. American Chemical Society: Washington, DC, 2000; Vol. 729, pp 503–515.
185. Kazakova, V. V.; Zhiltsov, A. S.; Gorbatsevitch, O. B.; Meshlov, I. B.; Pletneva, M. V.; Demchenko, N. V.; Cherkaev, G. V.; Muzafarov, A. M., Synthesis and Characterization of Hybrid Core-Shell Systems Based on Molecular Silicasols. *J. Inorg. Organomet. Polym. Mater.* **2012**, *22*, 564–576.

186. Patel, K.; Angelos, S.; Dichtel, W. R.; Coskun, A.; Yang, Y.-W.; Zink, J. I.; Stoddart, J.F., Enzyme-Responsive Snap-Top Covered Silica Nanocontainers. *J. Am. Chem. Soc.* **2008**, *130*, 2382–2383.
187. Cosgrove, T.; Turner, M. J.; Weatherhead, I.; Roberts, C.; Garasanin, T.; Schmidt, R. G.; Gordon, G. V.; Hannington, J. P., NMR Spin-Spin Relaxation Studies of Reinforced Poly(dimethylsiloxane) Melts. In *Silicones and Silicone-Modified Materials*, Clarson, S. J.; Fitzgerald, J. J.; Owen, M. J.; Smith, S. D., Eds. American Chemical Society: Washington, DC, 2000; Vol. 729, pp 204–213.
188. Dreiss, C. A.; Cosgrove, T.; Benton, N. J.; Kilburn, D.; Alam, M. A.; Schmidt, R. G.; Gordon, G. V., Effect of Crosslinking on the Mobility of PDMS Filled with Polysilicate Nanoparticles: Positron Lifetime, Rheology and NMR Relaxation Studies. *Polymer* **2007**, *48*, 4419–4428.
189. Mayer, B. P.; Reimer, J. A.; Maxwell, R. S., A Methodology for the Indirect Determination and Spatial Resolution of Shear Modulus of PDMS-Silica Elastomers. *Macromolecules* **2008**, *41*, 1323–1327.
190. Pinnavaia, T. J.; Beall, G., *Polymer-Clay Nanocomposites*. Wiley: New York, 2001.
191. Vu, Y. T.; Mark, J. E.; Pham, L. H.; Engelhardt, M., Clay Nanolayer Reinforcement of Cis-1,4-Polyisoprene and Epoxidized Natural Rubber. *J. Appl. Polym. Sci.* **2001**, *82*, 1391–1403.
192. Zhou, W.; Mark, J. E.; Unroe, M. R.; Arnold, F. E., Some Clay Nanocomposites Based on a High-Temperature, High-Performance Polymer. *J. Macromol. Sci., Pure Appl. Chem.* **2001**, *A38*, 1–9.
193. Wang, P.; Ding, T., Creep of Electrical Resistance Under Uniaxial Pressures for Carbon Black-Silicone Rubber Composite. *J. Mater. Sci.* **2010**, *45*, 3595–3601.
194. Zhang, Y.; Zhu, Y.; Lin, G.; Ruoff, R. S.; Hu, N.; Schaefer, D. W.; Mark, J. E., What Factors Control the Mechanical Properties of Poly (dimethylsiloxane) Reinforced with Nanosheets of 3-Aminopropyltriethoxysilane Modified Graphene Oxide. *Polymer* **2013**, *54*, 3605–3611.
195. Burnside, S. D.; Giannelis, E. P., Nanostructure and Properties of Polysiloxane-Layered Silicate Nanocomposites. *J. Polym. Sci., Polym. Phys. Ed.* **2000**, *38*, 1595–1604.
196. Sun, H.; Mark, J. E.; Tan, S. C.; Venkatasubramanian, N.; Houtz, M. D.; Arnold, F. E.; Lee, C. Y.-C., Microcellular Foams from Some High-Performance Thermoplastics and Their Composites. *Nonlinear Opt., Quantum Opt.* **2004**, *31*, 1–29.
197. Usuki, A.; Hasegawa, N.; Kato, M., Polymer-Clay Nanocomposites. *Adv. Polym. Sci.* **2005**, *179*, 135–195.
198. Xu, Y.; Ren, X.; Hanna, M. A., Chitosan/Clay Nanocomposite Film Preparation and Characterization. *J. Appl. Polym. Sci.* **2006**, *99*, 1684–1691.
199. Kader, M. A.; Lyu, M.-Y.; Nah, C., A Study on Melt Processing and Thermal Properties of Fluoroelastomer Nanocomposites. *Comp. Sci. Tech.* **2006**, *66*, 1431–1443.
200. Demirkol, E. A.; Kalyon, D. M., Batch and Continuous Processing of Polymer Layered Organoclay Nanocomposites. *J. Appl. Polym. Sci.* **2007**, *104*, 1391–1398.
201. Choudalakis, G.; Gotsis, A. D., Permeability of Polymer/Clay Nanocomposites; A Review. *Eur. Polym. J.* **2009**, *45*, 967–984.

202. Utracki, L. A.; Broughton, B.; Gonzalez-Rojano, N.; De Carvalho, L. H.; Achete, C. A., Clays for Polymeric Nanocomposites. *Polym. Eng. Sci.* **2011**, *51*, 559–572.
203. Guimont, A.; Beyou, E.; Martin, G.; Sonntag, P.; Cassagnau, P., Viscoelasticity of Graphite Oxide-Based Suspensions in PDMS. *Macromolecules* **2011**, *44*, 3893–3900.
204. Wang, S.; Long, C.; Wang, X.; Li, Q.; Qi, Z., Synthesis and Properties of Silicone Rubber/Organomontmorillonite Hybrid Nanocomposites. *J. Appl. Polym. Sci.* **1998**, *69*, 1557–1561.
205. Takeuchi, H.; Cohen, C., Reinforcement of Poly(dimethylsiloxane) Elastomers by Chain-End Anchoring to Clay Particles. *Macromolecules* **1999**, *32*, 6792–6799.
206. Osman, M. A.; Atallah, A.; Kahr, G.; Suter, U. W., Reinforcement of Poly(dimethylsiloxane) Networks by Montmorillonite Platelets. *J. Appl. Polym. Sci.* **2002**, *83*, 2175–2183.
207. Bourlinos, A. B.; Jiang, D. D.; Giannelis, E. P., Clay-Organosiloxane Hybrids: A Route to Cross-Linked Clay Particles and Clay Monoliths. *Chem. Mater.* **2004**, *16*, 2404–2410.
208. Ma, J.; Yu, Z.-Z.; Kuan, H.-C.; Dasari, A.; Mai, Y.-W., A New Strategy to Exfoliate Silicone Rubber/Clay Nanocomposites. *Macromol. Rapid Commun.* **2005**, *26*, 830–833.
209. Wang, J.; Chen, Y.; Jin, Q., Organic Montmorillonite as a Substitute for Aerosilica in Addition-Type Liquid Silicone Rubber Systems. *Macromol. Chem. Phys.* **2005**, *206*, 2512–2520.
210. Schmidt, D. F.; Clement, F.; Giannelis, E. P., On the Origins of Silicate Dispersion in Polysiloxane/Layered-Silicate Nanocomposites. *Adv. Funct. Mater.* **2006**, *16*, 417–425.
211. Feng, X.; Zhong, A.; Chen, D., Preparation and Properties of Poly(silicone-co-acrylate)/Montmorillonite Nanocomposite Emulsion. *J. Appl. Polym. Sci.* **2006**, *101*, 3963–3970.
212. Kim, E.-S.; Kim, H.-S.; Jung, S.-H.; Yoon, J.-S., Adhesion Properties and Thermal Degradation of Silicone Rubber. *J. Appl. Polym. Sci.* **2007**, *103*, 2782–2787.
213. Nagendiran, S.; Premkumar, S.; Alagar, M., Mechanical and Morphological Properties of Organic-Inorganic, Hybrid, Clay-Filled, and Cyanate Ester/Siloxane Toughened Epoxy Nanocomposites. *J. Appl. Polym. Sci.* **2007**, *106*, 1263–1273.
214. Kaneko, M. L. Q. A.; Yoshida, I. V. P., Effect of Natural and Organically Modified Montmorillonite Clays on the Properties of Polydimethylsiloxane Rubber. *J. Appl. Polym. Sci.* **2008**, *108*, 2587–2596.
215. Simon, M. W.; Stafford, K. T.; Ou, D. L., Nanoclay Reinforcement of Liquid Silicone Rubber. *J. Inorg. Organomet. Polym. Mater.* **2008**, *18*, 364–373.
216. Wang, J.; Chen, Y., Preparation of an Organomontmorillonite Master Batch and Its Application to High-Temperature-Vulcanized Silicone-Rubber Systems. *J. Appl. Polym. Sci.* **2008**, *107*, 2059–2066.
217. Fang, S.; Hu, Y.; Song, L.; Wu, J., Preparation and Investigation of Ethylene-Vinyl Acetate Copolymer/Silicone Rubber/Clay Nanocomposites. *J. Appl. Polym. Sci.* **2009**, *113*, 1664–1670.
218. Labruyere, C.; Gorrasi, G.; Monteverde, F.; Alexandre, M.; Dubois, P., Transport Properties of Organic Vapours in Silicone/Clay Nanocomposites. *Polymer* **2009**, *50*, 3626–3637.

219. Labruyere, C.; Monteverde, F.; Alexandre, M.; Dubois, P., Exfoliation of Clays in Poly(dimethylsiloxane) Rubber Using an Unexpected Couple: A Silicone Surfactant and Water *J. Nanosci. Nanotech.* **2009**, *9*, 2731–2738.
220. Kim, E.-S.; Shim, J. H.; Jung, S. H.; Joo, J. H.; Yoon, J.-S.; Lee, S. H., Influence of Clay Modification on the Reinforcement of Vinyl-Terminated Polydimethylsiloxane Networks. *Polym. Int.* **2009**, *59*, 479–485.
221. Kaneko, M. L. Q. A.; Romero, R. B.; Do Carmo Goncalves, M., High Molar Mass Silicone Rubber Reinforced with Montmorillonite Clay Masterbatches: Morphology and Mechanical Properties. *Eur. Polym. J.* **2010**, *46*, 881–890.
222. Schmidt, D. F.; Giannelis, E. P., Silicate Dispersion and Mechanical Reinforcement in Polysiloxane/Layered Silicate Nanocomposites. *Chem. Mater.* **2010**, *22*, 167–174.
223. Li, H.-Q.; Zeng, X.-R.; Deng, H.-P., Study on the Synthesis, Structure, and Properties of Polydimethylsiloxane-Acrylate/Organophilic Montmorillonite Nanocomposite Emulsions. *J. Appl. Polym. Sci.* **2010**, *118*, 63–68.
224. Voulomenou, A.; Tarantili, P. A., Preparation, Characterization, and Property Testing of Condensation-Type Sililcone/Montmorillonite Nanocomposites. *J. Appl. Polym. Sci.* **2010**, *118*, 2521–2529.
225. Vaulot, C.; Ziegler, P.; Haidar, B., Physiochemical Driving Forces Behind Exfoliation Process of a Synthetic Montmorillonite in PDMS Polymers. *Polymer* **2011**, *52*, 700–707.
226. Chen, L.; Chen, G.; Lu, L., Piezoresistive Behavior Study on Finger-Sensing Silicone Rubbe/Graphite Nanosheet Nanocomposites. *Adv. Funct. Mater.* **2007**, *17*, 898–904.
227. Bokobza, L.; Nugay, N., Orientational Effects of Mica in Fumed Silica Reinforced Composites. *J. Appl. Polym. Sci.* **2001**, *81*, 215–222.
228. Osman, M. A.; Atallah, A.; Muller, M.; Suter, U. W., Reinforcement of Poly(dimethylsiloxane) by Mica Flakes. *Polymer* **2001**, *42*, 6545–6556.
229. Park, E.-S., Mechanical Properties and Processibility of Glass-Fiber-, Wollastonite-, and Fluoro-Rubber-Reinforced Silicone Rubber Composites. *J. Appl. Polym. Sci.* **2007**, *105*, 460–468.
230. Lappas, A.; Zorko, A.; Wortham, E.; Das, R. N.; Giannelis, E. P.; Cevc, P.; Arcon, D., Clay-Low-Energy Magnetic Excitations and Morphology in Layered Hybrid Perovskite-Poly(dimethylsiloxane) Nanocomposites. *Chem. Mater.* **2005**, *17*, 1199–1207.
231. Roy, N.; Bhowmick, A. K., Novel In-Situ Polydimethylsiloxane-Sepiolite Nanocomposites: Structure-Property Relationship. *Polymer* **2010**, *51*, 5172–5185.
232. Bruzard, S.; Levesque, G., Polysiloxane-*g*-$TiNbO_5$ Nanocomposites: Synthesis via In Situ Intercalative Polymerization and Preliminary Characterization. *Chem. Mater.* **2002**, *14*, 2421–2426.
233. Beigbeder, A.; Bruzaud, S.; Spevacek, J.; Brus, J.; Grohens, Y., Dynamics and Crystallization in Polydimethylsiloxane Nanocomposites. *Macromol. Symp.* **2005**, *222*, 225–231.
234. Zhang, Y.; Gittins, D. I.; Skuse, D.; Cosgrove, T.; Van Duijneveldt, J. S., Composites of Kaolin and Polydimethylsiloxane. *Langmuir* **2008**, *24*, 12032–12039.
235. Pradhan, B.; Srivastava, S. K.; Ananthakrishnan, R.; Saxena, A., Preparation and Characterization of Exfoliated Layered Double Hydroxide Silicone Rubber Nanocomposites. *J. Appl. Polym. Sci.* **2011**, *119*, 343–351.
236. Monasterio, F.; Pita, V. R.; Dias, M. L.; Erdmann, E.; Destefanis, H., Thermal and Rheological Properties of Polyethylene Composites Based on

Poly(diphenylsiloxanes)/Organoclay Hybrids Obtained from Two different Silanes. *Macromol. Symp.* **2011**, *299/300*, 81–91.

237. Chrissopoulou, K.; Anastasiadis, S. H.; Giannelis, E. P., Quasielastic Neutron Scattering of Poly(methylphenylsiloxane) in the Bulk and Under Severe Confinement. *J. Chem. Phys.* **2007**, *127*, 144910.
238. Vaia, R. A.; Giannelis, E. P., Lattice Model of Polymer Melt Intercalation in Organically-Modified Layered Silicates. *Macromolecules* **1997**, *30*, 7990–7999.
239. Vaia, R. A.; Giannelis, E. P., Polymer Melt Intercalation in Organically-Modified Layered Silicates: Model Predictions and Experiment. *Macromolecules* **1997**, *30*, 8000–8009.
240. Hackett, E.; Manias, E.; Giannelis, E. P., Molecular Dynamics Simulations of Organically Modified Layered Silicates. *J. Chem. Phys.* **1998**, *108*, 7410.
241. Giannelis, E. P.; Krishnamoorti, R.; Manias, E., Polymer-Silicate Nanocomposites: Model Systems for Confined Polymers and Polymer Brushes. *Adv. Polym. Sci.* **1999**, *138*, 107–147.
242. Tobita, H.; Takekuma, K., Analytical Calculus and Monte Carlo Simulation of Crosslinked Polymer Formation in the Copolymerization of Tetraethoxysilane and Poly(dimethylsiloxane). *Macromol. Theory Simul.* **2000**, *9*, 181–187.
243. Toth, R.; Santese, E.; Pereira, S. P.; Nieto, D.; Prici, S.; Fermeglia, M.; Posocco, P., Size and Shape Matter! A Multiscale Molecular Simulation Approach to Polymer Nanocomposites. *J. Mater. Chem.* **2012**, *22*, 5398–5409.
244. Konduri, S.; Nair, S., A Computational Study of Gas Molecule Transport in a Polymer/Nanoporous Layered Silicate Nanocomposite Membrane Material. *J. Phys. Chem. C* **2007**, *111*, 2017–2024.
245. Ma, J.; Shi, L.; Yang, M.; Li, B.; Liu, H.; Xu, J., Synthesis and Structure of Polymethylsilsesquioxane-Clay Nanocomposite via In Situ Intercalative Polymerization. *J. Appl. Polym. Sci.* **2002**, *86*, 3708–3711.
246. Harada, M.; Minamigawa, S.; Tachibana, K.; Ochi, M., Flame Retardancy and Thermomechanical Properties of the Poly-(glcidyloxypropyl) Phenyl Silsesquioxane/Layered Titanate Nanocomposites. *J. Appl. Polym. Sci.* **2007**, *106*, 338–344.
247. Ozin, G. A.; Arsenault, A. C., *Nanochemistry. A Chemical Approach to Nanomaterials.* RSC Publishing: London, 2005.
248. Kubisz, L.; Skumiel, A.; Pankowski, E.; Hojan-Jezierska, D., Magnetically Induced Anisotropy of Electrical Permittivity in the PDMS Ferromagnetic Gel. *J. Non-Cryst. Solids* **2011**, *357*, 767–770.
249. Cho, H.-B.; Tokoi, Y.; Tanaka, S.; Suzuki, T.; Jiang, W.; Suematsu, H.; Niihara, K.; Nakayana, T., Facile Orientation of Unmodified BN Nanosheets in Polysiloxane/BN Composite Fillms Using a High Magnetic Field. *J. Mater. Sci.* **2011**, *46*, 2318–2323.
250. Sohoni, G. B.; Mark, J. E., Anisotropic Reinforcement in Elastomers Containing Magnetic Filler Particles. *J. Appl. Polym. Sci.* **1987**, *34*, 2853–2859.
251. Miles, W. C.; Goff, J. D.; Huffstetler, P. P.; Mefford, O. T.; Riffle, J. S.; Davis, R. M., The Design of Well-Defined PDMS-Magnetite Complexes. *Polymer* **2010**, *51*, 482–491.
252. Ding, T.; Wang, L.; Wang, P., Changes in Electrical Resistance of Carbon-Black-Filled Silicone Rubber Composite During Compressions. *J. Polym. Sci., Part B: Polym. Phys.* **2007**, *45*, 2700–2706.
253. Verdejo, R.; Barroso-Bujans, F.; Rodriguez-Perez, M. A.; De Sarja, J. A.; Lopez-Manchado, M. A., Functionalized Graphene Sheet Filled Silicone Foam Nanocomposites. *J. Mater. Chem.* **2008**, *18*, 2221–2226.

254. Jiang, M.-J.; Dang, Z.-M.; Xu, H.-P., Enhanced Electrical Conductivity in Chemically Modified Carbon Nanotube/Methylvinyl Silicone Rubber Nanocomposite. *Eur. Polym. J.* **2007**, *43*, 4924–4930.
255. Park, I.-S.; Kim, K. J.; Nam, J.-D.; Lee, J.; Yim, W., Mechanical, Dielectric, and Magnetic Properties of the Silicone Elastomer with Multi-Walled Carbon Nanotubes as a Nanofiller. *Polym. Eng. Sci.* **2007**, *47*, 1396–1405.
256. Vast, L.; Mekhalif, Z.; Fonseca, A.; Nagy, J. B.; Delhalle, J., Preparation and Electrical Characterization of a Silicone Elastomer Composite Charged with Multi-Wall Carbon Nanotubes Functionalized with 7-Octenyltrichlorosilane. *Comp. Sci. Tech.* **2007**, *67*, 880–889.
257. Wang, P.; Geng, S.; Ding, T., Effects of Carboxyl Radical on Electrical Resistance of Multi-Walled Carbon Nanotube Filled Silicone Rubber Composite Under Pressure. *Comp. Sci. Tech.* **2010**, *70*, 1571–1573.
258. Khosla, A.; Gray, B. L., Preparation, Micro-Patterning and Electrical Characterization of Functionalized Carbon-Nanotube Polydimethylsiloxane Nanocomposite Polymer. *Macromol. Symp.* **2010**, *297*, 210–218.
259. Nayak, G. C.; Rajesakar, R.; Das, C. K., Dispersion of *SiC* Coated MWCNTs in PEI/Silicone Rubber Blend and Its Effect on the Thermal and Mechanical Properties. *J. Appl. Polym. Sci.* **2011**, *119*, 3574–3581.
260. Chang, J.; Najeeb, C. K.; Lee, J.-H.; Kim, J.-H., Single-Walled Carbon Nanotubes/Polymer Composite Electrodes Patterned Directly from Solution. *Langmuir* **2011**, 7330–7338.
261. Geng, S.; Wang, P.; Ding, T., Impedance Characteristics and Electrical Modeling of Multi-Walled Carbon Nanotube/Silicone Rubber Composites. *Comp. Sci. Tech.* **2011**, *71*, 36–40.
262. Roy, N.; Bhowmick, A. K., In Situ Preparation, Morphology and Electrical Properties of Carbon Nanofiber/Polydimethylsiloxane Nanocomposites. *J. Mater. Sci.* **2012**, *47*, 272–281.
263. Roy, N.; Bhowmick, A., Novel *In-Situ* Carbon Nanofiber/Polydimethylsiloxane Nanocomposites: Synthesis, Morphology, and Physico-Mechanical Properties. *J. Appl. Polym. Sci.* **2012**, *123*, 3675–3687.
264. Beigbeder, A.; Linares, M.; Devalckenaere, M.; Degee, P.; Claes, M.; Beljonne, D.; Lazzaroni, R.; Dubois, P., CH-π Interactions as the Driving Force for Silicone-Based Nanocomposites with Exceptional Properties. *Adv. Mater.* **2008**, *20*, 1003–1007.
265. Zhao, Z. Y.; Khatri, N. D.; Nguyen, K.; Song, S. Q.; Sun, L., Synthesis and Mechanical Properties of Interconnected Carbon Nanofiber Network Reinforced Polydimethylsiloxane Composites. *J. Nanosci. Nanotech.* **2011**, *11*, 1092–1097.
266. Rigbi, Z.; Mark, J. E., Effects of a Magnetic Field Applied During the Curing of a Polymer Loaded with Magnetic Filler. *J. Polym. Sci., Polym. Phys. Ed.* **1985**, *23*, 1267–1269.
267. Sur, G. S.; Mark, J. E., Some Reinforcing Characteristics of Iron-Based Fillers. *Polym. Bull.* **1987**, *18*, 369–375.
268. Nikitin, L. V.; Mironova, L. S.; Kornev, K. G.; Stepanov, G. V., The Magnetic, Elastic, Structural and Magnetodeformational Properties of Magnetoelastics. *Polym. Sci. U.S.S.R., Ser. A* **2004**, *46*, 498–509.
269. Varga, Z.; Filipcsei, G.; Zrinyi, M., Magnetic Field Sensitive Functional Elastomers with Tunable Elastic Modulus. *Polymer* **2006**, *47*, 227–233.
270. Mefford, O. T.; Vadala, M. L.; Goff, J. D.; Carroll, M. R. J.; Mejia-Ariza, R.; Caba, B. L.; St. Pierre, T. G.; Woodward, R. C.; Davis, R. M.; Riffle, J. S.,

Stability of Polydimethylsiloxane-Magnetite Nanoparticle Dispersions Against Flocculation: Interparticle Interactions of Polydisperse Materials. *Langmuir* **2008**, *24*(5060-5069), 1489–1494.
271. Mefford, O. T.; Carroll, M. R. J.; Vadala, M. L.; Goff, J. D.; Mejia-Ariza, R.; Saunders, M.; Woodward, R. C.; St. Pierre, T. G.; Davis, R. M.; Riffle, J. S., Size Analysis of PDMS-Magnetite Nanoparticle Complexes: Experiment and Theory. *Chem. Mater.* **2008**, *20*, 2184–2191.
272. Lotonov, A. M.; Gavrilova, N. D.; Kramarenko, E. Y.; Alekseeva, E. I.; Popov, F. Y.; Stepanov, G. V., Efffect of Iron Particles on Dielectric Properties of Polydimethylsiloxane near Crystallization and Glass Transition Temperatures. *Polym. Sci., Ser. B* **2006**, *48*, 1898–1902.
273. Zhou, W.; Qi, S.; Tu, C.; Zhao, H., Novel Heat-Conductive Composite Silicone Rubber. *J. Appl. Polym. Sci.* **2007**, *104*, 2478–2483.
274. Acchar, W.; Diniz, M.; Fonesca, Y. A. A.; Sousa, C. R. C.; Lima, E. S., The Effect of Polysiloxane on the Pyrolysis of Al_2O_3-NbC Composite Material Produced by Pyrolysis Process. *J. Mater. Sci.* **2008**, *43*, 597–601.
275. Prasad, B. L. V.; Stoeva, S. I.; Sorensen, C. M.; Zaikovski, V.; Klabunde, K. J., Gold Nanoparticles as Catalysts for Polymerization of Alkylsilanes to Siloxane Nanowires, Filaments, and Tubes *J. Am. Chem. Soc.* **2003**, *125*, 10488–10489.
276. Kisailus, D.; Najarian, M.; Weaver, J. C.; Morse, D. E., Functionalized Gold Nanoparticles Mimic Catalytic Activity of a Polysiloxane-Synthesizing Enzyme. *Adv. Mater.* **2005**, *17*, 1234–1239.
277. Uhlenhaut, D. I.; Smith, P.; Caseri, W., Color Switching in Gold-Polysiloxane Elastomeric Nanocomposites. *Adv. Mater.* **2006**, *18*, 1653–1656.
278. Watanabe, M., Striped-Pattern Formation of a Thin Gold Film Deposited onto a Stretched Elastic Silicone Substrate. *J. Polym. Sci., Polym. Phys. Ed.* **2005**, *43*, 1532–1537.
279. Pastoriza-Santos, I.; Perez-Juste, J.; Kickelbick, G.; Liz-Marzan, L. M., Optically Active Poly(dimethylsiloxane) Elastomer Films Through Doping with Gold Nanoparticles. *J. Nanosci. Nanotech.* **2006**, *6*, 453–458.
280. Lichtenhan, J. D.; Schwab, J.; Reinerth, W. A., Sr., Nanostructured Chemicals: A New Era in Chemical Technology. *Chem. Innov.* **2001**, *31*, 3–5.
281. Laine, R. M.; Choi, J.; Lee, I., Organic/Inorganic Nanocomposites with Completely Defined Interfacial Interactions. *Adv. Mater.* **2001**, *13*, 800–803.
282. Loy, D. A.; Baugher, C. R.; Schnieder, D. A.; Sanchez, A.; Gonzalez, F., Sol-Gel Chemistry of 3-Isocyanatopropyltriethoxysilane. *Polym. Preprints* **2001**, *42*(1), 180–181.
283. Shea, K. J.; Loy, D. A., Bridged Polysilsesquioxanes: Molecular Engineering of Hybrid Organic-Inorganic Materials. *MRS Bull.* **2001**, *26*, 368–375.
284. Haddad, T. S.; Lee, A.; Phillips, S. H., Poly(dimethylsiloxanes) Modified with Inorganic Polyhedra. *Polym. Preprints* **2001**, *42*(1), 88–89.
285. Pan, G.; Mark, J. E.; Schaefer, D. W., Synthesis and Characterization of Fillers of Controlled Structure Based on Polyhedral Oligomeric Silsesquioxane Cages, and Their Use in Reinforcing Siloxane Elastomers. *J. Polym. Sci., Polym. Phys. Ed.* **2003**, *41*, 3314–3323.
286. Ryu, H.-S.; Kim, D.-G.; Lee, J.-C., Polysiloxanes Containing Polyhedral Oligomeric Silsesquioxane Groups in the Side Chains: Synthesis and Properties. *Polymer* **2010**, *51*, 2296–2304.

287. Schaefer, D. W.; Vu, B. T. N.; Mark, J. E., The Effect of Interphase Coupling on the Structure and Mechanical Properties of Silica-Siloxane Composites. *Rubber Chem. Technol.* **2002**, *75*, 795–810.
288. Vu, B. T. N.; Mark, J. E.; Schaefer, D. W., Interfacial Modification for Controlling Silica-Polysiloxane Interactions and Bonding in Some Elastomeric Composites. *Compos. Interfaces* **2003**, *10*, 451–473.
289. Liu, L.; Tian, M.; Zhang, W.; Zhang, L.; Mark, J. E., Crystallization and Morphology Study of Polyhedral Oligomeric Silsesquioxane (POSS)/Polysiloxane Elastomer Composites Prepared by Melt Blending. *Polymer* **2007**, *48*, 3201–3212.
290. Liu, L.; Tian, M.; Liang, G.; Chen, W.; Zhang, L.; Mark, J. E., Polyhedral Oligomeric Silsesquioxane (POSS) in a Polysiloxane Melt and Elastomer. Dependence of the Dispersion of the POSS on Its Solubility and Constraining Effects of a Network Structure. *J. Macromol. Sci., Pure Appl. Chem.* **2007**, *44*, 659–664.
291. Iyer, S.; Abu-Ali, A.; Detwiler, A.; Schiraldi, D. A., Transparent Polymer-Polyhedral Oligomeric Silsesquioxane Composites. In *Science and Technology of Silicones and Silicone-Modified Materials*, Clarson, S. J.; Fitzgerald, J. J.; Owen, M. J.; Smith, S. D.; Van Dyke, M. E., Eds. American Chemical Society: Washington, DC, 2007; Vol. 964, pp 313–325.
292. Mabry, J. M.; Vij, A.; Viers, B. D.; Grabow, W. W.; Marchant, D.; Iacono, S. C.; Ruth, P. N.; Vij, I., Hydrophobic Silsesquioxane Nanoparticles and Nanocomposite Surfaces: Overview of the Synthesis and Properties of Fluorinated Polyhedral Oligomeric Silsesquioxane (POSS) and Fluorinated POSS Nanocomposites. In *Science and Technology of Silicones and Silicone-Modified Materials*, Clarson, S. J.; Fitzgerald, J. J.; Owen, M. J.; Smith, S. D.; Van Dyke, M. E., Eds. American Chemical Society: Washington, DC, 2007; Vol. 964, pp 290–300.
293. Ouyang, Y.; Pan, Y.; Zhou, S.; Goh, S. H., Supramolecular Assembled Co_{60}-Containing Carboxylated Poly(dimethylsiloxane) Composites. *Polymer* **2006**, *47*, 6140–6148.
294. Nakamura, H.; Matsui, Y., Silica Gel Nanotubes Obtained by the Sol-Gel Method. *J. Am. Chem. Soc.* **1995**, *117*, 2651–2652.
295. Frisch, H. L.; Mark, J. E., Nanocomposites Prepared by Threading Polymer Chains through Zeolites, Mesoporous Silica, or Silica Nanotubes. *Chem. Mater.* **1996**, *8*, 1735–1738.
296. Tans, S. J.; Devoret, M. H.; Dai, H.; Thess, A.; Smalley, R. E.; Geerligs, L. J.; Dekker, C., Individual Single-Wall Carbon Nanotubes as Quantum Wires. *Nature* **1997**, *386*, 474–477.
297. Fu, B. X.; Hsiao, B. S.; White, H.; Rafailovich, M.; Mather, P. T.; Jeon, H. G.; Phillips, S.; Lichtenhan, J. D.; Schwab, J., Nanoscale Reinforcement of Polyhedral Oligomeric Silsesquioxane (POSS) in Polyurethane Elastomer. *Polym. Int.* **2000**, *49*, 437–440.
298. Jeon, H. G.; Mather, P. T.; Haddad, T. S., Shape Memory and Nanostructure in Poly(norbornyl-POSS) Copolymers. *Polym. Int.* **2000**, *49*, 453–457.
299. Costa, R. O. R.; Vasconcelos, W. L.; Tamaki, R.; Laine, R. M., Organic/Inorganic Nanocomposite Star Polymers via Atom Transfer Radical Polymerization of Methyl Methacrylate Using Octafunctional Silsesquioxane Cores. *Macromolecules* **2001**, *34*, 5398–5407.

300. Sharma, S.; Zhang, W.; Schrag, E.; Hsaio, B.; Mather, P. T.; Sokolov, J.; Rafailovich, M. H.; Gersappe, D.; Luo, H., Polymer Blend Thin Films with POSS: Compatibilization and Dewetting. *Chicago Am. Chem. Soc. Meeting* **2001**.
301. Pyun, J.; Matyjaszewski, K.; Kowalewski, T.; Mather, P. T.; Chun, S. B., Synthesis of Nanocomposite Silsesquioxane Materials Using Controlled Radical Polymerization. *Chicago Am. Chem. Soc. Meeting* **2001**.
302. Choi, J.; Harcup, J.; Yee, A. F.; Zhu, Q.; Laine, R. M., Organic/Inorganic Hybrid Composites from Cubic Silsesquioxanes. *J. Am. Chem. Soc.* **2001**, *123*, 11420–11430.
303. Tamaki, R.; Choi, J.; Laine, R. M., A Polyimide Nanocomposite from Octa(aminophenyl)silsesquioxane. *Chem. Mater.* **2003**, *15*, 793–797.
304. Ji, Y.; Huang, Y. Y.; Tajbakhsh, A. R.; Terentjev, E. M., Polysiloxane Surfactants for the Dispersion of Carbon Nanotubes in Nonpolar Organic Solvents. *Langmuir* **2009**, *25*, 12325–12331.
305. Kohlmeyer, B. R.; Javadi, A.; Pradhan, B.; Pilla, S.; Setyowati, K.; Chen, J.; Gong, S., Electrical and Dielectric Properties of Hydroxylated Carbon Nanotube-Elastomer Composites. *J. Phys. Chem. C* **2009**, *113*, 17626–17629.
306. Bayley, G. M.; Hedenquist, M.; Mallon, P. E., Large Strain and Toughness Enhancement of Poly(dimethylsiloxane) Composite Films Filled with Electrospun Polyacrylonitrile-graft-Poly(dimethylsiloxane) Fibers and Multi-Walled Carbon Nanotubes. *Polymer* **2011**, *52*, 4061–4172.
307. Nugay, N.; Erman, B., Property Optimization in Nitrile Rubber Composites via Hybrid Filler Systems. *J. Appl. Polym. Sci.* **2001**, *79*, 366–374.
308. Sen, S.; Nugay, N., Tuning of Final Performance of Unsaturated Polyester Composites with Inorganic Microsphere/Platelet Hybrid Reinforcers. *Eur. Polym. J.* **2001**, *37*, 2047–2053.
309. Konishi, Y.; Cakmak, M., Structural Hierarchy Developed in Injection Molding of Nylon 6/Clay/Carbon Black Nanocomposites. *Polymer* **2005**, *46*, 4811–4826.
310. Mora-Barrantes, I.; Ibarra, L.; Rodriguez, A.; Gonzales, L.; Valentin, J. I., Elastomer Composites Based on Improved Fumed Silica and Carbon Black. Advantages of Mixed Reinforcing Systems. *J. Mater. Chem.* **2011**, *21*, 17526–17533.
311. Mark, J. E., Polymer Chains in Constraining Environments. In *Molecular Catenanes, Rotaxanes and Knots*, Sauvage, J.-P.; Dietrich-Buchecker, C., Eds. Wiley-VCH: Weinheim, 1999; pp 223–246.
312. Kickelbick, G., Concepts for the Incorporation of Inorganic Building Blocks into Organic Polymers on a Nanoscale. *Prog. Polym. Sci.* **2003**, *28*, 83–114.
313. Okumura, H.; Okada, M.; Kawaguchi, Y.; Harada, A., Complex Formation Between Poly(dimethylsiloxane) and Cyclodextrins: New *Pseudo*-Polyrotaxanes Containing Inorganic Polymers. *Macromolecules* **2000**, *33*, 4297–4298.
314. Vu, B. T. N.; Mark, J. E.; Schaefer, D. W., Surface Modification of Silica Fillers Formed In-Situ for the Reinforcement of Polydimethylsiloxane Networks. *PSME Preprints, ACS Div. Polym. Mater. Sci. Eng.* **2000**, *83*, 411–412.
315. Vu, B. T. N., Interface Modification for the Reinforcement of Silicone Elastomer Composites. *M.S. Thesis in Chemistry, University of Cincinnati* **2001**.
316. Patwardhan, S. V.; Mukherjee, N.; Clarson, S. J., The Use of Poly-L-Lysine to Form Novel Silica Morphologies and the Role of Polypeptides in Biosilicification. *J. Inorg. Organomet. Polym.* **2001**, *11*, 193–198.
317. Patwardhan, S. V.; Clarson, S. J., Bioinspired Mineralization: Macromolecule Mediated Synthesis of Amorphous Germania Structures. *Polymer* **2005**, *46*, 4474–4479.

318. Patwardhan, S. V.; Clarson, S. J.; Perry, C. C., On the Role(s) of Additives in Bioinspired Silicification. *Chem. Commun.* **2005**, 1113–1121.
319. Patwardhan, S. V.; Clarson, S. J., Silicification and Biosilicification. Part 6. Poly-L-Histidine Mediated Synthesis of Silica at Neutral pH. *J. Inorg. Organomet. Polym.* **2003**, *13*, 49–53.
320. Patwardhan, S. V.; Clarson, S. J., Silicification and Biosilicification. Part 7. Poly-L-Arginine Mediated Bioinspired Synthesis of Silica. *J. Inorg. Organomet. Polym.* **2003**, *13*, 193–203.
321. Tilburey, G. E.; Patwardhan, S. V.; Huang, J.; Kaplan, D. L.; Perry, C. C., Are Hydroxyl-Containing Biomolecules Important in Biosilicification? A Model Study. *J. Phys. Chem. B* **2007**, *111*, 4630–4638.
322. Rovner, S. L., Emulating Nature's Silicon Skills. *Chem. Eng. News* **2009**, *14*, 30–33.
323. Belton, D. J.; Deschaume, O.; Patwardhan, S. V.; Perry, C. C., A Solution Study of Silica Condensation and Speciation with Relevance to in Vitro Investigation of Biosilicification. *J. Phys. Chem. B* **2010**, *114*, 9947–9955.
324. Colfen, H., Biomineralization: A Crystal-Clear View. *Nat. Mater.* **2010**, *9*, 960–961.
325. Patwardhan, S. V.; Tilbury, G. E.; Perrry, C. C., Interactions of Amines with Silicon Species in Undersaturated Solutions Leads to Dissolution and/or Precipitation of Silica. *Langmuir* **2011**, *27*, 15135–15145.
326. Stiubianu, G., New Materials Developed with Lignocellulose Siloxane Derivatives. In *Recent Developments in Silicone-Based Materials*, Cazacu, M., Ed. Nova Science: New York, **2010**; pp 135–178.
327. Deng, Q. Q.; Hahn, J. R.; Stasser, J.; Preston, J. D.; Burns, G. T., Reinforcement of Silicone Elastomers with Treated Silica Xerogels: Silica-Silicone IPNs. *Rubber Chem. Technol.* 2000, *73*, 647–665.
328. Kohls, D.; Beaucage, G.; Pratsinis, S. E.; Kammler, H.; Skillas, G., Carbon Coated Silica for Elastomer Reinforcement. In *Filled and Nanocomposite Polymer Materials*, Hjelm, R. J.; Nakatani, A. I.; Gerspacher, M.; Krishnamoorti, R., Eds. Materials Research Society: Warrendale, PA, 2001; Vol. 661.
329. Zou, H.; Zhang, L.; Tian, M.; Wu, S.; Zhao, S., Study on the Structure and Properties of Conductive Silicone Rubber Filled with Nickel-Coated Graphite. *J. Appl. Polym. Sci.* **2010**, *115*, 2710–2717.
330. Bourgeat-Lami, E.; Espiard, P.; Guyot, A.; Briat, S.; Gauthier, C.; Vigier, G.; Perez, J., Colloids from Functionalized Silica. In *ACS Symposium Series*, Mark, J. E.; Lee, C. Y.-C.; Bianconi, P. A., Eds. American Chemical Society: Washington, DC, 1995; Vol. 585, p 112.
331. Espiard, P.; Guyot, A.; Mark, J. E., Surface-Functionalized Colloidal Silica Particles from an Inverse Microemulsion Sol-Gel Process. *J. Inorg. Organomet. Polym.* **1995**, *5*, 391–407.
332. Luna-Xavier, J.-L.; Bourgeat-Lami, E.; Guyot, A., The Role of Initiation in the Synthesis of Silica/Poly(methyl methacrylate) Nanocomposite Latex Particles Through Emulsion Polymerization. *Coll. Polym. Sci.* **2001**, *279*, 947–958.
333. Perry, R. J., Applications for Cross-Linked Siloxane Particles. *Chem. Tech.* **1999**, *29*(2), 406–411.
334. Liles, D. T.; Morita, Y.; Kobayashi, K., Silicone Elastomeric Powders. *Polym. News* **2002**, *27*, 406–411.
335. Gillies, G.; Prestidge, C. A.; Attard, P., An AFM Study of the Deformation and Nanorheology of Cross-Linked PDMS Droplets. *Langmuir* **2002**, *18*, 1674–1679.

336. Bradley, C. A.; McMurdy, M. J.; Tilley, T. D., Selective Catalytic Cyclohexene Oxidation Using Titanium-Functionalized Silicone Nanospheres. *J. Phys. Chem. C* **2007**, *111*, 17570–17579.
337. Lu, X.; Xin, Z., Preparation and Characterization of Micron-Sized Polystyrene/Polysiloxane Core/Shell Particles. *Coll. Polym. Sci.* **2006**, *284*, 1062–1066.
338. Schmidt, R. G.; Badour, L. R.; Gordon, G. V., Modulus Reduction Mechanism of Trimethylsiloxy Silicates in a Polyorganosiloxane. In *Synthesis and Properties of Silicones and Silicone-Modified Materials*, Clarson, S. J.; Fitzgerald, J. J.; Owen, M. J.; Smith, S. D.; Van Dyke, M. E., Eds. American Chemical Society: Washington, DC, 2003; Vol. 838, pp 170–180.
339. Zhu, B.; Wu, Y.; Reese, H. H.; Katsoulis, D. E.; McGarry, F. J., Silicone-Organic Resin Hybrid Matrix Composites. *Macromol. Mater. Eng.* **2006**, *291*, 1052–1060.
340. Zaioncz, S.; Dahmouche, K.; Soares, B. G., SAXS Characterization of New Nanocomposites Based on Epoxy Resin/Siloxane/MMA/Acrylic Acid Hybrid Materials. *Macromol. Mater. Eng.* **2010**, *295*, 243–255.
341. Kang, B. M.; Park, Y. W., A Study on Phase Separatation of PA-PDMS Composites Dependent on Molecular Weight. *Polym. Plast. Technol. Eng.* **2010**, *49*(3), 223–237.
342. Chen, J.; Wu, W.; Zhou, T.; Fan, W.; Lu, Q., Thermal Properties of the PA66/PC/Silicone Rubber Composites. *J. Macromol. Sci., Part B: Phys.* **2011**, *50*, 111–122.
343. Liu, Y.; Zhang, L.-Q.; Wang, W.-C.; Yu, H.-T.; Lu, Y.-L., Preparation of Polydimethylsiloxane-Coated a-Alumina Filler with Cold Plasma for Elastomer Thermal Interface Materials. *J. Appl. Polym. Sci.* **2012**, *123*, 2875–2882.
344. Fang, C.; Shao, L.; Zhao, Y.; Wang, J.; Wu, H., A Gold Nanocrystal/Poly(dimethylsiloxane) Composite for Plasmonic Heating on Microfluidic Chips. *Adv. Mater.* **2012**, *24*, 94–98.
345. Esteves, A. C. C.; Brokken-Zijp, J.; Laven, J.; Huinink, H. P.; Reuvers, N. J. W.; Van, M. P.; De With, G., Garnet Particles Effect on the Cross-Linking of PDMS and the Network Structures Formed. *Polymer* **2010**, *51*, 136–145.
346. Cho, H.-B.; Konno, A.; Fujihara, T.; Suzuki, T.; Tanaka, S.; Jiang, W.; Suematsu, H.; Niihara, K.; Nakayama, T., Self-Assemblies of Linearly Aligned Diamond Fillers in Polysiloxane Diamond Composite Films with Enhanced Thermal Conductivity. *Compos. Sci. Tech.* **2011**, *72*, 112–118.
347. Raza, M. A.; Westwood, A. V. K.; Brown, A. P.; Stirling, C., Texture, Transport and Mechanical Properties of Graphite Nanoplatelet/Silicone Composites Produced by Three Roll Mill. *Compos. Sci. Tech.* **2012**, *72*, 112–118.
348. Del Castillo-Castro, T.; Castillo-Ortega, M. M.; Encinas, J. C.; Franco, P. J. H.; Carillo-Encalante, H. J., Piezo-Resistance Effect in Composite Based on Cross-Linked Polydimethylsiloxane and Polyaniline: Potential Pressure Sensor Application. *J. Mater. Sci.* **2012**, *47*, 1794–1802.
349. Cho, H.-B.; Najayama, T.; Tokoi, Y.; Endo, S.; Tanaka, S.; Suzuki, T.; Jiang, W.; Suematsu, H.; Niihara, K., Facile Preparation of a Polysiloxane-Based Hybrid Composite with Highly-Oriented Boron Nitride Nanosheets and an Unmodified Surface. *Compos. Sci. Tech.* **2010**, *70*, 1681–1686.
350. Karatas, Y.; Banhatti, R. D.; Kaskhedikar, N.; Burjanadze, M.; Funke, K.; Wiemhofer, H.-D., Synthesis and Modeling of Polysiloxane-Based Salt-in-Polymer Electrolytes with Various Additives. *J. Phys. Chem. B* **2009**, *113*, 15473–15484.

351. Quaranta, A.; Carturan, S.; Marchi, T.; Buffa, M.; Degerlier, M.; Cinausero, M.; Guastalla, G.; Gramegna, F.; Valotto, G.; Maggioni, G.; Kravchuk, V. L., Doped Polysiloxane Scintillators for Thermal Neutrons Detection. *J. Non-Cryst. Solids* **2011**, *357*, 1921–1925.
352. Kim, M. K.; Yi, D. K.; Paik, U., Tunable, Flexible Antireflection Layer of ZnO Nanowires Embedded in PDMS. *Langmuir* **2010**, *26*, 7552–7554.
353. Zhou, J.; Yan, H.; Zheng, Y.; Wu, H., Highly Fluorescent Poly(dimethylsiloxane) for On-Chip Temperature Measurements. *Adv. Funct. Mater.* **2009**, *19*, 324–329.
354. Yang, Y.; Li, W.-N.; Luo, Y.-S.; Xiao, H.-M.; Fu, S.-Y.; Mai, Y.-W., Novel Ultraviolet-Opaque, Visible-Transparent and Light-Emitting ZnO-QD/Silicone Composites with Tunable Luminescence Colors. *Polymer* **2010**, *51*, 2755–2762.
355. Lai, Q.; Lu, H.; Wang, D.; Wang, H.; Feng, S.; Zhang, J., Color-Tunable Luminescent Materials Based on Functional Polysiloxane and Lanthanide Ions. *Macromol. Chem. Phys.* **2011**, *212*, 1435–1442.
356. Yang, Y.; Shi, H.-Q.; Li, W.-N.; Xiao, H.-M.; Luo, Y.-S.; Fu, S.-Y.; Liu, T., Tunable Photo-Luminescent Properties of Novel Transparent CdSe-QD Silicone Nanocomposites. *Compos. Sci. Tech.* **2011**, *71*, 1652–1658.
357. Zhan, F.; Cheng, X. E.; Shi, W., Cationic UV Curing Behavior and Thermal Properties of Oxetane-Modified Polysiloxane Prepared from Tetraethyl Orthosilicate. *J. Appl. Polym. Sci.* **2012**, *123*, 717–724.
358. Naghash, H. J.; Hajati, F., Synthesis and Characterization of a Novel Silicone Containing Vinylic Macromonomer and Its Uses in the Preparation of Poly(silicone-*co*-styrene-*co*-butyl acrylate) with Montmorillonite Nanocomposite Emulsion. *J. Appl. Polym. Sci.* **2012**, *123*, 1227–1237.
359. Kim, E. S.; Lee, T. H.; Shin, S. H.; Yoon, J.-S., Effect of Incorporation of Carbon Fiber and Silicon Carbide Powders Into Silicone Rubber on the Ablation and Mechanical Properties of the Silicone Rubber-Based Ablation Material. *J. Appl. Polym. Sci.* **2011**, *120*, 831–838.
360. Zhang, Y.; Pang, M.; Xu, Q.; Lu, H.; Zhang, J.; Feng, S., The Curing Retardation and Mechanism of High Temperature Vulcanizing Silicone Rubber Filled with Superconductive Carbon Blacks. *Polym. Eng. Sci.* **2011**, *51*, 170–178.
361. Hu, S.-F.; Zhang, C.; Zhao, M.; Chen, W.; Peng, S.-X., Effect of Tensile Strain on the Electrical Resistivity of Silver-Coated Fly Ash Cenospheres/Sillicone Rubber Composites. *Polym. Bull.* **2011**, *66*, 955–961.
362. Im, H.; Kim, J., Enhancement of the Thermal Conductivity of Aluminum Oxide-Epoxy Terminated Poly(dimethyl silxane) with a Metal Oxide Polysiloxane. *J. Mater. Sci.* **2011**, *46*, 6571–6580.
363. Tran, T. H.; Lezhnina, M. A.; Kynast, U., Efficient Green Emission from Transparent Tb^{3+}-Silicone Hybrid Materials. *J. Mater. Chem.* **2011**, *21*, 12819–12823.
364. Tang, C. Y.; Chen, D.-Z.; Chan, K. Y. Y.; Chu, K. M.; Ng, P. C.; Yue, T. M., Fabrication of Antibacterial Silicone Composite by an Antibacterial Agent Deposition, Solution Casting and Cross-Linking Technique. *Polym. Int.* **2011**, *60*, 1461–1466.
365. Swart, M.; Olsson, R. T.; Hedenqvist, M. S.; Mallon, P. E., Organic-Inorganic Hybrid Copolymer Fibers and Their Use in Silicone Laminate Composites. *Polym. Eng. Sci.* **2010**, *50*, 2143–2152.
366. Goyal, A.; Mohl, M.; Kumar, A.; Puskas, R.; Kukovecz, A.; Konya, Z.; Kiricsi, I.; Ajayan, P. M., In Situ Synthesis of Catalytic Metal Nanoparticle-PDMS

Membranes by Thermal Decomposition Process. *Compos. Sci. Tech.* **2011**, *71*, 129–133.
367. Shimojima, A.; Liu, Z.; Ohsuna, T.; Terasaki, O.; Kuroda, K., Self-Assembly of Designed Oligomeric Siloxanes with Alkyl Chains into Silica-Based Hybrid Mesostructures. *J. Am. Chem. Soc.* **2005**, *127*, 14108–14116.
368. Yang, P.; Deng, T.; Zhao, D.; Feng, P.; Pine, D.; Chmelka, B. F.; Whitesides, G. M.; Stucky, G. D., Heirarchically Ordered Oxides. *Science* **1998**, *282*, 2244–2246.
369. Ilievski, F.; Mazzeo, A. D.; R. F., Shepherd; Chen, X.; Whitesides, G. M., Soft Robotics for Chemists. *Angew. Chem. Int. Ed. Engl.* **2011**, *50*, 1890–1895.
370. Trofin, L.; Leduc, P. R., Controlled Geometry Fabrication of Polydimethylsiloxane Nanofibers for Biomimetics. *J. Appl. Polym. Sci.* **2007**, *105*, 2549–2552.
371. Zoldesi, C. I.; Van Walree, C. A.; Imhof, A., Deformable Hollow Hybrid Silica/Siloxane Colloids by Emulsion Templating. *Langmuir* **2006**, *22*, 4343–4352.
372. Kepczynski, M.; Lewandowska, J.; Romek, M.; Zapotoczny, S.; Ganachaud, F.; Nowakowska, M., Silicone Nanocapsules Templated Inside the Membranes of Catanionic Vesicles. *Langmuir* **2007**, *23*, 7314–7320.
373. Kepczynski, M.; Bednar, J.; Lewandowska, J.; Staszewska, M.; Nowakowska, M., Hybrid Silica-Silicone Nanocapsules Obtained in Catanionic Vesicles. Cryo-TEM Studies. *J. Nanosci. Nanotech.* **2009**, *9*, 3138–3143.
374. Degen, P.; Chen, Z.; Rehage, H., Stimulated Deformation of Polysiloxane Capsules in External Electric Fields. *Macromol. Chem. Phys.* **2010**, *211*, 434–442.
375. Moon, M. W.; Sun, J. V.; Oh, K. H.; Vazin, A.; Hutchinson, J. W., Wrinkled Hard Skins on Polymers Created by Focused Ion Beam. *Proc. Nat. Acad. Sci.* **2007**, *104*, 1130–1133.
376. Py, C.; Reverdy, P.; Doppler, L.; Bico, J.; Roman, B.; Baroud, C. N., Capillary Origamic Spontaneous Wrapping of a Droplet with an Elastic Sheet. *Phys. Rev. Lett.* **2007**, *98*, 1561031–1561034.
377. Sozzani, P.; Bracco, S.; Comotti, A.; Simonutti, R.; Valsesia, P.; Sakamoto, Y.; Terasaki, O., Complete Shape Retention in the Transformation of Silica to Polymer Micro-Objects. *Nat. Mater.* **2006**, *5*, 545–551.
378. Bao, Z.; Weatherspoon, M. R.; Shian, S.; Cai, Y.; Graham, P. D.; Allan, S. M.; Ahmad, G.; Dickerson, M. B.; Church, B. C.; Kang, Z.; Abernathy, H. W., III; Summers, C. J.; Liu, M.; Sandhage, K. H., Chemical Reduction of Three-Dimensional Silica Micro-Assemblies into Microporous Silicon Replicas. *Nature* **2007**, *446*, 172–175.
379. Mark, J. E., Simulations on Aspects of Elastomer Structure That Provide Unusual Toughness. *J. Comput. Aided Mater. Design* **1996**, *3*, 311–318.
380. Yuan, Q. W.; Kloczkowski, A.; Mark, J. E.; Sharaf, M. A., Simulations on the Reinforcement of Poly(dimethylsiloxane) Elastomers by Randomly-Distributed Filler Particles. *J. Polym. Sci., Polym. Phys. Ed.* **1996**, *34*, 1647–1657.
381. Mark, J. E. In *Illustrative Modeling Studies on Elastomers and Rubberlike Elasticity*, 2001 International Conference on Computational Nanoscience, Laudon, M.; Romanowicz, B., Eds. Computational Publications, Boston: Hilton Head Island, SC, 2001; pp 53–56.
382. Mark, J. E., Some Simulations on Elastomers and Rubberlike Elasticity. *Makromol. Symp.* **2001**, *171*, 1–9.
383. Mark, J. E.; Curro, J. G., A Non-Gaussian Theory of Rubberlike Elasticity Based on Rotational Isomeric State Simulations of Network Chain

Configurations. I. Polyethylene and Polydimethylsiloxane Short-Chain Unimodal Networks. *J. Chem. Phys.* **1983**, *79*, 5705–5709.

384. Treloar, L. R. G., *The Physics of Rubber Elasticity*, 3rd ed. Clarendon Press: Oxford, 1975.
385. Sharaf, M. A.; Kloczkowski, A.; Mark, J. E., Simulations on the Reinforcement of Elastomeric Poly(dimethylsiloxane) by Filler Particles Arranged on a Cubic Lattice. *Comput. Polym. Sci.* **1994**, *4*, 29–39.
386. Sunkara, H. B.; Jethmalani, J. M.; Ford, W. T., Solidification of Colloidal Crystals of Silica. In *ACS Symposium Series*, Mark, J. E.; Lee, C. Y.-C.; Bianconi, P. A., Eds. American Chemical Society: Washington, DC, **1995**; Vol. 585, pp 181–191.
387. Pu, Z.; Mark, J. E.; Jethmalani, J. M.; Ford, W. T., Effects of Dispersion and Aggregation of Silica in the Reinforcement of Poly(methyl acrylate) Elastomers. *Chem. Mater.* **1997**, *9*, 2442–2447.
388. Rajan, G. S.; Mark, J. E.; Seabolt, E. E.; Ford, W. T., Thermal and Mechanical Properties of Some Poly(methyl acrylate)/Silica Composites with Intentionally Suppressed Interfacial Bonding. *J. Macromol. Sci., Pure Appl. Chem.* **2002**, *A39*, 39–51.
389. Nakatani, A. I.; Chen, W.; Schmidt, R. G.; Gordon, G. V.; Han, C. C., Chain Dimensions in Polysilicate-Filled Poly(dimethylsiloxane). *Polymer* **2001**, *42*, 3713–3722.
390. Sharaf, M. A.; Kloczkowski, A.; Mark, J. E., Monte Carlo Simulations on Reinforcement of an Elastomer by Oriented Prolate Particles. *Comput. Theor. Polym. Sci.* **2001**, *11*, 251–262.
391. Boonstra, B. B., Role of Particulate Filers in Elastomer Reinforcement: A Review. *Polymer* **1979**, *20*, 691–704.
392. Donnet, J.-B.; Vidal, A., Carbon Black: Surface Properties and Interactions with Elastomers. *Adv. Polym. Sci.* **1986**, *76*, 103–127.
393. Donnet, J.; Custodero, E., Reinforcement of Elastomers by Particulate Fillers. In *Science and Technology of Rubber*, 3rd ed.; Mark, J. E.; Erman, B., Eds. Elsevier: Amsterdam, **2005**; pp 367–400.
394. Mark, J. E.; Abou-Hussein, R.; Sen, T. Z.; Kloczkowski, A., Monte Carlo Simulations on Nanoparticles in Elastomers. Effects of the Particles on the Dimensions of the Polymer Chains and the Mechanical Properties of the Networks. *Macromol. Symp.* **2007**, *256*, 40–47.
395. Schmidt, H., Organically-Modified Silicates by the Sol-Gel Process. In *Better Ceramics Through Chemistry*, Brinker, C. J.; Clark, D. E.; Ulrich, D. R., Eds. Materials Research Society: Pittsburgh, PA, 1984; Vol. 32, pp 327–335.
396. Huang, H.-H.; Orler, B.; Wilkes, G. L., Hybrid Materials Incorporating Polymeric/Oligomeric Species with Inorganic Glasses by a Sol-Gel Process. *Polym. Bull.* **1985**, *14*(6), 557–564.
397. Mark, J. E.; Sun, C.-C., Polymer-Modified Silica Glasses. I. Control of Hardness. *Polym. Bull.* **1987**, *18*, 259–264.
398. Schmidt, H., Organically-Modified Silicates as Inorganic-Organic Polymers. In *Inorganic and Organometallic Polymers. Macromolecules Containing Silicon, Phosphorous, and Other Inorganic Elements*, Zeldin, M.; Wynne, K. J.; Allcock, H. R., Eds. American Chemical Society: Washington, DC, 1988; Vol. 360, pp 333–344.
399. Wang, B.; Huang, H.-H.; Brennan, A. B.; Wilkes, G. L., Synthesis of New Alumina-Containing Organic/Inorganic Hybrid Materials. *Prepr., Am. Chem. Soc., Div. Polym. Chem.* **1989**, *30*(2), 146.

400. Rodrigues, D. E.; Wilkes, G. L., Structure-Property Behavior of Sol-Gel Derived Hybrid Materials Using Microwave Processing. *Prepr., Am. Chem. Soc., Div. Polym. Chem.* **1989**, *30*(2), 227.
401. Wilkes, G. L.; Huang, H.-H.; Glaser, R. H., New Inorganic-Organic Hybrid Materials Through the Sol-Gel Approach. In *Silicon-Based Polymer Science. A Comprehensive Resource*, Zeigler, J. M.; Fearon, F. W. G., Eds. American Chemical Society: Washington, DC, 1990; Vol. 224, pp 207–226.
402. Wilkes, G. L.; Brennan, A. B.; Huang, H.-H.; Rodrigues, D.; Wang, B., The Synthesis, Structure, and Property Behavior of Inorganic-Organic Hybrid Network Materials Prepared by the Sol-Gel Process. In *Polymer-Based Molecular Composites*, Schaefer, D. W.; Mark, J. E., Eds. Materials Research Society: Pittsburgh, PA, 1990; Vol. 171, pp 15–29.
403. Wen, J.; Mark, J. E., Sol-Gel Preparation of Composites of Poly(dimethylsiloxane) with SiO_2 and SiO_2/TiO_2, and Their Mechanical Properties. *Polym. J.* **1995**, *27*, 492–502.
404. Chen, J.; Wu, W.; You, Y.; Fan, W.; Chen, Y., Morphology and Properties of the PA66/Silicone Rubber Composites. *J. Appl. Polym. Sci.* **2010**, *117*, 2964–2971.
405. Reed, P. E.; Squires, H. V., Application of Fracture Mechanics to Plastics Deformed at High Strain-Rates. *J. Mater. Sci.* **1974**, *9*(1), 129–135.
406. Meissner, B., Bound Rubber Theory and Experiment. *J. Appl. Polym. Sci.* **1993**, *50*, 285–292.
407. Leblanc, J., A Molecular Explanation for the Origin of Bound Rubber in Carbon Black Filled Rubber Compounds *J. Appl. Polym. Sci.* **1997**, *66*, 2257–2268.
408. Cohen-Addad, J. P.; Roby, C.; Sauviat, M., Characterization of Chain Binding to Filler in Silicone-Silica Systems. *Polymer* **1985**, *26*(8), 1231–1233.
409. Belousov, S. I.; Buzin, A. I.; Godovsky, Y. K., Self-Organization of Polysiloxanes into Mono- and Multilayer Structures at Interphase Boundaries. *Polym. Sci. U.S.S.R.* **1999**, *41*, 303–324.

CHAPTER 10

Applications

10.1 MEDICAL

Numerous medical applications have been developed for siloxane polymers.[1–4] Prostheses, artificial organs, objects for facial reconstruction, vitreous substitutes in the eyes, tubing and catheters, for example, take advantage of the inertness, stability, and pliability of polysiloxanes. Artificial skin, contact lenses, and drug delivery systems utilize their high permeability as well. Such biomedical applications have led to extensive biocompatability studies,[5–6] particularly on the interactions of polysiloxanes with proteins.[7] There has been considerable interest in modifying these materials to improve their suitability for biomedical applications in general.[8–9] Advances seem to be coming particularly rapidly in the area of high-tech drug-delivery systems.[10–11] Figure 10.1 shows the range of diameters of Silastic medical-grade siloxane tubing available for medical applications. The smallest tubing has an internal diameter of only 0.012 inches (0.031 cm) and an outer diameter of only 0.025 inches (0.064 cm).

Such materials must first be extensively tested (sensitization of skin, tissue cell culture compatibility, implant compatibility). There has been considerable controversy, for example, over the safety of using polysiloxanes in breast implants.[12–13] The major concern was "bleeding" of low molecular polysiloxanes out of the gels into the chest cavity, followed by transport to other parts of the body. The extent to which "bleeding" occurred and its possible systemic effects on the body were argued vigorously in the media and in the courts, and led to restrictions on the use of polysiloxanes.

In the case of controlled drug-delivery systems, the goal is to have the drug released at a relatively constant rate (zero-order kinetics) at a concentration

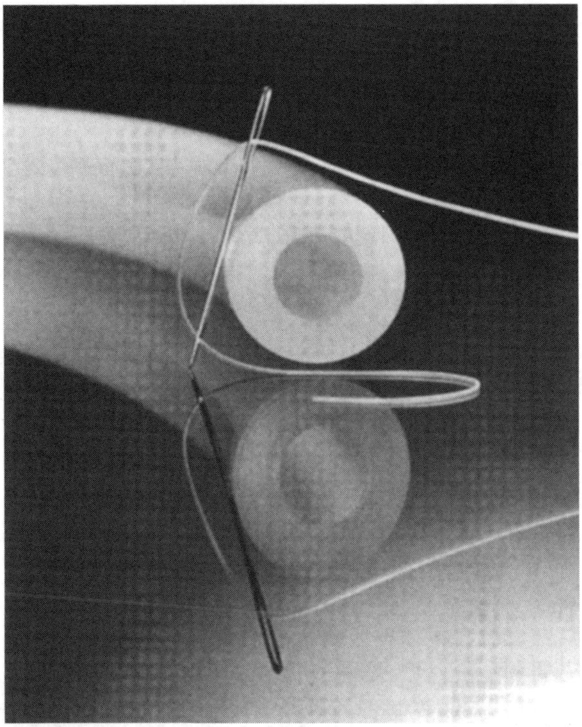

Figure 10.1:
Siloxane polymer used in tubing and catheters for medical applications. The photograph was provided by the Dow Corning Corporation of Midland, MI.

within the therapeutic range. It is obviously important to minimize the amount of time the concentration is in the low, ineffective range, and to eliminate completely the time it is in the high, toxic range (figure 10.2). Figure 10.3 illustrates the use of polysiloxanes in such drug-delivery systems. The goal mentioned is approached by placing the drug inside a siloxane elastomeric capsule, which is then implanted in an appropriate location in the body. The drug within the capsule can be in the free state, in a fluid suspension, or mixed or dissolved into an elastomeric matrix. Release rates for drugs that are much more hydrophilic than the polysiloxanes (e.g., melatonin and sulfanilamide) are frequently relatively slow. Rates can be increased by incorporating solvents, channeling agents, or fillers in the capsule.

10.2 NONMEDICAL

Typical nonmedical applications include high-performance elastomers, membranes, electrical insulators, water repellents, antifoaming agents,

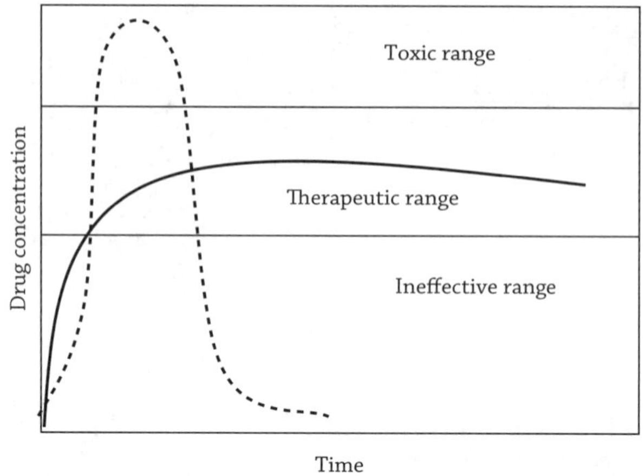

Figure 10.2:
Drug concentrations in the body as a function of time. The dashed line shows the uncontrolled dissolution of a pill, with two regions so low that the concentration would be ineffective and another region high enough to be toxic. The solid line shows the goal of controlled delivery, specifically a relatively constant delivery rate in the therapeutic range for an extended period of time.

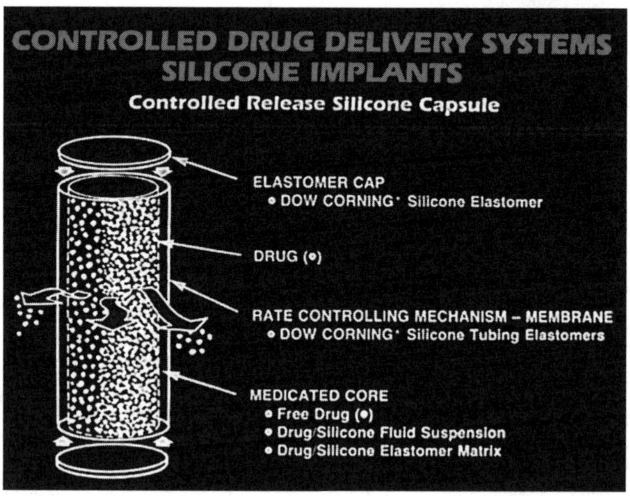

Figure 10.3:
A controlled release drug-delivery system that utilizes both siloxane elastomers and fluids. The photograph was provided by the Dow Corning Corporation of Midland, MI.

mold-release agents, adhesives, protective coatings,[14] release control agents for agricultural chemicals, encapsulation media, mold-making materials, coatings, layers in high-tech laminates, and hydraulic, heat-transfer, and dielectric fluids.[1, 3, 15–16] These applications are based on the same properties of polysiloxanes just mentioned and also their ability to modify surfaces and interfaces (e.g., as water repellents, antifoaming agents, and mold-release agents). In some cases the polysiloxane is blended into a material to improve its mechanical properties (e.g., its impact strength). For such applications, finely divided cross-linked powders are available commercially.[17–18] Two additional examples are the uses of polysiloxanes in polymer-electrolyte complexes [19] and in microlithographic applications. [20]

Figure 10.4 shows a typical nonmedical application of a siloxane polymer. In this case, a liquid, UV-curable polysiloxane is being used to protect a printed circuit board. The polymer being used was specifically chosen for its low viscosity, moisture resistance, and easy repairability. Also, some such materials can be treated so that thin spots in the coating are readily apparent. Although dip coating is illustrated in the figure, flow coating and spray coating can also be used.

One of the most impressive applications of polysiloxanes, particularly in the arts, is their use in making molds of intricate surfaces (figure 10.5). In this case, the surface to be copied was vertical and nonmovable, so a

Figure 10.4:
An electronic circuit board being given a protective polysiloxane coating. The photograph was provided by the Dow Corning Corporation of Midland, MI.

Figure 10.5:
High-transparency polysiloxane elastomer used as a stress-relieving interlayer in the type of safety glass or plastic used in windshields and canopies. The photograph was provided by the Dow Corning Corporation of Midland, MI.

"nonslumping" end-linkable paste had to be used rather than a liquid. After the surface was coated, it was cured to give a remarkably faithful reproduction that was easily removable from the original surface.

In the final example, a polysiloxane appears as an interlayer in the types of plastic and glass laminates used for increased safety in windshields and canopies in aircraft. The flexibility and thermal stability of these polymers are great advantages in the case of high-performance aircraft, which can experience wide variations in temperature. In this application, a polysiloxane is chosen to give the highest transparency and good adhesion to the plastic or glass outer layers. The excellent transparency of such composites is readily seen in the samples displayed in figure 10.6.

Polysiloxanes are the materials of choice in the new "soft-lithography" techniques.[21–25] Figure 10.7 briefly outlines the method. In the first step, linear (liquid) polydimethylsiloxane (PDMS) is poured over the surface of the master to be reproduced; it is then cross linked and peeled away from the master surface. The PDMS surface containing the pattern is then coated with a hydrophobic alkane thiol, and the pattern is transferred to a gold surface (to which the thiol strongly binds). The pattern placed onto the gold surface can then be developed using a variety of techniques, including washing with a hydrophilic thiol to coat those parts of the surface not covered by the hydrophobic thiol. The advantages of PDMS are (i) high fluidity (which is required to make good replicas of the

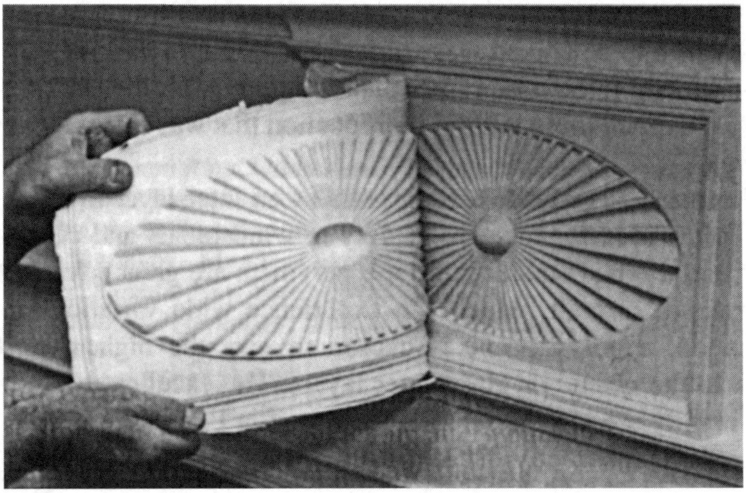

Figure 10.6:
A polysiloxane Room-Temperature Vulcanizate (RTV) being used to make a mold of an intricately carved surface. The photograph was provided by the Dow Corning Corporation of Midland, MI.

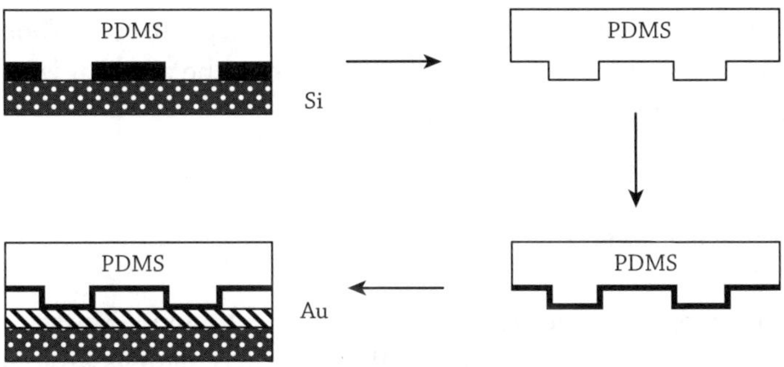

Figure 10.7:
Steps in soft lithography. The PDMS in the beginning sketch is first cross linked, then removed, and then coated with a thiol that transfers the original pattern to a gold layer.

master), (ii) ease of cross linking at ambient temperatures (to make the features of the stamp permanent), (iii) superb elasticity and robustness (for removing the PDMS stamp and using it in microcontact printing), and (iv) excellent stability (for long-term storage of patterns).

Mention should be made of the use of polysiloxanes in the area of microfluidics. [22,26] Devices of this type have channels or capillaries with dimensions 10–10,000 μm, and their preparation is relatively straightforward with PDMS, using the soft-lithography techniques just described to make the required molds.

10.3 CONCLUSIONS AND OUTLOOK

As documented in this review, the polysiloxanes have been of great interest and importance for a variety of reasons. The chains themselves have unusual structural features and flexibility and mobility unmatched by other polymer molecules. As a result, polysiloxanes have extraordinarily high permeabilities, very low viscosities, unusual surface properties, unexpected mesophases, and useful properties over an astonishingly wide range of temperatures. The fact that these novel properties are not well understood has attracted the attention of people doing analytical theory or computer simulations.

Networks produced from polysiloxane chains are also unique, because of the specific reactions that can be used to end link them into high-performance elastomeric materials of known network structures. Reinforcing these polysiloxane elastomers also provides some unusual opportunities in that they are compatible with fillers ranging from the commercially important silicas to new materials such as in situ–generated ceramic particles, ellipsoidal particles, clays and other layered materials, polysilsesquioxane molecular cubes, porous materials, and nanotubes. All indications are that the polysiloxanes will maintain their fascination to those doing basic research, and there seems to be no limit to the areas in which these materials are finding intriguing applications.

ACKNOWLEDGEMENTS

It is a pleasure to acknowledge the financial support provided by the National Science Foundation through Grant DMR-0314760 (Polymers Program, Division of Materials Research), and by the Dow Corning Corporation of Midland, MI. We appreciate the effort of Devesh Dadhich Shreeram in checking the references and Zheng Yan for preparing the index.

REFERENCES

1. Warrick, E. L.; Pierce, O. R.; Polmanteer, K. E.; Saam, J. C., Silicone Elastomer Developments 1967-1977. *Rubber Chem. Technol.* **1979**, *52*, 437–525.
2. Arkles, B.; Rediger, P., *Silicones in Biomedical Applications*. Technomic Publishing: Lancaster, PA, 1983.
3. Rochow, E. G., *Silicon and Silicones*. Springer-Verlag: Berlin, 1987.
4. Giordano, G. G.; Refojo, M. F., Silicone Oils as Vitreous Substitutes. *Prog. Polym. Sci.* **1998**, *23*, 509–532.

5. Shansong, M.; Mei, T.; Shunqing, T.; Changren, Z., Improvement of Blood Compatibility of Silicone Rubber by the Addition of Hydroxyapatite. *J. Mater. Sci. Lett.* **2003**, *22*, 343–344.
6. Fallahi, D.; Mirzadeh, H.; Khorasani, M. T., Physical, Mechanical, and Biocompatibility Evaluation of Three Different Types of Silicone Rubber. *J. Appl. Polym. Sci.* **2003**, *88*, 2522–2529.
7. Bartzoka, V.; McDermott, M. R.; Brook, M. A., Protein-Silicone Interactions. *Adv. Mater.* **1999**, *11*, 257–259.
8. Abbasi, F.; Mirzadeh, H.; Katbab, A.-A., Modification of Polysiloxane Polymers for Biomedical Applications: A Review. *Polym. Int.* **2001**, *50*, 1279–1287.
9. Gao, Z.; Nahrup, J. S.; Mark, J. E.; Sakr, A., Poly(dimethylsiloxane) Coatings for Controlled Drug Release. I. Preparation and Characterization of Pharmaceutically Acceptable Materials. *J. Appl. Polym. Sci.* **2003**, *90*, 658–666.
10. Langer, R., Where a Pill Won't Reach. *Sci. Am.* **2003**, *288*(4), 50–57.
11. Fischetti, M., Potent Patches. *Sci. Am.* **2003**, *288*(4), 92–93.
12. Garrido, L.; Kwong, K. K.; Pfleiderer, B.; Crawley, A. P.; Hulka, C. A.; Whitman, G. L.; Kopans, D. B., Echo-Planar Chemical Shift Imaging of Silicone Gel Prostheses. *Magn. Reson. Imaging* **1993**, *11*, 625–634.
13. El-Zaim, H. S.; Heggers, J. P., Silicones for Pharmaceutical and Biomedical Applications. In *Polymeric Biomaterials,* 2nd ed.; Dumitriu, S., Ed. Marcel Dekker: New York, 2002.
14. Zhang, S. L.; Tsou, A. H.; Li, C. M., Scratching Behavior of Elastomeric Poly(dimethylsiloxane) Coatings. *J. Polym. Sci., Part B: Polym. Phys.* **2002**, *40*, 1530–1537.
15. Andersson, L. H. U.; Hjertberg, T., Silicone Elastomers for Electronic Applications. I. Analyses of the Noncrosslinked Fractions. *J. Appl. Polym. Sci.* **2003**, *88*, 2073–2081.
16. Mark, J. E., Status and Trends in the Area of Elastomers and Rubberlike Elasticity. *Mol. Cryst. Liq. Cryst.* **2004**, *417*(1), 75–86.
17. Perry, R. J., Applications for Cross-Linked Siloxane Particles. *Chem. Tech.* **1999**, *29*(2), 406–411.
18. Liles, D. T.; Morita, Y.; Kobayashi, K., Silicone Elastomeric Powders. *Polym. News* **2002**, *27*, 406–411.
19. Smid, J.; Fish, D.; Khan, I. M.; Wu, E.; Zhou, G., Polysiloxane-Based Polymer-Electrolyte Complexes. In *Silicon-Based Polymer Science. A Comprehensive Resource*, Zeigler, J. M.; Fearon, F. W. G., Eds. American Chemical Society: Washington, DC, 1990; Vol. 224.
20. Reichmanis, E.; Novembre, A. E.; Tarascon, R. G.; Shugard, A.; Thompson, L. F., Organosilicon Polymers for Microlithographic Applications. In *Silicon-Based Polymer Science. A Comprehensive Resource*, Zeigler, J. M.; Fearon, F. W. G., Eds. American Chemical Society: Washington, DC, 1990; Vol. 224.
21. Kumar, A.; Abbott, N. L.; Kim, E.; Biebuyck, H. S.; Whitesides, G. M., Patterned Self-Assembled Monolayers and Meso-Scale Phenomena. *Acc. Chem. Res.* **1995**, *28*, 219–226.
22. Whitesides, G. M., Organic Materials Science. *MRS Bull.* **2002**, *27*(1), 56–65.
23. Paul, K. E.; Prentiss, M.; Whitesides, G. M., Patterning Spherical Surfaces at the Two-Hundred-Nanometer Scale Using Soft Lithography. *Adv. Funct. Mater.* **2003**, *13*, 259–263.

24. Choi, K. M.; Rogers, J. A., A Photocurable Poly(dimethylsiloxane) Chemistry Designed for Soft Lithographic Molding and Printing in the Nanometer Regime. *J. Am. Chem. Soc.* **2003**, *125*, 4060–4061.
25. Kim, D. H.; Lin, Z.; Kim, H.-C.; Jeong, U.; Russell, T. P., On the Replication of Block Copolymer Templates by Poly(dimethylsiloxane) Elastomers. *Adv. Mater.* **2003**, *15*, 811–814.
26. McDonald, J. C.; Whitesides, G. M., Poly(dimethylsiloxane) as a Material for Fabricating Microfluidic Devices. *Acc. Chem. Res.* **2002**, *35*(7), 491–499.

INDEX